Mathematics for Engineers

and Technologists

Mathematics for Engineers and Technologists

Huw Fox
Bill Bolton

OXFORD   AMSTERDAM   BOSTON   LONDON   NEW YORK   PARIS
SAN DIEGO   SAN FRANCISCO   SINGAPORE   SYDNEY   TOKYO

Butterworth-Heinemann
An imprint of Elsevier Science
Linacre House, Jordan Hill, Oxford, OX2 8DP
225 Wildwood Avenue, Woburn, MA 01801-2041

First published 2002

British Library Cataloguing in Publication Data
A catalogue record for this book is available from the British Library

ISBN 0 7506 5544 5

For information on all Butterworth-Heinemann publications
visit our website at www.bh.com

Printed and bound in Great Britain

Contents

Series Preface

'There is a time for all things: for shouting, for gentle speaking, for silence; for the washing of pots and the writing of books. Let now the pots go black, and set to work. It is hard to make a beginning, but it must be done' – Oliver Heaviside, *Electromagnetic Theory*, Vol 3 (1912), Ch. 9 'Waves from moving sources - Adagio, andante, Allegro Moderato.'

Oliver Heavyside was one of the greatest engineers of all time, ranking alongside Faraday and Maxwell in his field. As can be seen from the above excerpt from a seminal work, he appreciated the need to communicate to a wider audience. He also offered the advice 'So be rigorous: that will cover a multitude of sins. An do not frown.' The series of books that this prefaces takes up Heavyside's challenge but in a world which is quite different to that being experienced just a century ago.

With the vast range of books already available covering many of the topics developed in this series, what is this series offering which is unique? I hope the next few paragraphs help to answer that; certainly no one involved in this project would give up their time to bring these books to fruition if they had not thought that the series is both unique and valuable.

The motivation for this series of books was born out of the desire of the UK's Engineering Council to increase the number of incorporated engineers graduating from Higher Education establishments, and the Institution of Incorporated Engineers' (IIE) aim to provide enhanced services to those delivering Incorporated Engineering Courses. However, what has emerged from the project should prove of great value to a very wide range of courses within the UK and internationally – from Foundation Degrees or Higher Nationals through to first year modules for traditional 'Chartered' degree courses. The reason why these books will appeal to such a wide audience is that they present the core subject areas for engineering studies in a lively, student-centred way, with key theory delivered in real world contexts, and a pedagogical structure that supports independent learning and classroom use.

Despite the apparent waxing of 'new' technologies and the waning of 'old' technologies, engineering is still fundamental to wealth creation. Sitting alongside these are the new business focused, information and communication dominated, technology organisations. Both facets have an equal importance in the health of a nation and the prospects of individuals. In preparing this series of books, we have tried to strike a balance between traditional engineering and developing technology.

The philosophy is to provide a series of complementary texts which can be tailored to the actual courses being run – allowing the flexibility for course designers to take into account 'local' issues, such as areas of particular staff expertise and interest, while being able to demonstrate the depth and breadth of course material referenced to a common framework. The series is designed to cover material in the core texts which approximately corresponds to the first year of study with module texts focusing on individual topics to second and final year level. While the general structure of each of the texts is common, the styles are quite different, reflecting best practice in their areas.

Another set of factors which we have taken into account in designing this series is the reduction in contact hours between staff and students, the evolving responsibilities of both parties and the way in which advances in technology are changing the way study can be, and is, undertaken. As a result, the lecturers' support material which accompanies these texts, is paramount to delivering maximum benefit to the student.

It is with these thoughts of Voltaire that I leave the reader to embark on the rigours of study:

'Work banishes those three great evils: boredom, vice and poverty'

Alistair Duffy
Series Editor
De Montfort University, Leicester, UK

Further information on the IIE Textbook Series is available from bhmarketing@repp.co.uk
www.bh.com/iie

Please send book proposals to:
rachel.hudson@repp.co.uk

Other titles currently available in the IIE Textbook Series

Mechanical Engineering Systems	0 7506 5213 6
Business Skills for Engineers and Technologies	0 7506 5211 1
Design Engineering	0 7506 5211 X
Technology of Engineering Materials	0 7506 5643 3
Systems for Planning and Control	
in Manufacturing	0 7506 4977 1

Introduction: why mathematics?

Mathematics is an essential tool for the engineer and technologist. As an illustration, consider a number of simple situations:

- ## *A beam*

  A uniform horizontal beam rests on supports at each end and loads placed at its mid point. How does the deflection of the beam from the horizontal at its mid point depend on the applied load? Can we develop a mathematical relationship which will enable us to predict the deflection for a given load? We might need such a relationship in order to be able to consider the design of a simple plank bridge across a stream, or the elements in a more complex truss bridge.

See **Chapter 1** for a discussion of mathematical relationships; when one quantity is dependent on another it is said to be a function of it.

  We might develop such a relationship by conducting an experiment in which we measure the deflections for a number of loads and so develop an empirical relationship which fits the results. This could involve plotting a graph between the force and deflection and from the 'shape' of the graph determining a relationship. This requires an understanding of graphs and, in particular, straight line graphs. If the graph between two quantities is not a straight line then engineers use 'tricks' to persuade the graph to become straight line because straight line graphs enable relationships to be most easily discerned.

See **Chapter 3** for the determination of relationships from graphs by 'persuading' them to become straight line graphs.

  We might, however, develop the relationship from a consideration of how beams behave when subject to loads and so end up with a more general relationship which we can apply to other beams. In developing such a relationship we would use algebra, i.e. the quantities such as force and deflection are represented by letters such as F and y, and so we need to be able to manipulate algebraic expressions. In fact, the basic expression for elasticity involves a differential equation, i.e. an equation involving terms concerned with rates of change, and so to derive the relationship for the deflection we need to be able to solve such an equation.

See **Chapters 4 and 5** for a discussion of calculus and the solution of differential equations.

See **Chapter 1** for a discussion of logarithms and log graphs.

See **Chapter 1** for an introduction to phasors and **Chapter 2** for a more detailed consideration.

See **Chapter 6** for a discussion of how oscillations can be represented by differential equations and **Chapter 7** for how we can use the Laplace transforms to simplify the handling of such equations.

See **Chapter 1** for an introduction to phasors and **Chapter 2** for a more detailed consideration.

See **Chapter 7** for a discussion of the Fourier series.

See **Chapter 8** for a discussion of logic gates.

- *An oscillation*

  A loaded vertical spring is set in oscillation; what are the factors determining the frequency of the oscillation? We might carry out an experiment involving different loads on a spring, and also try a number of different springs. The relationship between frequency and load can be determined by plotting a graph. However, if we just plot frequency against load we obtain a non-linear graph and it is not easy to see the relationship. If we plot the logarithm of the frequency against the logarithm of the load then a straight line graph is obtained and the relationship can easily be discerned.

  Alternatively, a more general relationship might be found from the theory of oscillations. We might develop the theory by considering a model of oscillations in which the variation of displacement with time is represented by how the vertical height of a rotating radius varies with time when the radius rotates with a constant angular velocity.

  The relationships devised for this simple system of an oscillating spring can, however, provide a basic introduction to the consideration of much more complex oscillations.

- *An a.c. electrical circuit*

  A simple electrical circuit is set up with an a.c. voltage being applied to a series circuit of a capacitor and a resistor. To develop a theory which enables the circuit current to be determined by different values of capacitance and resistance and also for different frequency alternating currents, an approach based on a consideration of phasors is generally used. A phasor is used to model in its length the amplitude or root-mean-square value of a voltage or current and by its initial angle the phase. By using such models, the analysis of a.c. circuits is simplified and we do not need differential equations. The phasors can be drawn and added or subtracted graphically. However, an algebraic method is to use complex numbers.

  We might have the complication with such a circuit that the alternating voltage is not sinusoidal. In such a case we can deal with the circuit analysis by representing the signal as a series of sinusoidal terms.

- *Programmable logic controller (PLC)*

  PLCs are widely used for control systems. Such controllers can be easily programmed to carry out such operations as switching on motors when sensor A and sensor B both give ON signals or perhaps when either sensor A or sensor B gives an ON signal. Thus a central heating system controller needs to switch the pump motor on when either the temperature sensor on the hot water tank gives an ON signal or the room temperature sensor gives an ON signal. Such control systems require a consideration of basic logic systems.

See **Chapter 9** for a discussion of statistics and the handling of errors.

• *Measurements*

Engineering involves making measurements. With any measurement there is inevitably some associated error. To estimate and handle such errors in calculations based on the measurements, we need an understanding of basic statistics.

In considering the elasticity of a beam we form mathematical models of the real world situation. Thus, in the case of the beam, we make a number of simplifying assumptions, such as the deflections are small, the beam is thin and of uniform cross-section. With such simplifications we can produce a model. Real beams may behave differently because the assumptions are not valid but our mathematical model provides working relationship. In some situations we develop what are obviously models of systems. For example, we might represent the behaviour of a car suspension and wheel as a mass with a spring and a damper. In considering a motor we might be able to represent it by the model of an inductor in series with a resistor and a source of e.m.f. for the back e.m.f. of the motor.

See **Chapter 3** for an introductory discussion of mathematical models.

As the above examples illustrate, we need mathematics to be able to solve engineering problems. Mathematics, however, is not a tool that you can pick up and use without an understanding of the principles behind its development and its limitations. Thus, in this book, the principles behind the mathematics are explored and the book is more than just a collection of 'cookbook techniques' which can be used for particular situations. Such a 'cookbook' approach presents problems if you encounter, as engineers inevitably do, a new situation. The aim of this book is:

To enable the reader to understand the principles of the mathematics and acquire the ability to use it in engineering.

With that aim in mind, we hope you will enjoy the book and wish you well in your studies.

1 Functions

Summary

Engineers, whether electrical, electronic, mechatronic or mechanical, are concerned with expressing relationships between physical quantities clearly and unambiguously. This might be the relationship between the displacement of an oscillating object and time, or perhaps the amplitude of an a.c. voltage and time. This chapter is about how we can represent such relationships in mathematical terms, taking the opportunity to revise some basic mathematics in the process. This does not mean that it is not important to clearly explain in words what relationships there are between quantities but rather to supplement the written word by using a system that is both clear, unambiguous, and internationally understood, thereby removing the possibility of misinterpretation.

Objectives

By the end of this chapter, the reader should be able to:

- understand the concept of a function for relating quantities within engineering disciplines;
- use functions, and their notation, to describe relationships;
- manipulate and evaluate algebraically simple function expressions, including inverses;
- use graphs to express functions;
- express cyclic functions in terms of sine and cosine functions;
- know, and use, the relationships between sine, cosine and tangent ratios;
- describe waveforms in the general format $R \sin(\omega t \pm a)$;
- use exponential and logarithmic functions;
- use hyperbolic functions.

1.1 Introduction to functions

As you embark on studying engineering, whether electrical, electronic, mechatronic or mechanical, it will become apparent that equations are used to describe relationships between physical quantities and are more clear and unambiguous than the written word on its own. This does not mean that it is not important to clearly explain relationships in words but rather there is a need to supplement the written word by using equations. In any discussion

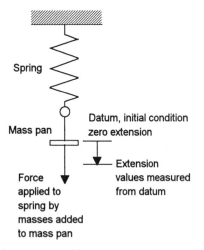

Spring

Mass pan

Datum, initial condition
zero extension

Extension
values measured
from datum

Force
applied to
spring by
masses added
to mass pan

Figure 1.1 *Simple spring with mass pan, the figure indicating the two physical quantities, namely the applied force and the resulting extension, that we are interested in determining the relationship between. Note, adding masses of 120 g results in force increments of about 1.2 N.*

of relationships between physical quantities, the term function is likely to be encountered. So what is a function?

So what is a function?

Let's commence our explanation of the term *function* by discussing a simple example.

We are all familiar with springs, whether they are those we find inside some pens, the springs holding the tremolo block in position on a classic Fender stratocaster guitar or the suspension springs on a car. Consider, therefore, a simple vertical spring dangling from a clamp with a mass pan attached to its lower end (Figure 1.1). Using such a system we can perform a simple experiment, adding masses to the pan and determining the relationship between the force exerted by the masses and the resulting extension of the spring. The extension is measured from the datum line of the system when in stable equilibrium before we start adding masses and recording extensions. The results of such an experiment might be of the form shown in Table 1.1.

Table 1.1 *Force–extension values for simple spring example*

Force F applied to spring, N	Corresponding extension e of spring, m
0.0	0.00
1.2	0.01
2.4	0.02
3.6	0.03
4.8	0.04
6.0	0.05

From Table 1.1, we can assume that there is some relationship or connection between the values of the force F acting on the spring and the corresponding observed extension e values. We can say:

the extension e depends on the corresponding applied force F [1]

Statement [1] is an inferred relationship between an observed measurement e and a varied quantity, in this case the applied force F. We can therefore call F the *independent* quantity, sometimes referred to as the *argument*, from which a *dependent* result is obtained. We can restate statement [1] as:

the extension e is a function of the applied force [2]

A *function* is a relationship which has for each value of the independent variable a unique value of the dependent variable. Statement [2] can be written as:

$e = f(F)$

This means exactly the same thing as statement [2] but is just easier and more concise to write. The 'f' simply is shorthand for 'function of'. Note that $f(F)$ does <u>not</u> mean a variable f multiplied by F. When we are dealing with a number of different functions it is customary to use different letters for the function label, e.g. we might use $y = f(x)$ and $z = g(x)$.

Tables, graphs and equations to define functions

The statement $e = f(F)$ merely tells us that the extension e is a function of the applied force F and does not describe the actual relationship between them. The data in Table 1.1 is one way we describe the relationship. If we know the force is 3.6 N then the extension must be 0.03 m.

However, a pattern can be seen from an inspection of the results in Table 1.1: if we double the applied force then the resulting extension doubles, if we treble the applied force the extension trebles and if we quadruple the applied force the extension quadruples. We can say, at least over the range of values we have observed:

the extension e is directly proportional to the applied force F

and we can write this as:

$$e \propto F \qquad\qquad [3]$$

The symbol $\propto$ simple means (or is shorthand for) 'proportional to'.

We can also see how the extension depends on the applied force by plotting a graph. If we plot a graph of the force values in Table 1.1 against the corresponding extension values we obtain the straight line graph shown in Figure 1.3. The straight line passes through the origin. This is a characteristic of relationships when one quantity is directly proportional to the other. The graph is thus one way of defining the functional relationship.

Key points

With $y = f(x)$ we have each value of y associated with an x value. We can plot such data points on a graph having x and y axes, the y-axis running vertically and the x-axis running horizontally and at 90° to the y-axis (Figure 1.2). The point of intersection of the two axes is called the *origin*; at this point y has the value 0 and x = 0. Values of x which are positive are plotted to the right of the origin, negative values to the left. Positive values of y are plotted upwards from the origin, negative values downwards.

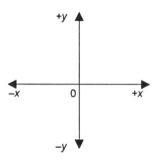

Figure 1.2 *Graph axes*

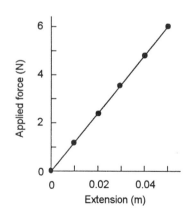

Figure 1.3 *Force–extension graph for the spring experiment values given in Table 1.1*

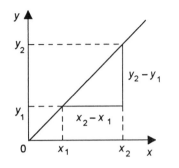

Figure 1.4 *Generalised straight line graph witth the slope being $(y_2 - y_1)/(x_2 - x_1)$*

We can use the graph in Figure 1.3 to determine an equation relating the extension to the applied force and so give another way of defining the functional relationship. The graph is a straight line and so has a constant slope (or gradient), the slope being defined in the same way as we define the slope of a road, i.e. as the change in vertical height of the line over a given horizontal distance (Figure 1.4). We can compute the slope by choosing a pair of values of force and extension, e.g. force 60 N, extension 0.05 m; force 1.2 N, extension 0.01 m, and so obtain:

$$\text{slope} = \frac{6.0 - 1.2}{0.05 - 0.01} = \frac{4.8}{0.04} = 120$$

Since the straight line passes through the origin, this tells us that the force in newtons is, over the interval that has been considered, 120 times the extension in metres. We can write this as:

$$F = 120e \tag{4}$$

The constant term, i.e. the 120 N/m, is called the *constant of proportionality* linking F to the corresponding e values.

To summarise: we can define the functional relationship between two variables by:

- a set of results (as in Table 1.1),

- a graph (as in Figure 1.3),

- an equation (as in equation [4]).

Note that to define a function $y = f(x)$ completely, we must define the range of values of x over which the definition is true. This is called the *domain* of the function.

Example

If y is a function of x and the relationship is defined by the equation $y = 4x^2$, what is the value of y when $x = 2$?

We have:

$y = f(x) = 4x^2$

and so:

$y = 4(2^2) = 4 \times 2^2 = 16$

Therefore, $y = 16$ when $x = 2$. Substituting $x = 2$ into the original functional equation, we can write:

$y = f(x) = f(2) = 16$.

> **Example**
>
> If y is a function of x and the relationship is defined by the equation $y = 12x^2 + 3x + 6$, what is the value of y when $x = 1$?
>
> We have:
>
> $y = f(x) = 12x^2 + 3x + 6$
>
> and so:
>
> $y = 12(1^2) + 3(1) + 6 = 12 + 3 + 6 = 21$
>
> Therefore, $y = 21$ when $x = 1$. Substituting $x = 1$ into the original functional equation, we can write:
>
> $y = f(x) = f(1) = 21$

Equations and functions

Functions may, as mentioned earlier, be defined using equations. Equations give the instructions for calculating the dependent variable of functions for values of the independent variable. For example, for an ohmic resistor the potential difference V across it is a function of the current I through it, i.e.

$V = f(I)$

The equation defining the functional relationship (Ohm's law) is $V = RI$, where R is the constant if proportionality connecting the variable V with the variable I, thereby defining their unique relationship. So given a value for the current we can use the equation to obtain a value of the potential difference. Thus, when $R = 10\ \Omega$ and $I = 2$ A we have:

$V = f(2) = 20$ V.

For an object freely falling from rest, the distance fallen s is a function of the time t for which it has been falling, i.e. $s = f(t)$. The defining equation is $s = \frac{1}{2}at^2$, where a, the acceleration, is a constant. The acceleration is the acceleration due to gravity g and so we can write the defining equation as $s = \frac{1}{2}gt^2$. Thus, given a value for the time we can use the equation to obtain a value for the distance fallen. If we assume that g has a value of 9.8 m/s^2, then for a time of 3 s;

$s = f(3) = \frac{1}{2} \times 9.8 \times 3^2 = 44.1$ m

Note that a function may be defined by several equations, with each giving the instructions for calculating the dependent variable for different values of the independent variable. For example, for

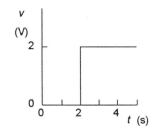

Figure 1.5   *Step voltage*

the voltage signal shown in Figure 1.5, a so-called *step voltage*, we have $v = f(t)$ and the relationship

$v = 0$ for t between 0 and 2 s, i.e. $0 \le t < 2$
$v = 2$ V for t greater than 2 s, i.e. $2 \le t$

The value of $v = f(1)$ is thus 0, while the value of $v = f(4)$ is 2 V.

Functions may of course get quite complex! For example, the natural frequency of transverse vibration of a cantilever is described by the equation:

$$\text{frequency} = \frac{1}{2\pi} \sqrt{\frac{3EI}{mL^3}}$$

The frequency of the cantilever is a function of E, I, m and L. So, in functional notation we can write:

$$\text{frequency} = f(E, I, m, L)$$

Example

If we have y as a function of x and described by the relationship $y = x^2$, what are the values of (a) $f(0)$, (b) $f(2)$?

(a) The function is described by $y = f(x) = x^2$. Thus $f(0)$ is the value of the function when $x = 0$ and so is 0.
(b) $f(2)$ is the value of the function when $x = 2$ and so is 4.

Example

Determine the values of (a) $f(2)$, (b) $f(4)$ if we have y as a function of x and defined by:

$y = 1$ for $0 \le x < 3$, $y = 2(x - 3) + 1$ for $3 \le x$

(a) The value of the function at $x = 2$ is given by the first relationship as 1.
(b) The value of the function at $x = 4$ is given by the second relationship as $y = 2(4 - 3) + 1 = 3$.

1.1.1 Combinations of functions

Many of the functions encountered in engineering and science can be considered to be combinations of other functions. Suppose we have the function $y = f(x) = x^2 + 2x$. We can think of the function $f(x)$ as resulting from the combination of two functions $g(x)$ and $h(x)$. One of the functions takes an input of x and gives an output of x^2 and the other takes an input of x and gives an output of $2x$.

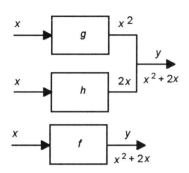

Figure 1.6 *Combination of functions g(x) and h(x)*

The two outputs are then added and we have $f(x) = g(x) + h(x)$. Figure 1.6 illustrates this.

Another way we can combine functions is by applying them in sequence. For example, if we have $h(x) = 2x$ and $g(x) = x^2$, then suppose we have the arrangement shown in Figure 1.7. The input of x to the g function box results in an output of x^2. The h function box takes its input and doubles it. Thus for an input of x^2 we have an output of $2x^2$, thus $f(x) = h\{g(x)\} = 2x^2$. Note that the order of the function boxes is important.

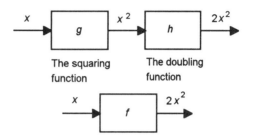

Figure 1.7 *Combination of functions to give the function f(x)*

Problems 1.1

1 If we have y as a function of x and defined by the equation $y = 2x + 3$, what are the values of (a) $f(0)$, (b) $f(1)$?

2 If we have y as a function of x and defined by the equation $y = x^2 + x$, what are the values of (a) $f(0)$, (b) $f(2)$?

3 If y is a function of x defined by the following equations, find the values of $f(0)$ and $f(1)$.

(a) $y = x^2 + 3$, (b) $y = x + 4$, (c) $y = (x + 1)^2 - 3$

4 Determine the values of (a) $f(0.5)$, (b) $f(2)$ if we have y as a function of x and defined by:

$y = 2$ for $0 \leq x < 1$, $y = 1$ for $1 \leq x$

5 Determine the values of (a) $f(1)$, (b) $f(3)$ if we have y as a function of t and defined by:

$y = 0$ for $0 \leq t < 2$, $y = 2(t - 2)$ for $2 \leq t$

6 The voltage in an electrical circuit is supplied by a constant voltage source of 10 V. If the voltage is switched on after time $t = 2$ s, state the equations defining the step voltage at any time t.

7 Sketch the periodic waveform described by the following equations:

$y = f(t) = t$ for $0 \leq t < 2$ and $y = f(t) = 2 - t$ for $2 \leq t < 4$

8 The period of oscillation T of a simple pendulum is a function of the length L of the pendulum, being defined by the equation

$$T = 2\pi \sqrt{\frac{L}{g}}$$

where g is the acceleration due to gravity. What are the values of (a) $f(1)$, (b) $f(10)$ if g can be taken as 10 m/s²?

9 The velocity v in metres per second of a moving object is a function of the time t in seconds, being defined by $v = 2 + 5t$. What are the values of (a) $f(0)$, (b) $f(1)$?

10 If $g(x) = 2x$ and $h(x) = x + 1$, what are (a) $g(x) + h(x)$, (b) $g\{h(x)\}$, (c) $h\{g(x)\}$?

11 If $f(x) = x^2 + 1$, $g(x) = 3x$ and $h(x) = 3x + 2$, determine: (a) $f(x) + g(x)$, (b) $f\{g(x)\}$, (c) $g\{f(x)\}$, (d) $f(x) - h(x)$, (e) $f\{h(x)\}$.

1.2 Linear functions

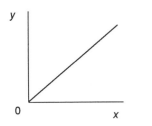

Figure 1.8 *Straight line graph with y directly proportional to x; the straight line thus passes though the origin*

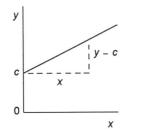

Figure 1.9 *Straight line graph not passing through the origin*

This section is about a form of functions that is very commonly encountered in engineering, namely linear functions. Quite simply, linear functions are ones that provide a linear or straight line relationship between two variables when plotted as a conventional graph of one variable against the other.

The potential difference V across a resistor is a function of the current I through it. If the resistor obeys Ohm's law then $V = RI$, the potential difference is proportional to the current. If the current is doubled then the potential difference is doubled, if the current is trebled the potential difference is trebled. This means that a graph of V plotted against I is a straight line graph passing through the origin. *Gradient* is defined as the change in y value divided by the change in x value. Thus, for all straight line graphs passing through the origin (Figure 1.8), the gradient is constant and given by gradient $m = y/x$. Hence the equation of such a straight line is of the form:

$$y = mx \qquad [5]$$

where m is the gradient of the line. *Only* when we have such a relationship is y directly proportional to x.

Straight line graphs which do not pass through the origin (Figure 1.9) have a gradient, change in y value divided by change in x value, given by $m = (y - c)/x$, where c is the value of y when $x = 0$, i.e. the intercept of the straight line with the y-axis. Thus, such lines have the equation:

$$y = mx + c \qquad [6]$$

This is the equation which defines a straight line and is termed a *linear equation*. It is important to realise that with $c \neq 0$ that y is *not proportional* to x.

The gradient m of a straight line graph may be positive or negative. The gradient may also have a value of zero and this is a line parallel to the x-axis. The intercept c may be positive or negative, or zero.

Example

State the gradients and intercepts of the graphs of the following equations: (a) $y = 2x + 3$, (b) $y = 2 - x$, (c) $y = x - 2$.

(a) This has a gradient of +2 and an intercept with the y-axis of +3. A positive gradient means that y increases as x increases.
(b) This has a gradient of –1 and an intercept with the y-axis of +2. A negative gradient means that y decreases as x increases.
(c) This has a gradient of +1 and an intercept with the y-axis of –2.

Example

During a test to find how the power of a CNC lathe varied with depth of cut, the following results were obtained:

Depth of cut d (mm)	0.51	1.02	1.52	2.03	2 54	3.00
Power P (W)	0.89	1.04	1.14	1.32	1.43	1.55

Use a graph to show that the function connecting the quantities d and P is of the form $y = mx + c$. Use this function to calculate the depth of cut when the power is 1 W.

Figure 1.10 shows the graph with P on the y-axis and d on the x-axis. The graph line represents the line of best fit through all the points and may therefore be prone to some error. Because it is a straight line, the function is of the form $y = mx + c$ and so we have:

$$P = md + c$$

The slope m of the graph is about 0.27 and the point where the line would intercept the P axis when $d = 0$ is about 0.76. The function is thus:

$$P = 0.27d + 0.76$$

We can check the integrity of the above equation by substituting values from the table of observed results, say $d = 2.03$ mm. This gives:

$$P = 0.27 \times 2.03 + 0.76 = 1.31$$

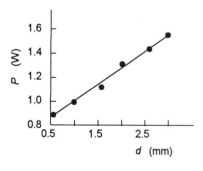

Figure 1.10 *Graph of power–depth of cut*

If we now refer back to the table of results we see that this is close to what is given there, i.e. 1.32 W.

To finally answer the question regarding the depth of cut to be expected when the power is 1 W:

$$1 = 0.27d + 0.76$$

Hence d is about 0.89 mm.

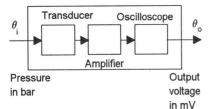

Pressure
in bar

Output
voltage
in mV

Figure 1.11 *Pressure measurement system*

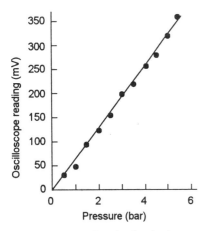

Figure 1.12 *Graph of output voltage–input pressure for pressure measurement system*

Example

A pressure measurement system using a piezoelectric transducer is set up as represented by the system diagram of Figure 1.11. As the input pressure signal is altered, corresponding output readings are taken off the oscilloscope screen and the results shown below were obtained:

Pressure (bar)	0.5	1.0	1.5	2.0	2.5	3.0
Output voltage (mV)	32	50	90	125	160	200

Pressure (bar)	3.5	4.0	4.5	5.0	5.5
Output voltage (mV)	220	260	280	320	350

Use a graph to show that the function connecting the quantities output voltage θ_o to the pressure θ_i is of the form $y = mx + c$. The static sensitivity of such a measurement system may be defined as the change in output signal divided by the change in the corresponding input signal. Determine the static sensitivity.

Figure 1.12 shows the graph obtained by plotting the above values. The graph line represents the line of best fit through all the points and may therefore be prone to some error. Because it is a straight line, the function is of the form $y = mx + c$ and so we have:

$$\theta_o = m\theta_i + c$$

From the graph, the approximate slope m is 350/5.5 = 63.6 mV/bar. The line passes through the origin and so c = 0. Hence:

$$\theta_o = 63.6\theta_i$$

The static sensitivity K is:

$$K = \frac{\Delta \theta_o}{\Delta \theta_i}$$

The symbol Δ in front of a quantity is used to indicate an increment of that quantity. But this is just the slope of the graph and so the static sensitivity is 63.6 mV/bar.

Problems 1.2

1 State which of the following will give a straight line graph and, if so, whether it passes through the origin:
(a) A graph of the extension of a spring plotted against the applied load when the extension is proportional to the applied load.
(b) A graph of the resistance R of a length of resistance wire plotted against the temperature t when $R = R_0(1 + at)$, with R_0 and a being constants.
(c) A graph of the distance d travelled by a car plotted against time t when $d = 10 + 4t^2$.
(d) A graph is plotted of the pressure p of a gas against its volume v, the pressure being related to the volume by Boyle's law, i.e. $pv = $ a constant.
2 Determine the straight line equations for the following data if linear functional relationships can be assumed:
(a) The current i and time t over a period of time if at the beginning of the time we have $i = 2$ A and $t = 0$ s and at the end we have $i = 3$ A and $t = 2$ s.
(b) The extension e of a strip of material as a function of its length L when subject to constant stress, given that:

| e in mm | 0.60 | 0.72 | 0.84 | 0.96 |
| L in m | 0.5 | 0.6 | 0.7 | 0.8 |

1.3 Quadratic functions

A *linear function* is one where the equation defining the function is of the form $y = mx + c$. The highest power of a variable is 1. This is only one type of function. Here we look at another form, the quadratic function, and examine its defining equation.

The term *quadratic function* is used for a function $y = f(x)$ where the defining equation has the general form:

$$f(x) = ax^2 + bx + c \qquad [7]$$

where a, b and c are constants. The highest power of the variable is 2.

Quadratic equations occur often in engineering. An example of such an equation in engineering occurs with the e.m.f. E of a thermocouple which can often be described by:

$$E = at + bt^2$$

Key point

Quadratic functions have defining equations in which the highest power of the variable is 2.

where t is the temperature and a and b are constants. Other examples occur in the relationships for the bending moment M for bending beams, such as that for a cantilever propped at its free end:

$$M = \tfrac{1}{2}wx^2 - \tfrac{5}{8}wLx + \tfrac{1}{8}wL^2$$

where x is the distance from the free end of a cantilever of length L and w the distributed load per unit length.

The linear equation and the quadratic equation are just two examples of what are termed *polynomials*. A polynomial is the term used for any equation involving powers of the variable which are positive integers. Such powers can be 1, 2, 3, 4, 5, etc. For example, $x^4 + 4x^3 + 2x^2 + 5x + 2 = 0$ is a polynomial with the highest power being 4.

1.3.1 Factors and roots

To factorise a number means to write it as the product of smaller numbers. Thus, for example, we can factor 12 to give $12 = 3 \times 4$. The 3 and 4 are factors of 12. If the 3 and the 4 are multiplied together then 12 is obtained. To *factorise a polynomial* means to write it as the product of simpler polynomials. Thus for the quadratic expression $x^2 + 5x + 6$ we can write:

$$x^2 + 5x + 6 = (x + 2)(x + 3)$$

$(x + 2)$ and $(x + 3)$ are factors. If the two factors are multiplied together then the $x^2 + 5x + 6$ is obtained. Note that, in general:

$$(a + b)(c + d) = a(c + d) + b(c + d) = ac + ad + bc + bd \qquad [8]$$

If we have $u \times v = 0$ then we must have either $u = 0$ or $v = 0$ or both u and v are 0. This is because 0 times any number is 0. Thus if we have the quadratic equation $x^2 + 5x + 6 = 0$ and rewrite it as $(x + 2)(x + 3) = 0$, then we must have either $x + 2 = 0$, or $x + 3 = 0$ or both equal to 0. This means that the solutions to the quadratic equation are the solutions of these two linear equations, i.e. $x = -2$ and $x = -3$. These values are called the *roots* of the equation. We can check these values by substituting them into the quadratic equation. Thus for $x = -2$ we have $4 - 10 + 6 = 0$ and thus $0 = 0$. For $x = -3$ we have $9 - 15 + 6 = 0$ and thus $0 = 0$.

Key point

We can solve a quadratic equation by:

1 Factoring the quadratic.

2 Setting each factor equal to 0.

3 Solving the resulting linear equations.

Example

Factorise and hence solve the quadratic equation $x^2 - 3x + 2 = 0$.

To factorise this equation we need to find the two numbers which when multiplied together will give 2 and which when added together will give -3.

If we multiply −1 and −2 we obtain 2 and the addition of −1 and −2 gives −3. Thus we can write:

$(x - 1)(x - 2) = 0$

The solutions are thus given by $x - 1 = 0$, i.e. $x = 1$, and $x - 2 = 0$, i.e. $x = 2$.

We can check these values by substituting them into the original equation, $x^2 - 3x + 2$. Thus, for $x = 1$ we have $1 - 3 + 2 = 0$ and so $0 = 0$. For $x = 2$ we have $4 - 6 + 2 = 0$ and so $0 = 0$.

Completing the square

Consider the equation $x^2 + 6x + 9 = 0$. This equation can be factorised to give $(x + 3)(x + 3) = 0$, i.e. $(x + 3)^2 = 0$. It is a perfect square, both the roots being the same. Now consider the equation $x^2 + 6x + 2 = 0$. What are the factors? We can rewrite the equation as:

$x^2 + 6x = -2$

If we add 9 to both sides of the equation then we obtain

$x^2 + 6x + 9 = -2 + 9 = 7$

The left-hand side of the equation has been made into a perfect square by the adding of the 9. Thus we can write:

$(x + 3)^2 = 7$

This means that $x + 3$ must be one of the square roots of 7, i.e.

$x + 3 = \pm\sqrt{7}$

The plus or minus is because every positive number has two square roots, one positive and one negative. Thus we have $(+\sqrt{7}) \times (+\sqrt{7}) = 7$ and $(-\sqrt{7}) \times (-\sqrt{7}) = 7$. Hence:

$x = -3 \pm \sqrt{7}$

The two solutions are thus $x = -3 + \sqrt{7}$ and $x = -3 - \sqrt{7}$.

   This method of determining the roots of a quadratic equation is known as *completing the square*. In the above discussion the left-hand side of the equation was made into a perfect square by the adding of 9. How do we determine what number to add in order to make a perfect square? Any expression of the form $x^2 + ax$ becomes a perfect square when we add $(a/2)^2$, since:

Key point

The procedure for determining the roots of a quadratic equation by completing the square can be summarised as:

1 Put the equation in the form $x^2 + ax = b$.

2 Determine the value of $(a/2)$.

3 Add $(a/2)^2$ to both sides of the equation to give:

$x^2 + ax + (a/2)^2 = b + (a/2)^2$

4 Hence obtain the equation:

$(x + a/2)^2 = b + (a/2)^2$

5 Determine the two roots by taking the square root of both sides of the equation, i.e.

$x + \dfrac{a}{2} = \pm\sqrt{b + \left(\dfrac{a}{2}\right)^2}$

$$x^2 + ax + \left(\frac{a}{2}\right)^2 = \left(x + \frac{a}{2}\right)^2 \qquad [9]$$

Thus for $x^2 + 6x$ we have $a = 6$ and so $(a/2) = 3$; we add $3^2 = 9$.

The above rule for completing the square only works if the coefficient of x^2, i.e. the number in front of x^2, is 1. However, if this is not the case we can simply divide throughout by that coefficient in order to make it 1.

Example

Use the method of completing the square to solve the quadratic equation $x^2 + 10x - 4 = 0$.

The quadratic equation can be written as:

$x^2 + 10x = 4$.

Adding $(10/2)^2 = 25$, to both sides of the equation gives:

$x^2 + 10x + 25 = 4 + 25$

Thus:

$(x + 5)^2 = 29$

Hence, $x + 5 = \pm\sqrt{29} = \pm 5.39$ and the solutions of the quadratic equation are $x = +5.39 - 5 = 0.39$ and $x = -5.39 - 5 = -10.39$.

We can check these values by substituting them in the equation $x^2 + 10x - 4 = 0$. Thus, for $x = 0.39$ we have $0.39^2 + 3.9 - 4 = 0.05$, which, because of the rounding used to limit the number of decimal places in determining the root, is effectively zero. For the other solution of $x = -10.39$ we have $(-10.39)^2 + 103.9 - 4 = 0.05$, which, because of the rounding used to limit the number of decimal places in determining the root, is effectively zero.

The quadratic formula

Consider the quadratic equation $ax^2 + bx + c = 0$. To obtain the roots by completing the square method, we divide throughout by a to give:

$$x^2 + \frac{b}{a}x + \frac{c}{a} = 0$$

This can be written as:

$$x^2 + \frac{b}{a}x = -\frac{c}{a}$$

To make the left-hand side of the equation a perfect square we must add $(b/2a)^2$ to both sides of the equation. Hence:

$$x^2 + \frac{b}{a}x + \left(\frac{b}{2a}\right)^2 = -\frac{c}{a} + \left(\frac{b}{2a}\right)^2$$

and so:

$$\left(x + \frac{b}{2a}\right)^2 = -\frac{c}{a} + \left(\frac{b}{2a}\right)^2 = \frac{-4ac + b^2}{4a^2}$$

Taking the square root of both sides of the equation gives

$$x + \frac{b}{2a} = \pm\frac{\sqrt{b^2 - 4ac}}{2a}$$

Thus:

$$x = -\frac{b}{2a} \pm \frac{\sqrt{b^2 - 4ac}}{2a}$$

and so we have the *general formula for the solution of a quadratic equation*:

$$x = \frac{-b \pm \sqrt{b^2 - 4ac}}{2a} \qquad [10]$$

Consider the following three situations:

- If we have $(b^2 - 4ac) > 0$, then the square root is of a positive number. There are then *two distinct roots* which are said to be *real*.

- If we have $(b^2 - 4ac) = 0$, then the square root is zero and the formula gives just one value for x. Since a quadratic equation must have two roots, we say that the equation has *two coincident real roots*.

- If we have $(b^2 - 4ac) < 0$, then the square root is of a negative number. A new type of number has to be invented to enable such expressions to be solved. The number is referred to as a *complex number* and the roots are said to be *imaginary* (the roots in 1 and 2 above are said to be *real*). Such numbers are discussed later in this book.

Key point

The general formula for the solution of a quadratic equations is:

$$x = \frac{-b \pm \sqrt{b^2 - 4ac}}{2a} \qquad [10]$$

Example

Determine, if they exist as real roots, the roots of the following quadratic equations:

(a) $4x^2 - 7x + 3$, (b) $x^2 - 4x + 4$, (c) $x^2 + 2x + 4$.

(a) Using the general quadratic formula [10], here we have $a = 4$, $b = -7$ and $c = 3$. Therefore:

$$x = \frac{+7 \pm \sqrt{49 - 48}}{8} = \frac{7 \pm 1}{8}$$

Therefore, $x = 1$ or $x = 0.75$ defines the two roots of the equation. We can now represent the function as:

$$(x - 1)(x - 0.75)$$

and this, when multiplied out, gives $x^2 - 1.75x + 0.76$ and which when multiplied by 4 gives the original equation $4x^2 - 7x + 3$.

(b) Using the general quadratic formula [10] gives:

$$x = \frac{-2 \pm \sqrt{16 - 16}}{2} = 2$$

Therefore, we have two roots with $x = 2$. We may now rewrite the equation in the form $(x - 2)(x - 2)$. This, when multiplied out, gives $x^2 - 4x + 4$.

(c) Using the general quadratic formula [10] gives:

$$x = \frac{-2 \pm \sqrt{4 - 16}}{2}$$

Since the term inside the square root is negative, we have *no real roots*.

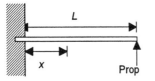

Figure 1.13   *Propped cantilever*

Example

Figure 1.13 shows a simple cantilever of length L, propped at its free end. It can be shown that the bending moment M of this type of cantilever is a function of the distance x measured from the fixed end of the beam, thus $M = f(x)$. The defining equation for the function is:

$$M = f(x) = \frac{Wx^2}{2} - \frac{5WLx}{8} + \frac{WL^2}{8}$$

where W is the distributed load in newtons per unit length. Using this quadratic formula, determine the positions along the beam at which the bending moment is zero (in engineering called the *points of contraflexure*).

When $M = 0$, we have:

$$\frac{Wx^2}{2} - \frac{5WLx}{8} + \frac{WL^2}{8} = 0$$

We can solve this by using the general quadratic formula [10]. Firstly, we can simplify the expression by multiplying through by 8 and taking the W term out as a factor:

$$4Wx^2 - 5WLx + WL^2 = W(4x^2 - 5Lx + L^2) = 0$$

and so the equation becomes $4x^2 - 5Lx + L^2 = 0$. Thus:

$$x = \frac{5L \pm \sqrt{(-5L)^2 - 4(4)(L^2)}}{2(4)}$$

$$x = \frac{5L \pm \sqrt{25L^2 - 16L^2}}{8}$$

Hence $x = L/4$ or $x = L$. The bending moment is thus zero at two locations (it has two points of contraflexure), i.e. at $L/4$ from the fixed end or at the extreme right-hand end of the beam when $x = L$.

Example

The distance s in metres moved by a vehicle over a period of time t seconds is defined by the equation $s = ut + \frac{1}{2}at^2$, with a being the constant acceleration and u the initial velocity. Assuming the vehicle commences motion with an initial velocity of 5 m/s and covers 84 m with a constant acceleration of 2 m/s^2, calculate the time over which this occurs.

Substituting the values in the equation gives:

$$84 = 5t + \frac{1}{2}(2)t^2$$

$$84 = 5t + t^2$$

Writing the equation in the general format:

$$t^2 + 5t - 84 = 0$$

Thus:

$$t = \frac{-5 \pm \sqrt{25 - 4(84)}}{2}$$

and so $t = -12$ s or $t = 7$ s. Since we cannot have a negative time, the only acceptable answer is $t = 7$ s.

The solution may be checked by substituting into the original equation $t^2 + 5t - 84 = 0$, when $t = 7$ s we have $7^2 + 5(7) - 84 = 0$. Since this is true, our solution holds.

Example

The total surface area A of a cylinder of radius r and height h is given by the equation $A = 2\pi r^2 + 2\pi rh$. If $h = 6$ cm, what will be the radius required to give a surface area of 88/7 cm^2? Take π as 22/7.

Putting the numbers in the equation gives

$$\frac{88}{7} = 2 \times \frac{22}{7} r^2 + 2 \times \frac{22}{7} \times 6r$$

Multiplying throughout by 7 and dividing by 44 gives

$$2 = r^2 + 6r$$

Hence we can write

$$r^2 + 6r - 2 = 0$$

and so:

$$r = \frac{-6 \pm \sqrt{36 - 4 \times 1 \times (-2)}}{2 \times 1} = \frac{-6 \pm \sqrt{44}}{2} = \frac{-6 \pm 6.63}{2}$$

Hence the solutions are $r = -6.32$ cm and $r = 0.32$ cm. The negative solution has no physical significance. Hence the solution is a radius of 0.32 cm.

We can check this value of 0.32 cm by substitution in the equation $2 = r^2 + 6r$. Hence $0.10 + 1.92 = 2.02$, which is effectively 2 bearing in mind the rounding of the root value to two decimal places that has occurred.

Problems 1.3

1   Determine, if they exist, the real roots of the following quadratic functions:

   (a) $x^2 + 2x - 4$, (b) $x^2 + 3x + 1$, (c) $x^2 - 2x - 1$, (d) $x^2 + x + 2$.

2   The e.m.f. E of a thermocouple is a function of the temperature T, being given by $E = -0.02T^2 + 6T$. The e.m.f. is in μV and the temperature in °C. Determine the temperatures at which the e.m.f. will be 200 μV.

3   When a ball is thrown vertically upwards with an initial velocity u from an initial height h_0, the height h of the ball is a function of the time t, being given by $h - h_0 = ut - 4.9t^2$. Determine the times for which the height is 1 m, if $u = 4$ m/s and $h_0 = 0.5$ m.

4   The deflection y of a simply supported beam of length L when subject to an impact load of mg dropped from a height h on its centre is obtained by equating the total energy released by the falling load with the strain energy acquired, i.e.

$$mgh + mgy = \frac{24EI}{L^3}$$

Hence obtain an expression for the deflection y.

5   The height h risen by an object, after a time t, when thrown vertically upwards with an initial velocity u is given by the equation $h = ut - \frac{1}{2}gt^2$, where g is the acceleration due to gravity. Solve the quadratic equation for t if $u = 100$ m/s, $h = 150$ m and $g = 9.81$ m/s^2.

6   A rectangle has one side 3 cm longer than the other. What will be the dimensions of the rectangle if the diagonals have to have lengths of 10 cm? Hint: let one of the sides have a length x, then the other side has a length of $3 + x$. The Pythagoras theorem can then be used.

1.4 Inverse functions

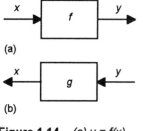

(a)

(b)

Figure 1.14   *(a) y = f(x),*
(b) x = g(y)

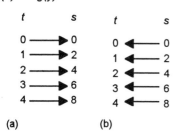

(a)   (b)

Figure 1.15 *For (a) function*
s = 2t, (b) function t = s/2

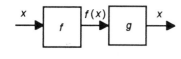

Figure 1.16   *x = g{f(x)}*

So far, in the treatment of a function we have started with a value of the independent variable x and used the function to find the corresponding value of the dependent variable y (Figure 1.14(a)). However, suppose we are given a value for y and want to find x (Figure 1.14(b)). For example, we might have distance s as a function of time t, e.g. $s = 2t$. Given a value of the independent variable t we can use the function to determine s. Suppose though that we are given a value of the dependent variable s and have to determine the corresponding t value? With the given equation we can rearrange it to give $t = s/2$. The function from t to s is $f(t)$, the function from s to t is a different function $g(s)$.

Figure 1.15(a) shows some values for the $s = f(t)$ function described by the equation $s = 2t$. Figure 1.15(b) shows the function obtained by reversing the arrows, i.e. starting with time values deducing the corresponding distance values. This figure represents the *inverse* relationship.

Note that there is a simple point of significance here: if we use $s = 2t$ to calculate a value for s given a value of t and then use the inverse by taking that value of t to calculate a value of s, we end up back where we started with our original value of s. This leads to a method of specifying an inverse function. Consider the arrangement shown in Figure 1.16. Here the g function system box operates on the output from the f function box in order to undo the work of the f box. Because the g function is undoing the work of the f function it is said to be the *inverse* of f. We may, therefore, write:

$$g\{f(x)\} = x \qquad \qquad [11]$$

This equation [11] is used to define an inverse function:

Key point

If f is a function of x then the function g which satisfies g{f(x)} = x for all values of x in the domain of f is called the inverse of f.

If f is a function of x then the function g which satisfies g{f(x)} = x for all values of x in the domain of f is called the inverse of f.

With regard to notation: the inverse of a function f of x is written as $f^{-1}(x)$. Note that $f^{-1}(x)$ does *not* mean $1/f(x)$, it is simply the notation to indicate the inverse function (the -1 not indicating a a power -1!). $f^{-1}(x)$ takes an input which is some function of x and inverts it to give an output of x. Thus the above definition gives:

$$f^{-1}\{f(x)\} = x \qquad\qquad [12]$$

As an illustration, consider a function f which adds 2, i.e. we have $f(x) = x + 2$ (Figure 1.17(a)). Then the inverse is a function that subtracts 2 in order to undo the action of the f function (Figure 1.17(b)). Thus, $f^{-1}(x) = x - 2$ and so if we put into the function $x + 2$ we obtain $(x + 2) - 2$.

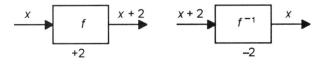

Figure 1.17 *(a) The function which adds 2, (b) the inverse function which subtracts 2*

As a further illustration, consider a function f which multiplies by 3, i.e. we have $f(x) = 3x$ (Figure 1.18(a)). Then the inverse is a function that divides by 3 in order to undo the action of the f function (Figure 1.18(b)). Thus $f^{-1}(x) = x/3$.

As another illustration, consider a function which multiplies by 3 and then adds 2, i.e. $f(x) = 3x + 2$. Here we have operated on the input x to the function twice, initially we multiplied the x by 3 and then we added the number 2 (Figure 1.19(a)). To arrive back at the original input, the inverse do two things (the reverse operation of those just detailed), namely initially subtract 2 and then secondly divided through by 3 (Figure 1.19(b)). Note that you must undo things in the reverse order to which they were done with the function f.

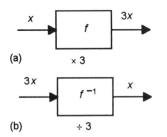

Figure 1.18 *(a) The function which multiplies by 3, (b) the inverse function which divides by 3*

Key point

A function can only have an inverse if there is a one-to-one relationship between the input to a function and its output. Some functions have inverses for just some part of their domain. For example, the function $y = f(x) = x^2$ with an input of $+1$ or -1 gives $y = +1$. Thus if we take the inverse we do not know whether the result should be $+1$ or -1. unless we place some resriction on the domain, e.g. restrict it to just positive values of x.

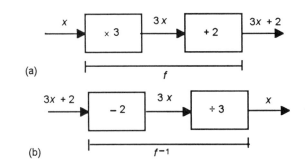

Figure 1.19 *(a) The functions which multiply by 3 and then add 2, (b) the inverse functions which subtract 2 and then divide by 3*

The above illustrations are rather basic functions. We will investigate more complex ones later. However, the basic rules still apply and, once understood, will provide a solid foundation from which to build more complex relationships.

Example

If $f(x) = 2x$, what is the inverse function?

The initial function $f(x) = 2x$ multiples by 2. Therefore, to reverse the process we simply divide by 2. Thus, $f^{-1}(x) = 1/2x$.

Example

If $f(x) = 2x + 3$, what is the inverse function?

$f(x) = 2x + 3$ involves doubling the input and then adding 3. The inverse is thus subtracting 3 from the input and then halving. Thus the inverse function is:

$$f^{-1}(x) = \frac{x-3}{2}$$

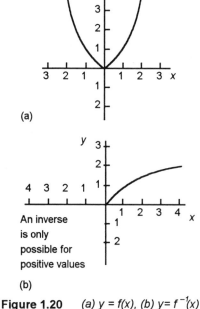

(a)

An inverse is only possible for positive values

(b)

Figure 1.20   (a) $y = f(x)$, (b) $y = f^{-1}(x)$

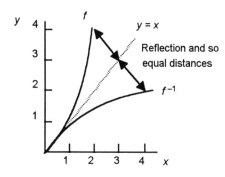

Figure 1.21   *The inverse as a reflection of function in line $y = x$*

1.4.1 Graphs of f and f^{-1}

We can use the above rules for a function and its inverse to find the graph of an inverse function from a graph of the function. Consider the graph of $y = f(x)$ shown in Figure 1.20(a). This is the graph described by the equation $y = x^2$. What is the graph of the inverse function $f^{-1}(x)$? This will be the graph of $y = \sqrt{x}$ (Figure 1.20(b)) since the function $\sqrt{x}$ is what we need to apply to undo the function x^2.

If we examine the two graphs we find that the inverse f^{-1} is just the reflection of the graph of f in the line $y = x$ (Figure 1.21). This is true for any function when it possesses an inverse.

Problems 1.4

1   Determine the inverses of:

   (a) $f(x) = 5x - 3$, (b) $f(x) = 4 + x$, (c) $f(x) = x^3$,
   (d) $f(x) = 2x^3 - 1$.

2   Does the function $f(x) = x^2$, have an inverse for all real values of x?

3   For each of the following functions, restrict the domain so that there is an inverse and then determine it:

   (a) $f(x) = (x - 1)^2$, (b) $f(x) = (x + 1)^2 - 4$.

1.5 Circular functions

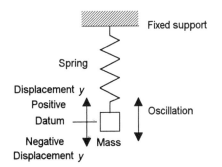

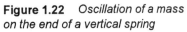

Figure 1.22 *Oscillation of a mass on the end of a vertical spring*

This section focuses on the so-called circular or trigonometric functions. Such functions are widely used in engineering. Thus in describing oscillations, whether mechanical of perhaps a vibrating beam or electrical and alternating current, the equations used to define the quantity which fluctuates with time is likely to involve a trigonometric function.

As an illustration, consider the mechanical oscillation of a mass on the end of a spring when is just vibrates up-and-down when the mass is given a vertical displacement (Figure 1.22). With very little damping, the mass will oscillate up-and-down for quite some time. Figure 1.23 shows how the displacement varies with time.

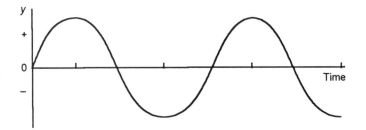

Figure 1.23 *Displacement y variation with time for the oscillating mass when damping is virtually absent*

If we look at the situation when there is noticeable damping present, then the displacement variation with time looks more like Figure 1.24. The difference between this graph and Figure 1.23 is that, though we have a similar form of graph, the effect of the damping is that the amplitude decreases with time.

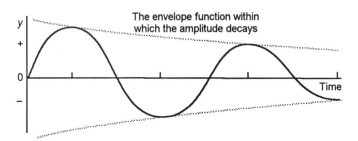

Figure 1.24 *Displacement y variation with time for the oscillating mass when damping is noticeable*

We can derive an equation to represent the variation of displacement with time in the absence of damping by using a simple model. Suppose we draw a circle with a radius OA equal to the amplitude of the oscillation, i.e. the maximum displacement, and consider a point P moving round the circle with a constant angular velocity ω (Figure 1.25) and starting from the horizontal. The vertical projection of the rotating radius OP gives a displacement–angle graph. Since the radius OP is rotating with a constant angular velocity, the angle rotated is proportional to the time. The result is a graph which replicates that of the undamped oscillating mass.

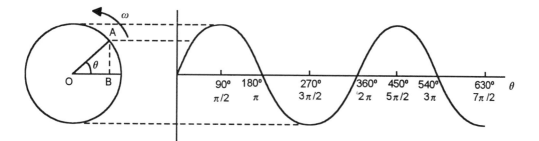

Figure 1.25 *The vertical projection AB of the rotating radius gives a displacement–time graph*

Key points

The convention used for angles is that they are referenced from zero degrees and when measured in an anticlockwise direction are termed positive angles. One unit for angles is degrees (Figure 1.26(a)). However, when angles occur in equations in engineering, it is usual to describe them in radians. One complete rotation of a radius is a rotation through 360°.

One radian is the angle swept so that the arc formed is the same length as the radius, hence since one complete sweep of a radius is an arc length of $2\pi r$, then one complete rotation is 2π radians (Figure 1.26(b)). Thus 360° = 2π radians (rad) and so 1 rad = 360/2π = 57.3°. Hence:

$$\theta_{radians} = \frac{\theta_{degrees}}{57.3}$$

$$\theta_{degrees} = \theta_{radians} \times 57.3$$

Since AB/OA = sin θ we can write:

$$AB = OA \sin \theta$$

where AB is the vertical height of the line at some instant of time, OA being its length. The maximum value of AB will be OA and occur when $\theta = 90°$. But a constant angular velocity ω means that in a time t the angle θ covered is ωt. Thus the vertical projection AB of the rotating line will vary with time and is described by the equation:

$$AB = OA \sin \omega t$$

If y is the displacement of the alternating mass and A the amplitude of its oscillation, the equation can be written as:

$$y = A \sin \omega t \qquad [13]$$

This type of oscillation is called *simple harmonic motion*.

It is usual to give angular velocities in units of radians per second, an angular rotation through 360° being a rotation through 2π radians. Since the periodic time T is the time taken for one cycle of a waveform, then T is the time taken for OA to complete one revolution, i.e. 2π radians. Thus:

$$T = \frac{2\pi}{\omega}$$

The frequency f is $1/T$ and so $\omega = 2\pi f$. Because ω is just 2π times the frequency, it is often called the *angular frequency*. The frequency f has units of hertz (Hz) or cycles per second and thus the angular frequency has units of s^{-1}. We can thus write the above equation as:

$$y = A \sin 2\pi f t \qquad [14]$$

We can use a similar model to describe alternating current; in this case the rotating radius OP is called a *phasor*. Thus the current i is related to its maximum value I by:

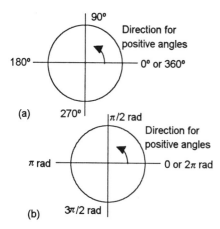

Figure 1.26 *Conventions for angles in degrees and radians*

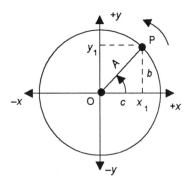

Figure 1.27 *Defining circular functions*

Figure 1.28 *Defining trig. ratios*

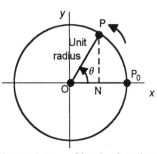

Figure 1.29 *Circular functions*

$i = I \sin \omega t$

For the damped oscillation, the amplitude decreases with time so we must figure out how to quantify this 'decay' and link it somehow with the basic sine wave function of the undamped system. As we will later discover, the damped oscillation is actually described by a combination of a sine function and an exponential function.

The circular functions

We can define the circular functions, i.e. the sine, cosine and tangent, in terms of the rotation of a radial arm of length A (it often represents the amplitude) in a circular path (Figure 1.27). Thus, we can define the sine of an angle as:

$$\sin \theta = \frac{b}{a}$$

But, with $b = y_i - 0$, then $\sin \theta = y_1/A$ and so:

$$y_1 = A \sin \theta \qquad [15]$$

This relationship now enables us to calculate the vertical side of the triangle Ox_1P, or the y-coordinate of point P.

Likewise, we can define the cosine of an angle as:

$$\cos \theta = \frac{c}{A}$$

But, with $c = x_1 - 0$, then $\cos \theta = x_1/A$ and so:

$$x_1 = A \cos \theta \qquad [16]$$

This relationship now enables us to calculate the horizontal side of the triangle Ox_1P, or the x-coordinate of point P.

We can define the tangent of an angle in terms of the gradient of the line OP as:

$$\tan \theta = \frac{b}{c} = \frac{y_1}{x_1} \qquad [17]$$

Using equations [15] and [16] we can write equation [17] as:

$$\tan \theta = \frac{A \sin \theta}{A \cos \theta} = \frac{\sin \theta}{\cos \theta} \qquad [18]$$

With reference to Figure 1.27, as the point P moves around the circle, so the angle θ changes. The trigonometrical ratios can be defined in terms of the angles in a right-angled triangle. However, the above definitions allow us to define them for all angles, not just those which are 90° or less. Because they are defined in terms of a circle, they are termed *circular functions*.

Consider the motion of a point P around a unit radius circle (Figure 1.29). P_0 is the initial position of the point and P the position to which it has rotated. The radial arm OP in moving

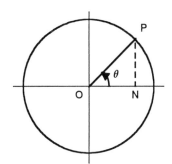

Figure 1.30 *First quadrant*

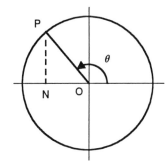

Figure 1.31 *Second quadrant*

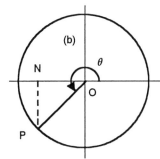

Figure 1.32 *Third quadrant*

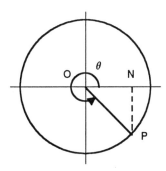

Figure 1.33 *Fourth quadrant*

from OP_0 has swept out an angle θ. The angle θ is measured between the radial arm and the OP_0 axis as a positive angle when the arm rotates in an anticlockwise direction. Since the circle has a unit radius, to obtain for angles up to 90° the same result as the trigonometric ratios defined in terms of the right-angled triangle, the perpendicular height NP defines the sine of the angle P_0OP and the horizontal distance ON defines the cosine of the angle P_0OP.

Consider the circular rotations for angles in each quadrant (note: the first quadrant is angles 0° to 90°, the second quadrant 90° to 180°, the third quadrant 180° to 270°, the fourth quadrant 270° to 360°):

1   *Angles between 0 and 90°*
    When the radial arm OP is in the first quadrant (Figure 1.30) with $0 \leq \theta < \pi/2$, $0 \leq \theta < 90°$, both NP and ON are positive. Thus both the sine and the cosine of angle θ are positive. Since the tangent is NP/ON then the tangent of angle θ is positive. For example, sin 30° = +0.5, cos 30° = +0.87 and tan 30° = +0.58.

2   *Angles between 90° and 180°*
    When the radial arm OP moves into the second quadrant (Figure 1.31) with $\pi/2 \leq \theta < \pi$, $90° \leq \theta < 180°$, NP is positive and ON negative. Thus the sine of angle θ is positive and the cosine negative. Since the tangent is NP/ON then the tangent of angle θ is negative. For example, sin 120° = 0.87, cos 120° = −0.50 and tan 120° = −1.73.

3   *Angles between 180° and 270°*
    When the radial arm moves into the third quadrant (Figure 1.32) with $\pi \leq \theta < 3\pi/2$, $180° \leq \theta < 270°$, NP is negative and ON negative. Thus the sine of angle θ is negative and the cosine negative. Since the tangent is NP/ON then the tangent of angle θ is positive. For example, sin 210° = −0.5, cos 210° = −0.87 and tan 210° = +0.58.

4   *Angles between 270° and 360°*
    When the radial arm is in the fourth quarter (Figure 1.33) with $3\pi/2 \leq \theta < 2\pi$, $270° \leq \theta < 360°$, NP is negative and ON positive. Thus the sine of angle θ is negative and the cosine positive. Since the tangent is NP/ON then the tangent of angle θ is negative. For example, sin 300° = −0.87, cos 300° = 0.5 and tan 300° = −1.73.

We can now summarise with Figure 1.34 as an aid to memory.

For angles greater than 2π rad (360°), the radial arm OP simply rotates more than one revolution. Negative angles are interpreted as a clockwise movement of the radial arm from OP_0.

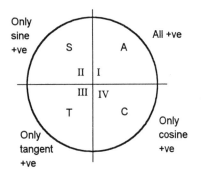

Figure 1.34 *Circular functions in the four quadrants*

Key point

Mathematicians call the sine function an 'odd' function. An *odd* function is defined as one which has:

$f(-x) = -f(x)$

An *even* function has:

$f(-x) = f(x)$

Cyclic functions

A cyclic function is one which repeats itself on a cyclic period. Thus, if we have a function $y = f(x)$ which is cyclic and repeats itself after a time T, then:

$$f(x) = f(x + T) = f(x + 2T) = \text{etc.} \qquad [19]$$

T is termed the *periodic time* and is the time taken to complete one cycle. Hence, if the frequency is f then f cycles are completed each second and so $T = 1/f$.

As the arm OP in Figure 1.35 rotates round-and-round its circular path, the values of its vertical projection NP is cyclic and generates the sine graph shown. Since the graph describes a periodic function of period 2π, then:

$$\sin \theta = \sin (\theta + 2\pi n) \qquad [20]$$

where $n = 0, \pm 1, \pm 2$, etc.

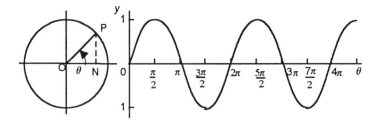

Figure 1.35 *Graph of $y = \sin \theta$*

Note that if OP rotates in a clockwise direction, i.e. the negative direction, then as θ is negative, this generates the sine function continued to the left of the origin O into the negative region (Figure 1.36). For negative values of θ, the sine function has the same values as the positive values except for a change in sign:

$$\sin (-\theta) = -\sin \theta$$

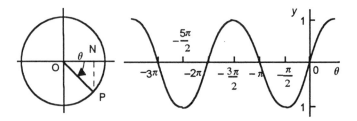

Figure 1.36 *Sine graph for negative angles*

To obtain the graph of $\cos \theta$ as the radial arm OP rotates round-and-round its circular path, we read off the values of its horizontal projection ON. Figure 1.37 shows the result. Since the graph describes a periodic function of period 2π, then:

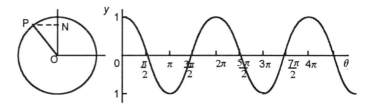

Figure 1.37 *Graph of y = cos θ*

$$\cos \theta = \cos (\theta + 2\pi n) \qquad [21]$$

where $n = 0, \pm 1, \pm 2$, etc.

Note that the graph of $y = \sin \theta$ is the same as that of $y = \cos \theta$ moved $\frac{1}{2}\pi$ to the right, while that of $y = \cos \theta$ is the same as $y = \sin \theta$ moved $\frac{1}{2}\pi$ to the left, i.e. $\sin \theta = \cos (\theta - \frac{1}{2}\pi)$ and $\cos \theta = \sin (\theta + \frac{1}{2}\pi)$.

In the projections of the radial arm OP to generate the sine or cosine graphs, we have let OP have the value of 1. If we consider a radial arm of length A, we have the same function but multiplied by A, i.e. $y = A \sin \theta$. The amplitude of the waveform is changed. To illustrate this look at the following functions and their graphs as plotted to the same scale and on the same axes (Figure 1.38): $y = 1 \sin \theta$ with amplitude $A = 1$, $y = 4 \sin \theta$ with amplitude $A = 4$ and $y = 0.5 \sin \theta$ with amplitude $A = 0.5$.

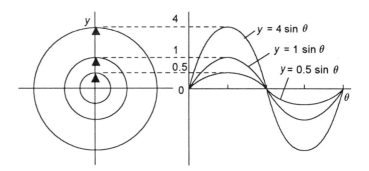

Figure 1.38 *Effect of changing A in y = A sin θ, only the amplitude of the graph waves is changed*

In engineering, we often encounter functions of the general form:

$$y = A \sin (\theta \pm \phi) \text{ or } y = A \cos (\theta \pm \phi) \qquad [22]$$

ϕ is the initial angle we start the rotating radial arm OP at and, as a consequence, ϕ is the angle by which the sine or cosine graph is moved to the left when positive and to the right when negative. It defines a phase shift of the complete waveform. Figure 1.39 illustrates this by showing the effect of a phase shift of $\pi/3$, i.e. 60°.

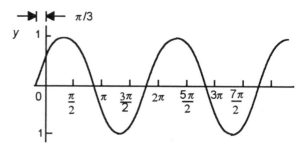

Figure 1.39 *Graph of y = sin (θ + π/3), showing the effect of a phase shift of π/3, i.e. 60°, as being to shift the graph to the left by that amount*

Now consider the graph of the function $y = \tan x$. For the radial arm OP rotating in a circle, the tangent is PN/OP (Figure 1.40). But if we draw a tangent to the circle at P_0 then, for a unit radius circle the tangent of the angle is P_0M. When the radius arm has moved to an angle between 90° and 180° then the tangent is P_0M_1. The graph describes a periodic function which repeats itself every period of π (not every 2π as for a sine or cosine function). Thus:

$$\tan \theta = \tan (\theta + \pi n) \tag{23}$$

for $n = 0, \pm1, \pm2$, etc.

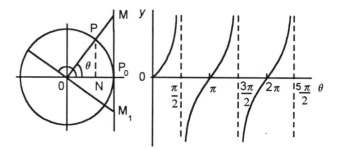

Figure 1.40 *y = tan θ*

Example

Draw graphs of $y = \cos \theta$ and $y = \cos 2\theta$ on the same axis and comment on how they differ.

A simple way to sketch the graphs is to formulate a table for values between $\theta = 0°$ and $\theta = 360°$ for $\cos \theta$ and $\cos 2\theta$, then plot the respective curves. So we have:

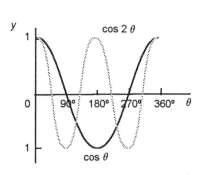

Figure 1.41 *Graphs of y = cos θ and y = cos 2 θ*

θ	0	30	45	60	90	120	135	150	180
2θ	0	60	90	120	180	240	270	300	360
$\cos \theta$	1.0	0.9	0.7	0.5	0	−0.5	−0.7	−0.9	−1.0
$\cos 2\theta$	1.0	0.5	0	−0.5	−1.0	−0.5	0	0.5	1.0

θ	210	225	240	270	300	315	330	360
2θ	420	450	480	540	600	630	660	720
$\cos \theta$	−0.9	−0.7	−0.5	0	0.5	0.7	0.9	1.0
$\cos 2\theta$	0.5	0	−0.5	−1.0	−0.5	0	0.5	1.0

Figure 1.41 shows the resulting graphs.

Example

Sketch the function y = 5 sin (θ + 30°) for values of θ between 0° and 360°.

The equation indicates that the waveform has an amplitude of 5 and a phase shift of +30°. Figure 1.42 shows the form of the function.

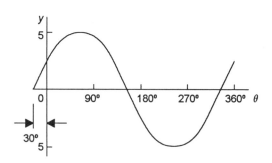

Figure 1.42 *Graph of y = 5 sin (θ + 30°)*

Maths in action

To illustrate a simple mechanical application, consider a piston head moving cyclically without damping, i.e. without friction, and represented by the spring–mass system shown in Figure 1.43. The spring represents the restoring or elastic driving force acting on the piston head.

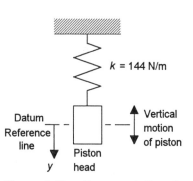

Figure 1.43 *Representation of a piston head*

We can find some basic properties of the system if we apply cyclic functions. We assume that the motion can be represented by the y-component of a rotating radial arm which rotates with a constant angular velocity and so is given by $y = A \sin \theta$, where y is the vertical displacement at a time t. Since $\theta = \omega t$, we can write:

$y = A \sin \omega t$

From mechanical theory, the angular frequency ω is:

$$\omega = \sqrt{\frac{k}{m}}$$

where k is the spring constant and m the mass. Hence $\omega = \sqrt{(144/4)} = 6$ rad/s and so $y = A \sin 6t$. We thus have y at a maximum when $\sin 6t = 1$, i.e. when $6t = 1.57$ and so $t = 0.26$ s.

If the maximum displacement, i.e. amplitude, of the piston is 0.1 m, we have $y = 0.1 \sin 6t$. Thus, at $t = 3$ s we have $y = 0.1 \sin 6(3) = 0.1 \sin 18 = 0.1$. Since 18 rad = 1031° then $y = 0.1(-0.75) = -0.075$ m. The minus sign indicates that the displacement is in the upward direction from the datum line.

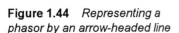

Figure 1.44 *Representing a phasor by an arrow-headed line*

Phasors

A sinusoidal alternating current can be represented by the equation $i = I \sin \omega t$, where i is the current at time t and I the maximum current. In a similar way we can write for a sinusoidal alternating voltage $v = V \sin \omega t$, where v is the voltage at time t and V the maximum voltage. Thus we can think of an alternating current and voltage in terms of a model in which the instantaneous value of the current or voltage is represented by the vertical projection of a line rotating in an anticlockwise direction with a constant angular velocity. The term *phasor*, being an abbreviation of the term phase vector, is used for such rotating lines. The length of the phasor can represent the maximum value of the sinusoidal waveform (or the generally more convenient root-mean-square value, the maximum value is proportional to the root-mean-square value). The line representing a phasor is drawn with an arrowhead at the end that rotates and is drawn in its position at time $t = 0$, i.e. the phasor represents a frozen view of the rotating line at one instant of time of $t = 0$ (Figure 1.44).

Alternating currents or voltages which do not always start with zero values at time $t = 0$ and can be represented in general by:

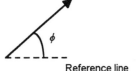

Reference line

Figure 1.45 *Phasor with phase angle ϕ*

$$i = I \sin (\omega t + \phi) \text{ or } v = V \sin (\omega t + \phi) \qquad [24]$$

The phasor for such alternating currents or voltages is represented by a phasor (Figure 1.45) at an angle ϕ to the reference line, this line being generally taken as being the horizontal. The angle ϕ is termed the *phase angle*. We can describe such a phasor by merely stating its magnitude and phase angle (the term used is polar coordinates). Thus $2\angle 40°$ A describes a phasor with a magnitude, represented by its length, of 2 A and with a phase angle of 40°.

In discussing alternating current circuits we often have to consider the relationship between an alternating current through a

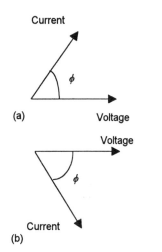

Current

(a) Voltage

Voltage

ϕ

Current

(b)

Figure 1.46   *(a) Current leading voltage, (b) current lagging voltage*

component and the alternating voltage across it. If we take the alternating voltage as the reference and consider it to be represented by a horizontal voltage phasor, then the current may have some value at that time and so be represented by another phasor at some angle ϕ. There is said to be a *phase difference* of ϕ between the current and the voltage. If ϕ has a positive value then the current is said to be *leading* the voltage, if a negative value then *lagging* the voltage (Figure 1.46).

Example

A sinusoidal voltage has a maximum value of 10 V and a frequency of 100 Hz. If the voltage has a phase angle of 30°, what will be the instantaneous voltage at times of (a) $t = 0$, (b) $t = 0.5$ ms?

The equation for the sinusoidal voltage will be:

$v = V_m \sin (2\pi f t + \phi)$

The term $2\pi f t$, i.e. ωt, is in radians. Thus, for consistency, we should express ϕ in radians. An angle of 30° is $\pi/6$ radians. Thus:

$v = 10 \sin (2\pi \times 100t + \pi/6)$ volts

It should be noted that it is quite common in engineering to mix the units of radians and degrees in such expressions. Thus you might see:

$v = 10 \sin (2\pi \times 100t + 30°)$ volts

However, when carrying out calculations involving the terms in the bracket there must be consistency of the units.

(a) When $t = 0$ then: $v = 10 \sin \pi/6 = 5$ V.

(b) When $t = 0.5$ ms then:

$v = 10 \sin (2\pi \times 100 \times 0.5 \times 10^{-3} + \pi/6)$ volts

and so $v = 10 \sin 0.838 = 7.43$ V.

1.5.2 Manipulating circular functions

Often in working through engineering problems, it is necessary to rearrange circular functions in a different format. This section looks at how we can do this.

Key points

The cosecant, secant and cotangent ratios are defined as the reciprocals of the sine, cosine and tangent:

$$\csc \theta = \frac{1}{\sin \theta}$$

$$\sec \theta = \frac{1}{\cos \theta}$$

$$\cot \theta = \frac{1}{\tan \theta}$$

Figure 1.47   Right-angled triangle

Key points

$\sin (A + B) = \sin A \cos B + \cos A \sin B$

$\sin (A - B) = \sin A \cos B - \cos A \sin B$

$\cos (A + B) = \cos A \cos B - \sin A \sin B$

$\cos (A - B) = \cos A \cos B + \sin A \sin B$

$\tan (A + B) = \frac{\tan A + \tan B}{1 - \tan A \tan B}$

$\tan (A - B) = \frac{\tan A - \tan B}{1 + \tan A \tan B}$

$2 \sin A \cos B = \sin (A + B) + \sin (A - B)$

$2 \cos A \cos B = \cos (A + B) + \cos (A - B)$

$2 \sin A \sin B = \cos (A - B) - \cos (A + B)$

$2 \cos A \sin B = \sin (A + B) - \sin (A - B)$

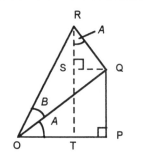

Figure 1.48   Compound angle

Trigonometric relationships

For the right-angled triangle shown in Figure 1.47, the *Pythagoras theorem* gives $AB^2 + BC^2 = AC^2$. Dividing both sides of the equation by AC^2 gives:

$$\left(\frac{AB}{AC} \right)^2 + \left(\frac{BC}{AC} \right)^2 = 1$$

Hence:

$$\cos^2 \theta + \sin^2 \theta = 1 \qquad [25]$$

Dividing this equation by $\cos^2 \theta$ gives:

$$1 + \tan^2 \theta = \sec^2 \theta \qquad [26]$$

and dividing equation [25] by $\sin^2 \theta$ gives:

$$\cot^2 \theta + 1 = \csc^2 \theta \qquad [27]$$

Example

Simplify $\dfrac{\cos \theta}{1 - \sin \theta} + \dfrac{\cos \theta}{1 + \sin \theta}$.

$$\frac{\cos \theta}{1 - \sin \theta} + \frac{\cos \theta}{1 + \sin \theta} = \frac{\csc \theta (1 + \sin \theta) + \cos \theta (1 - \sin \theta)}{(1 - \sin \theta)(1 + \sin \theta)}$$

$$= \frac{2 \cos \theta}{1 - \sin^2 \theta} = \frac{2 \cos \theta}{\cos^2 \theta} = 2 \sec \theta$$

Trigonometric ratios of sums of angles

It is often useful to express the trigonometric ratios of angles such as $A + B$ or $A - B$ in terms of the trigonometric ratios of A and B. In such situations the relationships shown in Key points prove useful.

As an illustration of how we can derive such relationships, consider the two right-angled triangles OPQ and OQR shown in Figure 1.48:

$$\sin(A + B) = \frac{TR}{OR} = \frac{TS + SR}{OR} = \frac{PQ + SR}{OR} = \frac{PQ}{OQ} \frac{OQ}{OR} + \frac{SR}{QR} \frac{Q}{O}$$

Hence:

$$\sin (A + B) = \sin A \cos B + \cos A \sin B \qquad [28]$$

If we replace B by $-B$ we obtain:

$$\sin (A - B) = \sin A \cos B - \cos A \sin B \qquad [29]$$

Key points

$\sin 2A = 2 \sin A \cos A$

$\cos 2A = \cos^2 A - \sin^2 A$

$\quad = 1 - 2\sin^2 A = 2\cos^2 A - 1$

$\tan 2A = \dfrac{2\tan A}{1 - \tan^2 A}$

If in equation [28] we replace A by $(\pi/2 - A)$ we obtain:

$$\cos (A + B) = \cos A \cos B - \sin A \sin B \qquad [30]$$

If in equation [29] we replace B by $-B$ we obtain:

$$\cos (A - B) = \cos A \cos B + \sin A \sin B \qquad [31]$$

We can obtain tan $(A + B)$ by dividing sin $(A + B)$ by cos $(A + B)$ and likewise tan $(A - B)$ by dividing sin $(A - B)$ by cos $(A - B)$. By adding or subtracting equations from above we obtain the relationships such as $2 \sin A \cos B$.

If, in the above relationships for the sums of angles A and B we let $B = A$ we obtain the double-angle equations shown in Key points.

Example

Solve the equation cos $2x$ + 3 sin x = 2.

Using equation [46] for cos $2x$ gives:

$1 - 2\sin^2 x + 3 \sin x = 2$

This can be rearranged as:

$2 \sin^2 x - 3 \sin x + 1 = 0$

$(2 \sin x - 1)(\sin x - 1) = 0$

Hence sin x = ½ or 1. For angles between 0° and 90°, x = 30° or 90°.

Example

In an alternating current circuit, the instantaneous voltage v is given by $v = 5 \sin \omega t$ and the instantaneous current i by $i = 10 \sin (\omega t - \pi/6)$. Find an expression for the instantaneous power P at a time t given $P = vi$.

As $P = vi$ we have:

$P = 5 \sin \omega t \{10 \sin (\omega t - \pi/6)\} = 50 \sin \omega t \{\sin (\omega t - \pi/6)\}$

Using $2 \sin A \sin B = \cos (A - B) - \cos (A + B)$ gives:

$P = 25\{\cos (\pi/6) - \cos (2\omega t - \pi/6)\}$

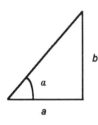

Figure 1.49 *Right-angled triangle*

Key points

$a \sin \theta + b \cos \theta = R \sin (\theta + a)$

$a \sin \theta - b \cos \theta = R \sin (\theta - a)$

with:

$R = \sqrt{a^2 + b^2}$ and $\tan a = \frac{a}{b}$

$a \cos \theta + b \sin \theta = R \cos (\theta - a)$

$a \cos \theta - b \sin \theta = R \cos (\theta + a)$

with:

$R = \sqrt{a^2 + b^2}$ and $\tan a = \frac{b}{a}$

---

a cos θ + b sin θ

Sometimes it is useful to write an equation of the form $a \sin \theta + b \cos \theta$ in the form $R \sin (\theta - a)$. We can do this by using the trigonometric formula for compound angles, e.g. equation [29] for $\sin (A - B)$. Thus:

$R \sin (\theta - a) = R(\sin \theta \cos a + \cos \theta \sin a)$

Hence, we require:

$R(\sin \theta \cos a + \cos \theta \sin a) = a \sin \theta + b \cos \theta$

Therefore, comparing coefficients of the $\sin \theta$ terms:

$R \cos a = a$

and comparing coefficients of the $\cos \theta$ terms:

$R \sin a = b$

Dividing these two equations gives

$$\tan a = \frac{a}{b} \qquad [32]$$

This leads us to be able to describe the angle a by the right-angled triangle shown in Figure 1.49. Hence:

$$R = \sqrt{a^2 + b^2} \qquad [33]$$

Thus:

$$a \sin \theta + b \cos \theta = R \sin (\theta + a) \qquad [34]$$

The Key points show other relationships which can be derived in a similar way.

Example

Express $3 \cos \theta + 4 \sin \theta$ in the form (a) $R \cos (\theta - a)$, (b) $R \sin (\theta + a)$.

(a) We can derive it by using the double-angle formula $\cos (A - B) = \cos A \cos B + \sin A \sin B$. Thus:

$3 \cos \theta + 4 \sin \theta = R (\cos \theta \cos a + \sin \theta \sin a)$

Thus $3 = R \cos a$ and $4 = R \sin a$. Hence $\tan a = 4/3$ and so $a = 53.1°$ or 0.93 rad. $R = \sqrt{(3^2 + 4^2)} = 5$. Hence:

$3 \cos \theta + 4 \sin \theta = 5 \cos (\theta - 0.93)$

(b) We can derive it by directly using the double-angle relationship $\sin(A + B) = \sin A \cos B + \cos A \sin B$. Thus:

$3 \cos \theta + 4 \sin \theta = R(\sin \theta \cos a + \cos \theta \sin a)$

Thus $3 = R \sin a$ and $4 = R \cos a$. Hence $\tan a = 3/4$ and so $a = 36.9°$ or 0.64 rad. $R = \sqrt{(3^2 + 4^2)} = 5$. Hence:

Example

Express $6 \sin \omega t - 2.5 \cos \omega t$ in the form $R \sin(\omega t + a)$.

Using the double angle formula $\sin(A + B) = \sin A \cos B + \cos A \sin B$:

$6 \sin \omega t - 2.5 \cos \omega t = R(\sin \omega t \cos a + \cos \omega t \sin a)$

Comparing coefficients of $\sin \omega t$ gives $6 = R \cos a$ and of $\cos \omega t$ gives $-2.5 = R \sin a$. Thus $\tan a = -2.5/6$. The negative sign for the sine and the tangent means that the angle must be in the fourth quadrant (see Figure 1.34). Hence $a = -0.39$ rad. $R = \sqrt{(6^2 + 2.5^2)} = 6.5$ and so:

$6 \sin \omega t - 2.5 \cos \omega t = 6.5 \sin(\omega t - 0.39)$

Thus, by subtracting the waveform $2.5 \cos \omega t$ from $6 \sin \omega t$ we end up with a waveform of amplitude 6.5 and a phase shift of -0.39 rad.

Example

Two sinusoidal alternating voltages of $v_1 = 1.25 \sin \omega t$ and $v_2 = 1.60 \cos \omega t$ are combined. Show that the result is a voltage of $v = 2 \sin(\omega t + 52°)$.

$v = v_1 + v_2 = 1.25 \sin \omega t + 1.60 \cos \omega t$

Using $\sin(A + B) = \sin A \cos B + \cos A \sin B$, then:

$2 \sin(\omega t + 52°) = 2(\sin \omega t \cos 52° + \cos \omega t \sin 52°)$

$= 1.23 \sin \omega t + 1.57 \cos \omega t$

With the accuracy to which the result was quoted, the case is proved.

Adding phasors

Often in alternating current circuits we need to add the voltages across two components in series. We must take account of the possibility that the two voltages may not be in phase, despite having the same frequency since they are supplied by the same source. This means that if we consider the phasors, they will rotate with the same angular velocity but may have different lengths and start with a phase angle between them. Consider one of the voltages to have an amplitude V_1 and zero phase angle (Figure 1.50(a)) and the other an amplitude V_2 and a phase difference of ϕ from the first voltage (Figure 1.50 (b)). We can obtain the sum of the two by adding the two graphs, point-by-point, to obtain the result shown in Figure 1.50(c). Thus at the instant of time indicated in the figures, the two voltages are v_1 and v_2. Hence the total voltage is $v = v_1 + v_2$. We can repeat this for each instant of time and hence end up with the graph shown in Figure 1.50(c).

However, exactly the same result is obtained by adding the two phasors by means of the *parallelogram rule* of vectors. If we place the tails of the arrows representing the two phasors together and complete a parallelogram, then the diagonal of that parallelogram drawn from the junction of the two tails represents the sum of the two phasors. Figure 17.16(c) shows such a parallelogram and the resulting phasor with magnitude V.

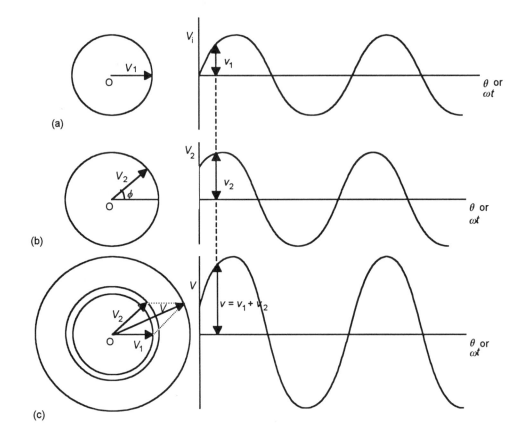

Figure 1.50 *Adding two sinusoidal signals of the same frequency*

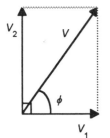

Figure 1.51   *Adding two phasors*

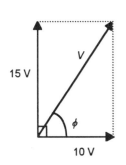

Figure 1.52   *Example*

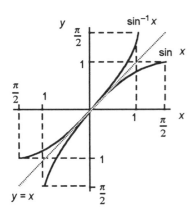

Figure 1.53   *sin x and its inverse*

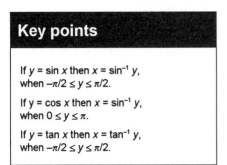

Key points

If $y = \sin x$ then $x = \sin^{-1} y$, when $-\pi/2 \le y \le \pi/2$.

If $y = \cos x$ then $x = \sin^{-1} y$, when $0 \le y \le \pi$.

If $y = \tan x$ then $x = \tan^{-1} y$, when $-\pi/2 \le y \le \pi/2$.

If the phase angle between the two phasors of sizes V_1 and V_2 is 90°, as in Figure 1.51, then the resultant can be calculated by the use of the Pythagoras theorem as having a size V of:

$$V^2 = V_1^2 + V_2^2 \qquad [35]$$

and is at a phase angle ϕ relative to the phasor for V_2 of:

$$\tan \phi = \frac{V_2}{V_1} \qquad [36]$$

Example

Two sinusoidal alternating voltages are described by the equations of $v_1 = 10 \sin \omega t$ volts and $v_2 = 15 \sin (\omega t + \pi/2)$ volts. Determine the sum of these voltages.

Figure 1.52 shows the phasor diagram for the two voltages. The angle between the phasors is $\pi/2$, i.e. 90°. We could determine the sum from a scale drawing or by calculation using the Pythagoras theorem. Thus:

$$(\text{sum})^2 = 10^2 + 15^2$$

Hence the magnitude of the sum of the two voltages is 18.0 V. The phase angle is given by:

$$\tan \phi = \frac{15}{10}$$

Hence $\phi = 56.3°$ or 0.983 rad. Thus the sum is an alternating voltage described by a phasor of amplitude 18.0 V and phase angle 56.3° (or 0.983 rad). This alternating voltage is thus described by:

$v = 18.0 \sin (\omega t + 0.983)$ volts.

1.5.4 The inverse circular functions

If $\sin x = 0.8$ what is the value of x? This requires the inverse being obtained. There is an inverse if the function is one-to-one or restrictions imposed to give this state of affairs. However, the function $y = \sin x$ gives many values of x for the same value of y. To obtain an inverse we have to restrict the domain of the function to $-\pi/2$ to $+\pi/2$. With that restriction $y = \sin x$ has an inverse. The inverse function is denoted as $\sin^{-1} x$ (sometimes also written as arcsin x). Note that the -1 is *not* a power here but purely notation to indicate the inverse. If $\sin x = 0.8$ then the value of x that gives this sine is the inverse and so $x = \sin^{-1} 0.8$, i.e. $x = 53°$. Figure 1.53 shows the graphs for $\sin x$ and its inverse function. In a similar way we can define inverses for cosines and tangents.

Problems 1.5

1 State the amplitude and phase angle (with respect to $y = 5 \sin \theta$) of the function $y = 5 \sin (\theta + 30°)$.

2 A cyclic function used to describe a rotating radius (phasor) is defined by the equation $y = 4 \sin 3t$. What is the amplitude and the angular frequency of the function?

3 State the amplitude, period and phase angle for the following cyclic functions:

(a) $2 \sin (5t + 1)$, (b) $6 \cos 3t$, (c) $5 \cos\left(\dfrac{2t+1}{3}\right)$,

(d) $2 \cos (t - 0.6)$

4 State the amplitude, period and phase angle for the following cyclic functions:

(a) $6 \sin (2t + 1)$, (b) $2 \cos 9t$, (c) $5 \cos\left(\dfrac{2t-1}{5}\right)$,

(d) $2 \cos (t - 0.2)$, (e) $5 \sin (4t + \pi/8)$, (f) $½ \sin (t - \pi/0.6)$

5 The potential difference across a component in an electrical circuit is given by the equation $v = 40 \sin 40\pi t$. Deduce the maximum value of the potential difference and its frequency.

6 A sinusoidal voltage has a maximum value of 1 V and a frequency of 1 kHz. If the voltage has a phase angle of 60°, what will be the instantaneous voltage at times of (a) $t = 0$, (b) $t = 0.5$ ms?

7 A sinusoidal alternating current has an instantaneous value i at a time t, in seconds, given by $i = 100 \sin (200\pi t - 0.25)$ mA. Determine (a) the maximum current, (b) the frequency, (c) the phase angle.

8 A sinusoidal alternating voltage has an instantaneous value v at a time t, in seconds, given by $v = 12 \sin (100\pi t + 0.5)$ volts. Determine (a) the maximum voltage, (b) the frequency, (c) the phase angle.

9 What is the value of v, when $t = 30$ μs, for an amplitude-modulated radio wave with a voltage v in volts which varies with time t in seconds and is defined by the equation $v = 50(1 + 0.02 \sin 2400\pi t) \sin (200 \times 10^3 \pi t)$.

10 Show that $\sin (A + B + C) = \cos A \cos B \cos C (\tan A + \tan B + \tan C - \tan A \tan B \tan C)$.

11 Find the values of x between 0 and 360° which satisfy the condition $8 \cos x + 9 \sin x = 7.25$.

12 Write $5 \sin \theta + 4 \cos \theta$ in the forms (a) $R \sin (\theta - a)$, (b) $R \cos (\theta + a)$.

13 Express $W (\sin a + \mu \cos a)$ in the form $R \cos (a - \beta)$ giving the values of R and $\tan \beta$. Also show that the maximum value of the expression is $W\sqrt{(1 + \mu^2)}$ and that this occurs when $\tan a = 1/\mu$.

14 Write the following functions in the form $R \sin (\omega t + a)$:
(a) $3 \sin \omega t + 4 \cos \omega t$, (b) $4.6 \sin \omega t - 7.3 \cos \omega t$,

(c) $-2.7 \sin \omega t - 4.1 \cos \omega t$

15 Express $3 \sin \theta + 5 \cos \theta$ in the form $R \sin (\theta + a)$ with a measured in degrees.

16 Write the following functions in the form $R \sin (\omega t \pm a)$:
(a) $4 \sin \omega t - 3 \cos \omega t$, (b) $-7 \sin \omega t + 4 \cos \omega t$,
(c) $-3 \sin \omega t - 6 \cos \omega t$

17 The currents in two parallel branches of a circuit are $10 \sin \omega t$ milliamps and $20 \sin (\omega t + \pi/2)$ milliamps. What is the total current entering the parallel arrangement?

18 The voltage across a component in a circuit is $5.0 \sin \omega t$ volts and across another component in series with it $2.0 \sin (\omega t + \pi/6)$ volts. Determine the total voltage across both components.

19 The sinusoidal alternating voltage across a component in a circuit is $50\sqrt{2} \sin (\omega t + 40°)$ volts and across another component in series with it $100\sqrt{2} \sin (\omega t - 30°)$ volts. What is the total voltage across the two components?

20 The currents in two parallel branches of a circuit are $4\sqrt{2} \sin \omega t$ amps and $6\sqrt{2} \sin (\omega t - \pi/3)$ amps. What is the total current entering the parallel arrangement?

21 Determine the value in radians of:
(a) $\sin^{-1} 0.74$, (b) $\cos^{-1} 0.10$, (c) $\tan^{-1} 0.80$, (d) $\sin^{-1} 0.40$

1.6 Exponential functions

There are many situations in engineering where we are concerned with functions which grow or decay with time, e.g.

- The variation with time of the temperature of a cooling object.

- The variation with time of the charge on a capacitor when it is being charged and when it is being discharged.

- The variation with time of the current in a circuit containing inductance when the current is first switched on and then when it is switched off.

- The decay with time of the radioactivity of a radioactive isotope.

This section is about the equations we can use to describe such growth or decay.

Exponentials

In general, we can describe growth and decay processes by an equation of the form:

$$y = a^t \qquad\qquad [37]$$

where a is some constant called the *base*, and y the value of the quantity at a time t. Thus, for growth, we might have 2^t, 3^t, 4^t, etc. and for decay 2^{-t}, 3^{-t}, 4^{-t}, etc. We could write equations for growth or decay processes with different values of the base. However, we usually standardise the base to one particular value. The most widely used form of equation is e^x, where e is a constant with the value 2.728 281 828 ... Whenever an engineer refers to an exponential change he or she is almost invariably referring to an equation written in terms of e^x. Why choose this strange number 2.718... for the base? The reason is linked to the properties of expressions written in this way. For $y = e^x$, the rate of change of y with x, i.e. the slope of a graph of y against x, is equal to e^x (this is discussed in more detail in the chapter concerned with differentiation):

$$\text{slope of graph of } y \text{ against } x = y = e^x \qquad [38]$$

and there are many engineering situations where this property occurs.

A simple illustration of the above is given if we take a strip of paper and cut it into half, throwing away one of the halves. We then take the half strip and cut it into half, throwing away one of the halves. If we keep on repeating this procedure we obtain the graph shown in Figure 1.54(a). This is an exponential decay in the length of the paper. Now look at the change in length per tear, i.e. the 'gradient' of the graph, Figure 1.54(b). We have the same exponential function. A similar type of relationship exists in the discharge of a charged capacitor. The charge on the capacitor decreases exponentially with time and the rate of change of charge, i.e. the current, follows the same exponential decay.

The following shows the values of e^x and e^{-x} for various values of x and Figure 1.55 the resulting graphs

$y = e^x$	0.14	0.37	1	2.72	7.39	20.09	... infinite
x	−2	−1	0	1	2	3	... infinite

$y = e^{-x}$	7.39	2.72	1	0.37	0.14	0.05	... 0
x	−2	−1	0	1	2	3	... infinite

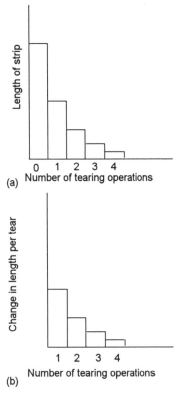

(a)

Length of strip

Number of tearing operations

(b)

Change in length per tear

Number of tearing operations

Figure 1.54 *An 'exponential decay'*

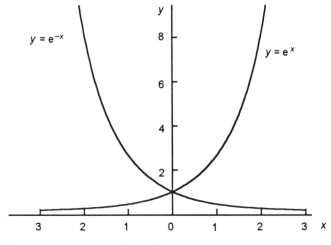

Figure 1.55 $y = e^x$ and $y = e^{-x}$

The e^x graph describes a growth curve, the e^{-x} a decay curve. Note that both graphs have $y = 1$ when $x = 0$.

In a more general form we can write the exponential equation in the form $y = e^{kx}$, or $y = e^{-kx}$, where k is some constant. This constant k determines how fast y changes with x. The following data illustrates this:

x	0	1	2	3	... infinite
$y = e^{-1x}$	1	0.368	0.135	0.050	... 0
$y = e^{-2x}$	1	0.135	0.018	0.003	... 0

x	0	1	2	3	... infinite
$y = e^{1x}$	1	2.718	7.389	20.086	... infinite
$y = e^{2x}$	1	7.389	54.598	403.429	... infinite

The bigger k is the faster y decreases, or increases, with x.

When $x = 0$ then for $y = e^{kx}$, or $y = e^{-kx}$, $y = e^0$ and so $y = 1$. This is thus the value of y that occurs when x is zero. Since we may often have an initial value other than 1, we write the equation in the form:

$$y = A\, e^{kx} \qquad [39]$$

where A is the initial value of y at $x = 0$. For example, for the discharging of a capacitor in an electrical circuit we have, for the charge q on the capacitor at a time t, the equation:

$$q = Q_0\, e^{-t/CR} \qquad [40]$$

When $t = 0$ then $q = Q_0$. The constant k is $1/CR$. The bigger the value of CR the smaller the value of $1/CR$ and so the slower the rate at which the capacitor becomes discharged.

One form of equation involving exponentials that is quite common is of the form:

$$y = A - A\, e^{-kx} \qquad [41]$$

When $x = 0$ then $e^0 = 1$ and so $y = A - A = 0$. The initial value is thus 0. As x increases then e^{-kx} decreases from 1 towards 0, eventually becoming zero when x is infinite. Thus the value of y increases as x increases. When x is very large then e^{-kx} becomes virtually 0 and so y becomes equal to A. Figure 8.4 shows the graph. It shows a quantity y which increases rapidly at first and then slows down to become eventually A.

For example, for a capacitor which starts with zero charge on its plates and is then charged we have the equation:

$$q = Q_0 - Q_0\, e^{-t/CR} \qquad [42]$$

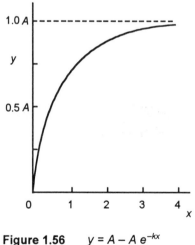

Figure 1.56 $y = A - A\, e^{-kx}$

When $t = 0$ then $e^0 = 1$ and so $q = Q_0 - Q_0 = 0$. As t increases, so the value of $e^{-t/CR}$ decreases and so q becomes more and more equal to Q_0.

Example

For an object cooling according to Newton's law, the temperature θ of the object varies with time t according to the equation $\theta = \theta_0\, e^{-kt}$, where θ_0 and k are constants. (a) Explain why this equation represents a quantity which is decreasing with time. (b) What is the value of the temperature at $t = 0$? (c) How will the rates at which the object cools change if in one instance $k = 0.01$ and in another $k = 0.02$ (the units of k are per °C)?

(a) If we assume that t and k will only have positive values, then the $-kt$ means that the power is negative and so the temperature decreases with time.
(b) When $t = 0$ then $e^{-kt} = 1$ and so $\theta = \theta_0$. Thus θ_0 is the initial value at the time $t = 0$.
(c) Doubling the value of k means that the object will cool faster, in fact it will cool twice as fast.

Example

The current i in amperes in an electrical circuit varies with time t according to the equation $i = 10(1 - e^{-t/0.4})$. What will be (a) the initial value of the current when $t = 0$, (b) the final value of the current at infinite time?

(a) When $t = 0$ then $e^{-t/0.4} = e^0 = 1$. Thus $i = 10(1 - 1) = 0$.
(b) When t becomes very large then $e^{-t/0.4}$ becomes 0. Thus we have $i = 10(1 - 0)$ and so the current becomes 10 A.

Maths in action

Time constant

Consider the discharging of a charged capacitor through a resistance (Figure 1.57). The voltage v_C across the capacitor varies with time t according to the equation $v_C = V\, e^{-RC}$, where V is the initial potential difference across the capacitor at time $t = 0$. Suppose we let $\tau = RC$, calling τ the *time constant* for the circuit. Thus, $v_C = V\, e^{-t/\tau}$. The time taken for v_C to drop from V to $0.5V$ is thus given by:

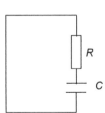

Figure 1.57 *Discharge of a charged capacitor*

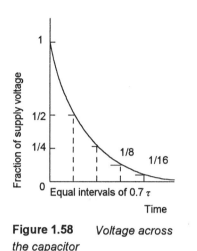

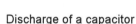

Figure 1.58 caption area:

Figure 1.58 *Voltage across the capacitor*

Discharge of a capacitor

Time	v_C
0	V
0.7T	0.5V
1.4T	0.25V
2.1T	0.125V
2.8T	0.0625V
3.5T	0.03125V

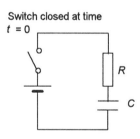

Figure 1.59 *Charging a capacitor*

$$0.5V = V\,e^{-t/\tau}$$

$$e^{-t/\tau} = 0.5$$

$$-\frac{t}{\tau} = \ln 0.5 = -0.693$$

Thus in a time of 0.693τ the voltage will drop to half its initial voltage. The time taken to drop to $0.25V$ is given by:

$$0.25V = V\,e^{-t/\tau}$$

$$e^{-t/\tau} = 0.25$$

$$-\frac{t}{\tau} = \ln 0.25 = -1.386$$

Thus in a time of 1.386τ the voltage will drop to one-quarter of its initial voltage. This is twice the time taken to drop to half the voltage. This is a characteristic of a decaying exponential graph: if t is the time taken to reach half the steady-state value, then in $2t$ it will reach one-quarter, in $3t$ it will reach one-eighth, etc. In each of these time intervals it reduces its value by a half (Figure 1.58).

When $t = 1\tau$ then $v_C = V\,e^{-1} = 0.632V$. Thus in a time equal to the time constant the voltage across the capacitor drops to 63.2% of the initial voltage. When $t = 2\tau$ then $v_C = V\,e^{-2} = 0.135V$. Thus the voltage across the capacitor drops to 13.5% of the initial voltage. When $t = 3\tau$ then $v_C = V\,e^{-3} = 0.050V$. Thus the voltage across the capacitor drops to 5.0% of the initial voltage.

Now consider the growth of the charge on an initially uncharged capacitor when a voltage is switched across it (Figure 1.59). The time constant τ is RC. Thus:

$$v_C = V(1 - e^{-t/RC}) = V(1 - e^{-t/\tau})$$

What time will be required for v_C to reach $0.5V$?

$$0.5V = V(1 - e^{-t/\tau})$$

$$e^{-t/\tau} = 0.5$$

$$-\frac{t}{\tau} = \ln 0.5 = -0.693$$

Thus in a time of 0.693τ the voltage will reach half its steady-state voltage. The time taken to reach $0.75V$ is given by:

$$0.75V = V(1 - e^{-t/\tau})$$

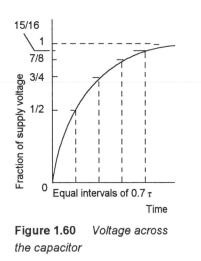

Figure 1.60 *Voltage across the capacitor*

Growth of the p.d. across *C*

Time	v_C
0	0
0.7T	0.5V
1.4T	0.75V
2.1T	0.875V
2.8T	0.938V
3.5T	0.969V

$$e^{-t/\tau} = 0.25$$

$$-\frac{t}{\tau} = \ln 0.25 = -1.386$$

Thus in a time of 1.386τ the voltage will reach three-quarters of its steady-state value. This is twice the time taken to reach half the steady-state voltage. This is a characteristic of exponential graphs: if *t* is the time taken to reach half the steady-state value, then in 2*t* it will reach three-quarters, in 3*t* it will reach seven-eighths, etc. In each successive time interval of 0.7τ the p.d. across the capacitor reduces its value by a half (Figure 1.60).

When $t = 1\tau$ then $v_C = V(1 - e^{-1}) = 0.632V$. Thus in a time equal to the time constant the voltage across the capacitor rises to 63.2% of the steady-state voltage. When $t = 2\tau$ then $v_C = V(1 - e^{-2}) = 0.865V$. Thus the voltage across the capacitor rises to 86.5% of the steady-state voltage. When $t = 3\tau$ then $v_C = V(1 - e^{-3}) = 0.950V$. Thus the voltage across the capacitor rises to 95.0% of the steady-state voltage.

Damped oscillations

In Section 1.5 we considered the vertical oscillations of a mass on the end of a spring (Figure 1.22) with Figure 1.24 showing how the vertical displacement of the mass can be described by a sinusoidal oscillation with an amplitude which decays with time. In the absence of damping the displacement is described by:

$$y = A \sin \omega t$$

where the amplitude *A* is a constant. With the damped oscillation we replace the constant *A* by a term involving exponential decay, i.e.

$$y = C e^{-\zeta \omega t} \sin \omega t \qquad [43]$$

with *C* being a constant and ζ a damping term called the *damping factor*. At zero time the exponential term has the value 1 and so *C* is the initial amplitude. As the time increases so the exponential term becomes smaller and smaller and the amplitude term thus decreases.

1.6.1 Manipulating exponentials

The techniques used for the manipulation of exponentials are the same as those for manipulating powers. The following examples illustrate this.

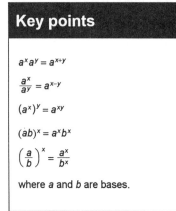

Key points

$a^x a^y = a^{x+y}$

$\dfrac{a^x}{a^y} = a^{x-y}$

$(a^x)^y = a^{xy}$

$(ab)^x = a^x b^x$

$\left(\dfrac{a}{b}\right)^x = \dfrac{a^x}{b^x}$

where *a* and *b* are bases.

Example

Simplify the following:

(a) $e^{2t}e^{4t}$, (b) $(e^{2t})^{-3}$, (c) $\dfrac{e^{5x}}{e^{2x}}$, (d) $\dfrac{10\,e^{t/2}}{2\,e^{t/3}}$, (e) $\dfrac{1}{e^{2t}}+\dfrac{2}{e^{3t}}$

(a) $e^{2t}e^{4t}=e^{2t+4t}=e^{6t}$

(b) $(e^{2t})^{-3}=e^{-6t}$

(c) $\dfrac{e^{5x}}{e^{2x}}=e^{5x-2x}=e^{3x}$

(d) $\dfrac{10\,e^{t/2}}{2\,e^{t/3}}=5\,e^{\frac{t}{2}-\frac{t}{3}}=5\,e^{t/6}$

(e) Bringing the fraction to a common denominator:

$\dfrac{1}{e^{2t}}+\dfrac{2}{e^{3t}}=\dfrac{e^{3t}+2\,e^{3t}}{e^{2t}e^{3t}}=\dfrac{e^{3t}+2\,e^{3t}}{e^{5t}}=e^{-2t}+2\,e^{-3t}$

Alternatively we could take the reciprocals of each term and write the equation as:

$\dfrac{1}{e^{2t}}+\dfrac{2}{e^{3t}}=e^{-2t}+2\,e^{-3t}$

Problems 1.6

1 The number N of radioactive atoms in a sample of radioactive material decreases with time t and is described by the equation $N = N_0\,e^{-\lambda t}$, where N_0 and λ are constants. (a) Explain why this equation represents a quantity which is decreasing with time. (b) What will be the number of radioactive atoms at time $t = 0$? (c) For a radioactive material that decreases only very slowly with time, will λ have a large or smaller value than with a radioactive material which decreases quickly with time?

2 The length L of a rod of material increases from some initial length with the temperature θ above that at which the initial length is measured and is described by the equation $L = L_0\,e^{\alpha\theta}$, where L_0 and a are constants. (a) Explain why the equation represents a quantity which increases with time. (b) What will be the length of the rod when h = 0? (c) What will be the effect of a material having a higher value of a than some other material?

3 For an electrical circuit involving inductance, the current in amperes is related to the time t by the equation $i = 3(1 - e^{-10t})$. What is the value of the current when (a) $t = 0$, and (b) t is very large?

4 What are the values of y in the following equations when (i) $x = 0$, (ii) x is very large, i.e. infinite?

(a) $y = 2\,e^{3x}$, (b) $y = 10\,e^{-5x}$, (c) $y = 2(1 - e^{-2x})$,

(d) $y = 2\,e^{-0.2x}$, (e) $y = -4\,e^{-x/3}$, (f) $y = 0.5(1 - e^{-x/5})$,

(g) $y = 4(1 - e^{-x/2})$, (h) $y = 10\,e^{4x}$, (i) $y = 0.2 - 0.2\,e^{-3x}$

5 The voltage, in volts, across a capacitor is given by $20\,e^{-0.1t}$, where t is the time in seconds. Determine the voltage when t is (a) 1 s, (b) 10 s.

6 The atmospheric pressure p is related to the height h above the ground at which it is measured by the equation $p = p_0\,e^{-h/c}$, where c is a constant and p_0 the pressure at ground level where $h = 0$. Determine the pressure at a height of 1000 m if p_0 is 1.01×10^5 Pa and $c = 70\,000$ (unit m^{-1}).

7 The current i, in amperes, in a circuit involving an inductor in series with a resistor when a voltage is E is applied to the circuit at time $t = 0$ is given by the equation

$$i = \frac{E}{R}(1 - e^{-Rt/L})$$

If R/L has the value 2 Ω/H (actually the same unit as seconds), what is the current when (a) $t = 0$, (b) $t = 1$ s?

8 The voltage v across a resistor in series with an inductor when a voltage E is applied to the circuit at time $t = 0$ is given by the equation $v = E(1 - e^{-t/T})$, where T is the so-called time constant of the circuit. If $T = 0.5$ s, what is the voltage when (a) $t = 0$, (b) $t = 1$ s?

9 The charge q on a discharging capacitor is related to the time t by the equation $q = Q_0\,e^{-t/CR}$, where Q_0 is the charge at $t = 0$, R is the resistance in the circuit and C the capacitance. Determine the charge on a capacitor after a time of 0.2 s if initially the charge was 1 μC (1 μC $= 10^{-6}$ C), R is 1 MΩ and C is 4 μF. Note that with the units in seconds (s), coulombs (C), ohms (Ω) and farads (F), the resulting charge will be in coulombs.

10 The current i, in amperes, in a circuit with an inductor in series with a resistor is given by the equation $i = 4(1 - e^{-10t})$, where the time t is in seconds. Determine the current when (a) $t = 0$, (b) $t = 0.05$ s, (c) $t = 0.10$ s, (d) $t = 0.15$ s, (e) $t = $ infinity.

11 The voltage v, in volts, across a capacitor after a time t, in seconds, is given by the equation $v = 10\,e^{-t/3}$. Determine the value of the voltage v after 2 s.

12 The resistance R, in ohms, of an electrical conductor at a temperature of θ°C is given by the equation $R = R_0\,e^{a\theta}$. Determine the resistance at a temperature of 1000°C if R_0 is 5000 Ω and a is 1.2×10^{-4} (unit per °C).

13 The current i, in amperes, in an electrical circuit varies with time t and is given by the equation $i = 2(1 - e^{-10t})$. Determine the current after times of (a) 0.1 s, (b) 0.2 s, (c) 0.3 s.

14 The amount N of a radioactive material decays with time t and is given by the equation $N = N_0 e^{-0.7t}$, where t is in years. If at time $t = 0$ the amount of radioactive material is 1 g, what will be the amount after five years?

15 The atmospheric pressure p, in pascals, varies with the height h, in kilometres, above sea level according to the equation $p = p_0 e^{-0.15h}$. If the pressure at sea level is 10^5 Pa, what will be the pressure at heights of (a) 1 km, (b) 2 km?

16 The voltage v, in volts, across an inductor in an electrical circuit varies with time t, in milliseconds, according to the equation $v = 200 e^{-t/10}$. Determine the voltage after times of (a) 0.1 ms, (b) 0.5 ms.

17 When the voltage E to a circuit consisting of an inductor in series with a resistor is switched off, the voltage across the inductor varies with time t according to the equation $v = -E e^{-t/T}$, where T is the time constant of the circuit. If $T = 2$ s, determine the voltage when (a) $t = 0$, (b) $t = 1$ s.

18 When a voltage E is applied to a circuit consisting of a capacitor in series with a resistor at time $t = 0$, the voltage v across the capacitor varies with time according to the equation $v = E(1 - e^{-t/T})$, where T is the time constant of the circuit. If $T = 0.1$ s, determine the voltage when (a) $t = 0$, (b) $t = 0.1$ s.

19 The temperature θ, in °C, of a cooling object varies with time t, in minutes, according to $\theta = 200 e^{-0.04t}$. Determine the temperature when (a) $t = 0$, (b) $t = 10$ minutes, (c) t is infinite.

20 Under one set of conditions the amplitude A of the oscillations of a system varies with time t according to the equation $A = A_0 e^{kt}$. Under other conditions the amplitude varies according to the equation $y = A_0 e^{-kt}$. If k is a positive number, how do the oscillations differ?

21 Simplify the following:

(a) $e^3 e^5$, (b) $e^{3t} e^{5t}$, (c) $e^{-5t} e^{3t}$, (d) $(e^{-4t})^3$, (e) $(1 + e^{2t})^2$,

(f) $\dfrac{1}{e^{3t}}$, (g) $\dfrac{e^{3t}}{e^{5t}}$, (h) $\left(\dfrac{1}{e^{4t}}\right)^2$, (i) $\dfrac{10 e^{4t}}{2 e^t}$.

1.7 Log functions

Consider the function $y = 2^x$. If we are given a value of x then we can determine the corresponding value of y. However, suppose we are given a value of y and asked to find the value of x that could have produced it. The inverse function is called the *logarithm function* and is defined, for $y = a^x$ and $a > 0$, as:

$$x = \log_a y \qquad\qquad [44]$$

This is stated as 'log to base a of y equals x'. Thus, if we take an input of x to a function $f(x) = a^x$ and then follow it by the inverse

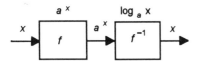

Figure 1.61 $f(x)f^{-1}(x) = x$

Key points

The defining equations for logs is:

$\log_a a^x = x$

Most logarithms use base 10 or base e. Logarithms to base 10 are often just written as log or lg, the base 10 being then understood. Logarithms to base e are termed *natural logarithms* and often just written as ln.

---

function $f^{-1}(x) = \log_a(x)$, as in Figure 1.61, then because it is an inverse we obtain x and so:

$$\log_a a^x = x \tag{45}$$

While logarithms can be to any base, most logarithms use base 10 or base e. Logarithms to base 10 are often just written as log or lg, the base 10 being then understood. Logarithms to base e are termed *natural logarithms* and often just written as ln. Figure 1.61 shows the graph of $y = e^x$ and its inverse of the natural logarithm function.

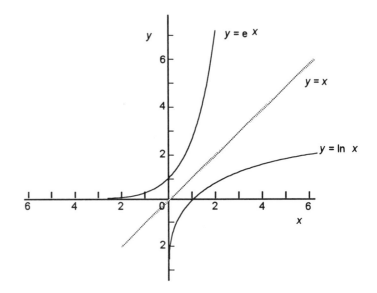

Figure 1.62 *The exponential and its inverse of the natural logarithm function*

Since $a^{A+B} = a^A a^B$ then:

$$\log_a A + \log_a B = \log_a AB \tag{46}$$

$$n \log_a A = \log_a(A^n) \tag{47}$$

Since $a^{A-B} = \dfrac{a^A}{a^B}$ then:

$$\log_a A - \log_a B = \log_a \frac{A}{B} \tag{48}$$

Since $a^1 = a$ then $\log_a a = 1$.

Sometimes there is a need to change from one base to another, e.g. $\log_a x$ to $\log_b x$. Let $u = \log_b x$ then $b^u = x$ and so taking logarithms to base a of both sides gives $\log_a b^u = \log_a x$ and so $u \log_a b = \log_a x$. Since $u = \log_b x$ then $(\log_b x)(\log_a b) = \log_a x$ and so:

$$\log_b x = \frac{\log_a x}{\log_a b}$$

[49]

Example

Write $\lg\left(\dfrac{\sqrt{a}}{bc^3}\right)$ in terms of $\lg a$, $\lg b$ and $\lg c$.

We have:

$$\lg\left(\frac{\sqrt{a}}{bc^3}\right) = \lg \sqrt{a} - \lg(bc^3)$$

Hence:

$$\lg\left(\frac{\sqrt{a}}{bc^3}\right) = \tfrac{1}{2}\lg a - \lg b - 3\lg c$$

Example

Simplify (a) $\lg x + \lg x^3$, (b) $3\ln x + \ln(1/x)$.

(a) $\lg x + \lg x^3 = \lg(x \times x^3) = \lg x^4$
(b) $3\ln x + \ln(1/x) = \ln x^3 + \ln(1/x) = \ln(x^3/x) = \ln x^2$

Example

Solve for x the equation $2^{2x-1} = 12$.

Taking logarithms of both sides of the equation gives:

$$(2x - 1)\lg 2 = \lg 12$$

Hence:

$$2x - 1 = \frac{\lg 12}{\lg 2} = 3.58$$

Thus $x = 2.29$.

The decibel

The power gain of a system is the ratio of the output power to the input power. If we have, say, three systems in series (Figure 1.63) then the power gain of each system is given by:

$$G_1 = \frac{P_2}{P_1},\ G_2 = \frac{P_3}{P_2},\ G_3 = \frac{P_4}{P_3}$$

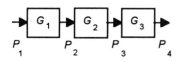

Figure 1.63 *Systems in series*

The overall power gain of the system is P_4/P_1 and is the product of the individual gains, i.e.

$$G = \frac{P_4}{P_1} = \frac{P_2}{P_1} \times \frac{P_3}{P_2} \times \frac{P_4}{P_3} = G_1 \times G_2 \times G_3 \qquad [50]$$

Taking logarithms gives:

$$\lg G = \lg G_1 + \lg G_2 + \lg G_3 \qquad [51]$$

We thus can add the log ratio of the powers. This log of the power ratio was said to be the power ratio in units of the *bel*, named in honour of Alexander Graham Bell:

$$\text{Power ratio in bels} = \lg \frac{\text{power out}}{\text{power in}} \qquad [52]$$

Thus the overall power gain in bels can be determined by simply adding together the power gains in bels of each of the series systems. The bel is an inconveniently large quantity and thus the *decibel* is used:

$$\text{Power ratio in decibels} = 10 \lg \frac{\text{power out}}{\text{power in}} \qquad [53]$$

A power gain of 3 dB is thus a power ratio of 2.0.

Log graphs

When a graph is a straight line then the relationship between the two variables can be stated as being of the form $y = mx + c$ and we can easily determine the constants m and c from the graph and hence obtain the relationship. However, if we have a relationship of the form $y = ax^b$, where a and b are constants, then a plot of y against x gives a non-linear graph from which it is not easy to determine a and b. However, we can write the equation as:

$$\lg y = \lg x^b + \lg a = b \lg x + \lg a \qquad [54]$$

A graph of $\lg y$ against $\lg x$ will thus be a straight line graph with a gradient of b and an intercept of $\lg a$. Likewise, if we have the relationship $y = a\, e^{bx}$ then, taking logarithms to base e:

$$\ln y = \ln e^{bx} + \ln a = bx + \ln a \qquad [55]$$

A graph of $\ln y$ against x will give a straight line graph with a gradient of b and an intercept of $\ln a$.

To avoid having to take the logarithms of quantities, it is possible to use special graph paper which effectively takes the logarithms for you. Figure 1.64 shows the form taken by log-linear and log-log graph paper. On a logarithmic scale, the distance between 1 and 10 is the same as between 10 and 100, each of these distances being termed a cycle.

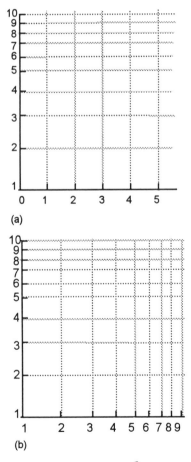

(a)

(b)

Figure 1.64 *(a) Log-linear and (b) log-log graph paper*

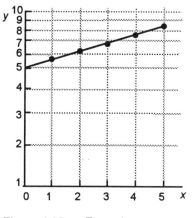

Figure 1.65 *Example*

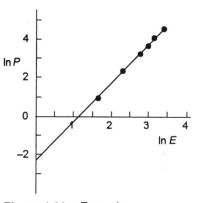

Figure 1.66 *Example*

Example

It is believed that the relationship between y and x for the following data is of the form $y = a\,e^{bx}$. Show that this is the case and determine, using log-linear graph paper, the values of a and b.

y	5.53	6.11	6.75	7.46	8.24
x	1	2	3	4	5

Taking logarithms to base e gives $\ln y = bx + \ln a$. We thus require log-linear graph paper. The y-axis, which is the ln axis, has to range from $\ln 5.53 = 1.7$ to $\ln 8.24 = 2.1$ and so just one cycle from 1 to 10 is required. Figure 1.65 shows the resulting graph. The graph is straight line and so the relationship is valid. The gradient is

$$\text{gradient} = \frac{\ln 8.44 - \ln 5.53}{5 - 1} = \frac{2.13 - 1.71}{4} = 0.10$$

The intercept with the y-axis, i.e. $x = 0$ ordinate, is at 5. Thus the required equation is $y = 5\,e^{0.10x}$.

Key point

Before actually plotting graphs, or creating spreadsheets to plot graphs, it is useful to first sketch the 'form' that the graph might be expected to have in order to get the 'feel' of what the actual plotted graph should look like.

Example

The relationship between power P (in watts), the e.m.f. E (in volts) and the resistance R (in ohms) is thought to be of the form $P = E^n/R$. In a test in which R was kept constant, the following measurements were recorded:

E (volts)	5	10	15	20	25	30
P (watts)	2.5	10	22.5	40	62.5	90

Determine whether the above relationship is true (or approximately so) and determine the values for n and R.

Taking ln of both sides of the equation gives:

$$\ln P = \ln\left(\frac{E^n}{R}\right) = \ln E^n - \ln R = n\ln E - \ln R$$

So, if the relationship is true, a graph of $\ln P$ against $\ln E$ should be a straight line. The values of $\ln P$ and $\ln E$ are:

$\ln E$	1.61	2.30	2.71	3.00	3.22	3.40
$\ln P$	0.92	2.30	3.11	3.69	4.14	4.50

Figure 1.66 shows the plot. From the graph we obtain an intercept on the y-axis of -2.3 and a gradient of about 2.

We thus have $-2.3 = -\ln R$ and so:

$\ln R = 2.3$

$R = e^{2.3}$

and $R = 9.9$, or 10 when rounded up. With $n = 2$ we thus have:

$P = \dfrac{E^2}{10}$

We can test that this is valid by choosing any two results from the test, e.g. $E = 5$ V, $P = 2.5$ W and substituting them into the equation. With $E = 5$ V the equation gives $P = 23/10 = 2.5$ V and so the test confirms the equation.

Maths in action

Radioactive materials, e.g. uranium 235, decay and the mass of that isotope decreases with time. The rate of decay of the isotope is proportional to the mass of isotope present:

rate of decay $= -\lambda m$

where λ is a constant called the decay constant. If m_0 is the mass at time $t = 0$ and mass m the mass at time t, then the following relationship can be derived from the above equation:

$m = m_0\, e^{-\lambda t}$

Taking ln gives:

$\ln m = -\lambda t + \ln m_0$

A graph of $\ln m$ plotted against t will be a straight line graph of slope $-\lambda$ and intercept $+\ln m_0$.

Problems 1.7

1 Simplify (a) $2 \lg x + \log x^2$, (b) $\ln 2x^3 - \ln(4/x^2)$.

2 Write the following in terms of $\lg a$, $\lg b$ and $\lg c$:

(a) $\lg\left(\dfrac{b\sqrt{2}}{ac}\right)$, (b) $\lg\left(\dfrac{ab}{\sqrt{c}}\right)^3$

3 Solve for x the equations: (a) $3^x = 300$, (b) $10^{2-3x} = 6000$, (c) $7^{2x+1} = 4^{3-x}$.

4 The following data indicates how the voltage v across a component in an electrical circuit varies with time t. It is considered that the relationship between V and t might be of the form $v = V e^{-bt}$. Show that this is so and determine the values of V and b.

v in volts	3.75	1.38	0.51	0.19	0.07
t in s	10	20	30	40	50

5 A hot object cools with time. The following data shows how the temperature θ of the object varies with time t. The relationship between θ and t is expected to be of the form $\theta = a e^{-bx}$. Show that this is so and determine the values of a and b.

θ in °C	536	359	241	162	108
t in min	2	4	6	8	10

6 The rate of flow Q of water over a V-shaped notch weir was measured for different heights h of the water above the point of the V and the following data obtained. The relationship between Q and h is thought to be of the form $Q = ah^b$. Show that this is so and determine the values of a and b.

Q in m³/s	0.13	0.26	0.46	2.12	1.07
h in m	0.3	0.4	0.5	0.6	0.7

7 The amplitude A of oscillation of a pendulum decreases with time t and gives the following data. Show that the relationship is of the form $A = a e^{bt}$ and determine the values of a and b.

A in mm	268	180	120	81	54
t in s	20	40	60	80	100

8 The tension T and T_0 in the two sides of a belt driving a pulley and in contact with the pulley over an angle of q is given by the equation $T = T_0 e^{\mu \theta}$. Determine the values of T_0 and μ for the following data:

T in N	69.5	80.8	91.1	109.1	126.7
θ in radians	1.1	1.6	2.0	2.6	3.1

9 In an electrical circuit, the current i in mA occurring when an 8.3 μF capacitor is being discharged varies with time t in ms as shown in the following table:

i (mA)	50.0	17.0	5.8	1.7	0.58	0.24
t (ms)	200	255	310	375	425	475

If I and T are constants, with I being the initial current in mA, show that the above results are connected by the equation $i = I\,e^{t/T}$ and determine I and T.

10 The pressure P at a height h above ground level is given by $P = P_0\,e^{-h/c}$, where P_0 is the pressure at ground level and c is a constant. When P_0 is 1.013×10^5 Pa and the pressure at a height of 1570 m is 9.871×10^4 Pa, determine graphically the value of c.

1.8 Hyperbolic functions

When we want to describe the curve a rope hangs in we use, what is termed, an hyperbolic function. The sine, cosine and tangent are termed circular functions because their definition is associated with a circle. In a similar way, the sinh (pronounced sinch or shine), cosh (pronounced cosh) and tanh (pronounced than or tanch) are *hyperbolic functions* associated with a hyperbola. Sinh is a contracted form of 'hyperbolic sine', cosh of 'hyperbolic cosine' and tanh of 'hyperbolic tangent'. Figure 1.67 shows the comparison of the circular and hyperbolic functions. The hyperbolic functions are defined as:

$$\sinh x = \tfrac{1}{2}(e^x - e^{-x}) \qquad [56]$$

$$\cosh x = \tfrac{1}{2}(e^x + e^{-x}) \qquad [57]$$

$$\tanh x = \frac{\sinh x}{\cosh x} = \frac{e^x - e^{-x}}{e^x + e^{-x}} \qquad [58]$$

Also we have sech x = 1/cosh x, cosech x = 1/sinh x and coth x = 1/tanh x.

Key points

$\sinh x = \tfrac{1}{2}(e^x - e^{-x})$

$\cosh x = \tfrac{1}{2}(e^x + e^{-x})$

$\tanh x = \dfrac{\sinh x}{\cosh x} = \dfrac{e^x - e^{-x}}{e^x + e^{-x}}$

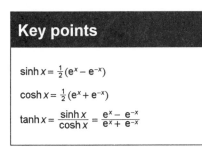

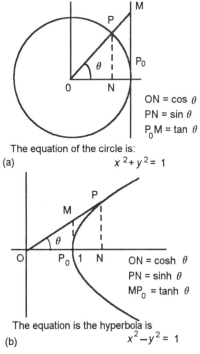

$ON = \cos \theta$
$PN = \sin \theta$
$P_0M = \tan \theta$

The equation of the circle is:
(a) $\quad x^2 + y^2 = 1$

$ON = \cosh \theta$
$PN = \sinh \theta$
$MP_0 = \tanh \theta$

The equation is the hyperbola is
(b) $\quad x^2 - y^2 = 1$

Figure 1.67 *(a) Circular functions, (b) hyperbolic functions*

Example

Determine, using a calculator, the values of (a) cosh 3, (b) sinh 3.

Some calculators have hyperbolic functions so that they can be evaluated by the simple pressing of a key, with others you will have to evaluate the exponentials.

(a) Evaluating the exponentials:

$$\cosh 3 = \tfrac{1}{2}(e^3 + e^{-3}) = 10.07.$$

(b) Evaluating the exponentials:

$$\sinh 3 = \tfrac{1}{2}(e^3 - e^{-3}) = 10.02.$$

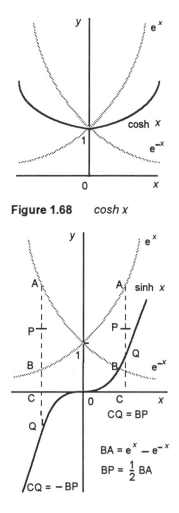

Figure 1.68   *cosh x*

Figure 1.69   *sinh x*

1.8.1 Graphs of hyperbolic functions

Since cosh x is the average value of e^x and e^{-x} we can obtain a graph of cosh x as a function of x by plotting the e^x and e^{-x} graphs and taking the average value. Figure 1.68 illustrates this. Note that unlike cos x, cosh x is not a periodic function. At $x = 0$, cosh $x = 1$. The curve is symmetrical about the y-axis, i.e. cosh $(-x) =$ cosh x and is termed an even function.

To obtain the graph of sinh x from those of e^x and e^{-x}, at a particular value of x we subtract the second from the first and then take half the resulting value. Figure 1.69 illustrates this. Note that unlike sin x, sinh x is not a periodic function. When $x = 0$, sinh $x = 0$. The curve is symmetrical about the origin, i.e. sinh$(-x) = -$sinh x, and is said to be an odd function.

Figure 1.70 shows the graph of tanh x, obtained by taking values of e^x and e^{-x} and calculating values of tanh x for particular values of x. Unlike tan x, tanh x is not periodic. When $x = 0$, tanh $x = 0$. All the values of tanh x lie between -1 and $+1$. As x tends to infinity, tanh x tends to 1. As x tends to minus infinity, tanh x tends to -1. The curve is symmetrical about the origin, i.e. tanh$(-x) = -$tanh x, and is said to be an odd function.

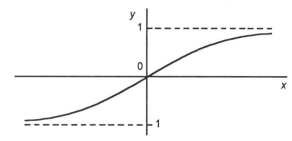

Figure 1.70 *y = tanh x*

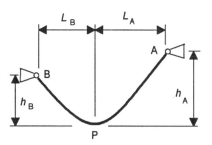

Figure 1.71   *Sagging cable*

Maths in action

Hyperbolic functions and suspended cables

Often it is necessary for engineers to analyse frameworks in order to test for their integrity, i.e. safety and ability to function as designed under a range of conditions. The design engineer needs to formulate a mathematical 'model' which will accurately represent the real system when built. Such problems often involve hyperbolic functions as the following example shows.

Consider a uniform cable which is suspended from two fixing points A and B and which hangs under its own weight (Figure 1.71). Point A is higher than point B and the cable has a uniform weight μ per unit length.

By drawing free-body diagrams for the forces involved on an element of the cable and considering its equilibrium we can arrive at a differential equation (see Chapter 4 for a discussion of such equations), which when solved leads to the equation for the gradient a distance x from P:

$$\text{gradient} = \frac{e^{\mu x/T_0} - e^{-\mu x/T_0}}{2} = \sinh\left[\frac{\mu x}{T_0}\right]$$

T_0 is the horizontal component of the tension in the cable at P. This equation can then give (by integration, see Chapter 4) the height y above P of the cable at distance x as:

$$y = \frac{T_0}{\mu}\cosh\left(\frac{\mu x}{T_0}\right) + k$$

where k is a constant. Since $x = 0$ when $y = 0$, we can put these values in the equation and obtain $k = -T_0/\mu$. Thus:

$$y = \frac{T_0}{\mu}\left[\cosh\left(\frac{\mu x}{T_0}\right) - 1\right]$$

This is the equation of the curve of the cable, known as a catenary. For a full analysis of the system, see the companion book in this series: *Mechanical Engineering Systems* by R. Gentle, P. Edwards and W. Bolton

Problems 1.8

1 Determine, using a calculator, the values of (a) sinh 2, (b) cosh 5, (c) tanh 2, (d) sinh(–2), (e) cosech 1.4, (f) sech 0.8.
2 A flexible cable suspended between two horizontal points hangs in the form of a catenary (Figure 1.72), the equation of the curve being given by $y = c[\cosh(x/c) - 1]$, where y is the sag of the cable, x the horizontal distance from the midpoint to one end of the cable and c is a constant. Determine the sag of a cable when $c = 20$ and $2x = 16$ m.
3 The speed v of a surface wave on a liquid is given by:

$$v = \sqrt{\left[\left(\frac{g\lambda}{2\pi} + \frac{2\pi\gamma}{\rho\lambda}\right)\tanh\frac{2\pi h}{\lambda}\right]}$$

where g is the acceleration due to gravity, λ the wavelength of the waves, γ the surface tension, ρ the density and h the depth of the water. What will the speed approximately be for (a) shallow water waves when h/k tends to zero, (b) deep water waves when h/k tends to infinity?

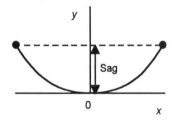

Figure 1.72 *Problem 2*

2 Vectors, phasors and complex numbers

Summary

Vectors are means by which engineers describe quantities which need both a direction and a magnitude specified if their effects are to be ascertained. Vectors play a strong part in the formulation and analysis of mechanical systems, both static and dynamic. Phasors are a means by which sinusoidal alternating voltages and currents can be specified in terms of a rotating radius and an angle, they behaving like vectors. This chapter looks at how we can work with such quantities, considering both vector algebra and complex numbers.

Objectives

By the end of this chapter, the reader should be able to:

- add and subtract vector quantities;
- use vector components to add and subtract vectors;
- use phasors to describe sinusoidal alternating voltages and currents;
- represent phasors by polar notation and be able to work with quantities expressed in this way;
- represent phasors by complex numbers and work with quantities expressed in this way.

2.1 Vectors

Key points

A scalar quantity is defined by purely its magnitude; a vector quantity has to have both its magnitude and direction defined.

If we talk of the mass of this book then we quote just a number, this being all that is needed to give a specification of its mass. However, if we quote a force then in order to fully describe the force we need to specify both its size and the direction in which it acts. Quantities which are fully specified by a statement of purely size are termed *scalars*. Quantities for which we need to specify both size and direction in order to give a full specification are termed *vectors*. Examples of scalar quantities are mass, distance, speed, work and energy. Examples of vector quantities are displacement, velocity, acceleration and force.

To specify a vector we need to specify its magnitude and direction. Thus, we can represent it by a line segment AB (Figure 2.1) with a length which represents the magnitude of the vector and a direction, indicated by the arrow on the segment, which

Figure 2.1   *Representing a vector*

represents the direction of the vector. We can denote this vector representation as

$$\overrightarrow{AB}$$

the arrow indicating the direction of the line segment being from A to B. Note that:

$$\overrightarrow{AB} \neq \overrightarrow{BA}$$

One of the vectors is directed from A to B while the other is directed from B to A. An alternative notation is often used, lower case bold notation **a** being used in print, or underlining $\underline{a}$ in writing. With this notation, if we write **a** or $\underline{a}$ from the vector from A to B then the vector from B to A is represented as −**a** or −$\underline{a}$, the minus sign being used to indicate the vector is in the opposite direction.

The length of the line segment represents the *magnitude* of the vector. This is indicated by the notation:

$$\overline{|AB|} \quad \text{or} \quad |\mathbf{a}| \quad \text{or} \quad a$$

Unit vector

A vector which is defined as having a magnitude of 1 is termed a *unit vector*, such a vector often being denoted by the symbol **â**.

Like vectors

Two vectors are equal if they have the same magnitude and direction. Thus the vectors in Figure 2.2 are equal, even if their locations differ. A vector is only defined in terms of its magnitude and direction, its location is not used in its specification. Thus, for Figure 2.2, we can write:

$$\overrightarrow{AB} = \overrightarrow{CD} \quad \text{or} \quad \mathbf{a} = \mathbf{c} \quad \text{or} \quad \underline{a} = \underline{c}$$

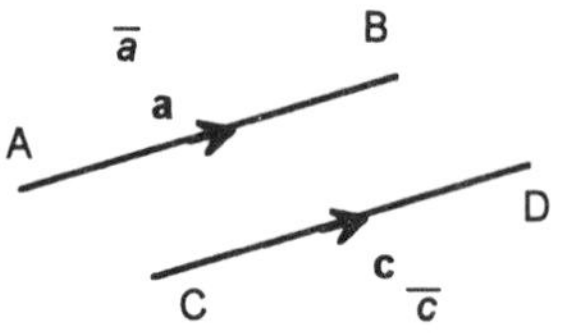

Figure 2.2   *Equal vectors*

Multiplication of vectors by a number

If a vector is multiplied by a positive real number k then the result is another vector with the same direction but with a magnitude that is k times the original magnitude. This is multiplication of a vector by a scalar.

$$k \times \mathbf{a} = k\mathbf{a} \tag{1}$$

We can consider a vector **a** with magnitude $|\mathbf{a}|$ as being a unit vector, i.e. a vector with a magnitude 1, multiplied by the magnitude $|\mathbf{a}|$ (note that the magnitude $|\mathbf{a}|$ is a scalar), i.e.

$$\mathbf{a} = |\mathbf{a}|\hat{\mathbf{a}} \tag{2}$$

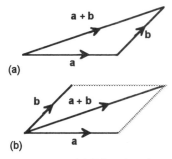

Figure 2.3 *(a) Triangle rule,*
(b) parallelogram rules

Key points

The *triangle rule* can be stated as: to add two vectors **a** and **b** we place the tail of the line segment representing one vector at the head of the line segment representing the other and the line that forms the third side of the triangle represents the vector sum of **a** and **b**.

The *parallelogram* rule can be stated as: to add two vectors **a** and **b** we place the tails of the line segments representing the vectors together and then draw lines parallel to them to complete a parallelogram, the diagonal of the parallelogram drawn from the initial junction of the two tails represents the vector sum of a and b

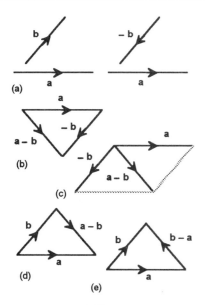

Figure 2.4 *(a) The vectors,*
(b) subtraction by the triangle rule,
(c) subtraction by the parallelogram rule
(d) **a** – **b**, *(e)* **b** – **a**

Maths in action

In the vector analysis of a mechanical system, we can write Newton's 2nd Law of Motion as a vector equation:

F = *m***a**

where **F** is the resultant force acting on a system and **a** is the resulting acceleration. The equation is a vector equation since the direction of the acceleration must be in the same direction as the force; both the force and the acceleration are vector quantities. Newton's 1st Law contains the principle of equilibrium of forces and is used in the following section concerned with the addition and subtraction of vectors; we have to bother about both magnitude and direction to consider equilibrium. Newton's 3rd Law is basic to our understanding of force, stating that forces always occur in pairs with equal in magnitude but opposite in direction forces.

2.1.1 Adding and subtracting vectors

Consider the following situation involving displacement vectors. An aeroplane flies 100 km due west, then 60 km in a north-westerly direction. What is the resultant displacement of the aeroplane from its start point? If the initial displacement vector is **a** and the second displacement vector is **b**, then what is required is the vector sum **a** + **b**.

One way we can determine the sum of two vectors involves the *triangle rule* and is shown in Figure 2.3(a). Note that **a** and **b** have directions that go in one sense round the triangle and the sum **a** + **b** has a direction in the opposite sense. An alternative way of determining the sum involves the *parallelogram rule* and is shown in Figure 2.3(b).

Subtraction of vector **b** from **a** is carried out by adding −**b** to **a**:

$$\mathbf{a} - \mathbf{b} = \mathbf{a} + (-\mathbf{b}) \tag{3}$$

The addition of **a** and −**b** is carried out using the triangle (Figure 2.4(b)) or parallelogram rules (Figure 2.4(c)). Note that, whatever rule we use, the vector **a** – **b** can be represented by the vector from the end point of **b** to the end point of **a** (Figure 2.4(d)), the vector from the end point of **a** to the end point of **b** being **b** – **a** (Figure 2.4(e)).

The triangle rule for the addition of vectors can be extended to the addition of any number of vectors. If the vectors are represented in magnitude and direction by the sides of a *polygon* then their sum is represented in magnitude and direction by the

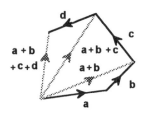

Figure 2.5 *Polygon of vectors*

line segment used to close the polygon (Figure 2.5). Essentially what we are doing is determining the sum of vector 1 and vector 2 using the triangle, then adding to this sum vector 3 by a further triangle and repeating this for all the vectors.

If we have a number of vectors and the vectors give a closed triangle or polygon, then, since the line segment needed to close the figure has zero length, the sum of the vectors must be a vector with no magnitude. This is a statement of equilibrium.

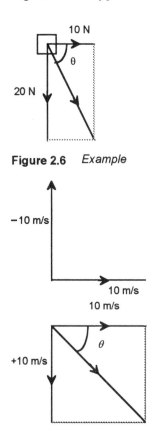

Figure 2.6 *Example*

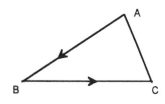

Figure 2.7 *Example*

Example

An object is acted on by two forces, one of which is has a size of 10 N and acts horizontally and the other a size of 20 N which acts vertically. Determine the resultant force.

Figure 2.6 shows the vectors and the use of the parallelogram rule to determine the sum. We can calculate, using the Pythagoras theorem, the diagonal as a having a size of $\sqrt{(20^2 + 10^2)} = 22.4$ N. It is at an angle θ to the horizontal force, with $\theta = \tan^{-1}(20/10) = 63.4°$.

Example

Determine the resultant velocity if we have velocities of 10 m/s acting horizontally to the right and –10 m/s acting vertically upwards.

This problem requires the addition of two vectors, Figure 2.7 showing the vectors and the use of the parallelogram rule to determine the sum. A –10 m/s vector upwards is the same as a +10 m/s vector downwards. Hence, the magnitude of the sum, i.e. the diagonal of the parallelogram, is given by the Pythagoras theorem as $\sqrt{(10^2 + 10^2)} = 14.1$ m/s and it is at an angle below the horizontal of θ where $\theta = \tan^{-1}(10/10) = 45°$.

Figure 2.8 *Example*

Example

For the triangle ABC (Figure 2.8) if **a** is the vector from A to B and **b** the vector from B to C, express the vector from C to A in terms of **a** and **b**.

Using the triangle rule: $\overrightarrow{AC} = \overrightarrow{BC} + \overrightarrow{AB}$.

Since $\overrightarrow{CA} = -\overrightarrow{AC}$, then we have: $\overrightarrow{CA} = -(\mathbf{a} + \mathbf{b})$.

2.1.2 Components

In mechanics a common technique to aid in the solution of problems is to replace a single vector by two components which are at right angles to each other, generally in the horizontal and the vertical directions. Then we can sum all the horizontal components, sum all the vertical components, and are then left with the simple problem of determining the resultant of two vectors at right angles to each other.

For the vector **a** in Figure 2.9 we have **h** and **v** as the horizontal and vertical components. Thus for the magnitudes we must have:

$$|\mathbf{h}| = |\mathbf{a}| \cos \theta \qquad\qquad [4]$$

$$|\mathbf{v}| = |\mathbf{a}| \sin \theta \qquad\qquad [5]$$

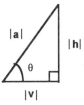

Figure 2.9 *Resolution of a vector into two components*

Example

Express a force of 10 N at 40° to the horizontal in terms of horizontal and vertical components.

Horizontal component = 10 cos 40° = 7.7 N
Vertical component = 10 sin 40° = 6.4 N

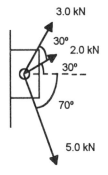

Figure 2.10 *Example*

Example

Determine the resultant force acting on the bracket shown in Figure 2.10 due to the three forces indicated.

For the 3 kN force we have:

horizontal component = 3.0 cos 60° = 1.5 kN
vertical component = 3.0 sin 60° = 2.6 kN

For the 2.0 kN force we have:

horizontal component = 2.0 cos 30° = 1.7 kN
vertical component = 2.0 sin 30° = 1.0 kN

For the 5.0 kN force we have:

horizontal component = 5.0 cos 70° = 1.7 kN
vertical component = –5.0 sin 70° = –4.7 kN

The minus sign is because this force is acting downwards and in the opposite direction to the other vertical components which we have taken as being positive. All the horizontal components are in the same direction. Thus:

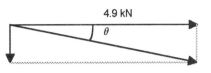

4.9 kN

θ

1.1 kN

Figure 2.11   *Example*

sum of horizontal components = 1.5 + 1.7 + 1.7 = 4.9 kN
sum of vertical components = 2.6 + 1.0 − 4.7 = −1.1 kN

Figure 2.11 shows how we can use the parallelogram rule to find the resultant with these two components. Since the two components are at right angles to each other, the resultant can be calculated using the Pythagoras theorem. Thus, the magnitude of the resultant is:

resultant = $\sqrt{(4.9^2 + 1.1^2)}$ = 5.0 kN

The resultant is at an angle θ downwards from the horizontal given by:

$$\tan \theta = \frac{1.1}{4.9}$$

Thus θ = 12.7°.

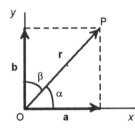

Figure 2.12   *Components*

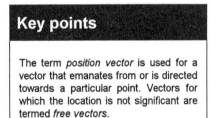

Key points

The term *position vector* is used for a vector that emanates from or is directed towards a particular point. Vectors for which the location is not significant are termed *free vectors*.

A unit vector may be formed by dividing a vector by its magnitude.

Components in terms of unit vectors

A useful way of tackling problems involving summing vectors by considering their components is to write them in terms of unit vectors. Consider the *x-y* plane shown in Figure 2.12. Point P has the coordinates (*x*, *y*) and is joined to the origin O by the line OP. This line from O to P can be considered to be a vector **r** anchored at O and specifying a position, being defined by its two components **a** and **b** along the *x* and *y* directions with:

$$\mathbf{r} = \mathbf{a} + \mathbf{b}$$

If we define **i** to be a unit vector along the *x*-axis then **a** = *a***i**, where *a* is the magnitude of the **a** vector. If we define **j** to be a unit vector along the *y*-axis then **b** = *b***j**, where *b* is the magnitude of the **b** vector. Thus:

$$\mathbf{r} = a\mathbf{i} + b\mathbf{j}$$

But *a* is the *x*-coordinate of P and *b* the *y*-coordinate of P. Thus we can write:

$$\mathbf{r} = x\mathbf{i} + y\mathbf{j} \qquad [6]$$

For example, we might specify a position vector as 3**i** + 2**j**. This would mean a position vector from the origin to a point with the coordinates (3, 2).

The magnitude of the vector **r** is given by the Pythagoras theorem as:

$$|\mathbf{r}| = \sqrt{x^2 + y^2} \qquad [7]$$

If α and β are the angles the vector **r** makes with the *x*- and *y*-axes, then:

$$\cos\alpha = \frac{x}{|\mathbf{r}|} \quad \text{and} \quad \cos\beta = \frac{y}{|\mathbf{r}|} \qquad \text{[8]}$$

These are known as the *direction cosines* of **r**.

Example

If **r** = 4**i** + 7**j** determine |**r**| and the angle **r** makes with the *x*-axis.

$$|\mathbf{r}| = \sqrt{4^2 + 7^2} = 8.1$$

The angle with the *x*-axis is given by:

$$\cos\alpha = \frac{4}{8.1}$$

Thus the angle is 60.4°.

Example

Figure 2.13 shows three forces F_1, F_2 and F_3 all acting at a single point A on a wall bracket. In order to calculate the pulling force on the bracket at the wall, so that it can be safely connected to the wall when under load, determine the size of the force components of F_1, F_2 and F_3 in the *x* and *y* directions.

The components of F_1 in the *x* and *y* directions are:

$$F_{1x} = 600 \cos 35° = 491 \text{ N}$$

$$F_{1y} = 600 \sin 35° = 344 \text{ N}$$

The components of F_2 in the *x* and *y* directions are (the vector forms the hypotenuse of a 3-4-5 triangle):

$$F_{2x} = -500 \,(4/5) = 400 \text{ N}$$

$$F_{2y} = 500 \,(3/5) = 300 \text{ N}$$

The components of F_3 in the *x* and *y* directions are, with α = $\tan^{-1}$ (0.2/0.4) = 26.6°:

$$F_{3x} = 800 \sin 26.6° = 358 \text{ N}$$

$$F_{3y} = -800 \cos 26.6° = 715 \text{ N}$$

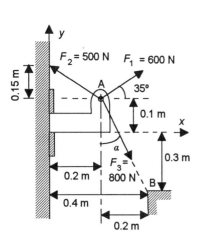

Figure 2.13 *Example*

Alternatively, the size of the components of F_3 may be obtained by writing F_3 at a magnitude times a unit vector r_{AB} in the direction of A to B. The position vector is 0.2i – 0.4j and its magnitude is $\sqrt{(0.2^2 + 0.4^2)} = 0.447$. Thus the unit vector is (0.2i – 0.4j)/0.447 and so:

$F_3 = 800(0.447i – 0.896j) = 358i – 715j$ N

and so $F_{3x} = 358$ N and $F_{3y} = –715$ N.

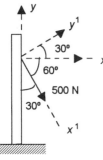

Figure 2.14 *Example*

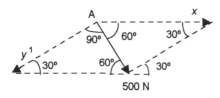

Figure 2.15 *Example*

Key point

The *sine rule*: For a triangle, the length of a side *a* divided by the sine of the opposite angle *A* equals the length of side *b* divided by the sine of its opposite angle *B*.

Example

In a structural test, a 500 N force was applied to a vertical pole, as shown in Figure 2.14. (a) Write the 500 N force in terms of the unit vectors **i** and **j** and identify its *x* and *y* components. (b) Determine the components of the 500 N force along the x^1 and y^1 directions. (c) Determine the components of the 500 N force along the *x* and y^1 directions.

(a) **F** = (500 cos 60°)**i** – (500 sin 60°)**j** = 250**i** – 433**j**

Thus the vector components are **F**x = 250**i** N and **F**y = –433**j** N.

(b) Axis y^1 is at 90° to x^1 and so, since the 500 N is in the x^1 direction we have the component in the x1 direction as 500 N and in the y^1 direction as 0.

(c) Here the required directions are not at right angles to each other and so we determine them by using the parallelogram rule. Figure 2.15 shows the parallelogram. If we use the sine rule:

$$\frac{|\mathbf{F}_x|}{\sin 90°} = \frac{500}{\sin 30°}$$

Hence the size of the *x* component is 1000 N.

$$\frac{|\mathbf{F}_{y^1}|}{\sin 60°} = \frac{500}{\sin 30°}$$

Hence the size of the y^1 component is 866 N. The two components are thus 1000 N and –866 N.

Addition and subtraction of vectors

Consider the addition of the two position vectors $\overrightarrow{OP}$ and $\overrightarrow{OQ}$ shown in Figure 2.16, P having the coordinates (x_1, y_1) and Q the coordinates (x_2, y_2). Thus:

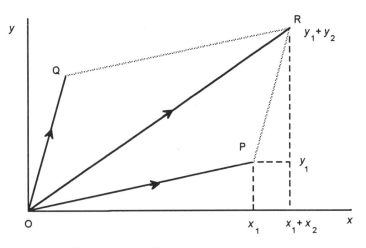

Figure 2.16 *Adding position vectors*

$$\overrightarrow{OP} = x_1\mathbf{i} + y_1\mathbf{j} \quad \text{and} \quad \overrightarrow{OQ} = x_2\mathbf{i} + y_2\mathbf{j}$$

We can obtain the sum by the use of the parallelogram rule as $\overrightarrow{OR}$. R has the coordinates $(x_1 + x_2, y_1 + y_2)$. Thus:

$$\overrightarrow{OR} = (x_1 + x_2)\mathbf{i} + (y_1 + y_2)\mathbf{j} \qquad [9]$$

Key point

Adding or subtracting position vectors is achieved by adding or subtracting their respective co-ordinates.

Example

If **a** = 2**i** + 4**j** and **b** = 3**i** + 5**j**, determine (a) **a** + **b**, (b) **a** − **b**, (c) **a** + 2**b**.

(a) **a** + **b** = (2 + 3)**i** + (4 + 5) **j** = 5**i** + 8**j**
(b) **a** − **b** = (2 − 3)**i** + (4 − 5) **j** = −1**i** + −1**j**
(c) **a** + 2**b** = (2 + 6)**i** + (4 + 10) **j** = 8**i** + 14**j**

Example

ABCD is a square. If forces of magnitudes 1 N, 2 N and 3 N act parallel to AB, BC and CD respectively, in the directions indicated by the order of the letters, determine the magnitude and direction of the resultant force.

Figure 2.17 shows the directions of the forces. Expressing the forces in terms of unit vector components then the force parallel to AB is 1**i**, parallel to BC is 2**j** and that parallel to CD is −3**i**. Thus the resultant is 1**i** + 2**j** − 3**i** = −2**i** + 2**j** N. This will have a magnitude $\sqrt{[(-2)^2 + 2^2]}$ = 2.8 N at an angle of $\tan^{-1}(2/-2)$ = 135° to AB.

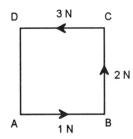

Figure 2.17 *Example*

Example

Forces of 5i − 5j N and −1i + 3j N act on a particle of mass 2 kg. Determine the resulting acceleration.

The resultant force is 5i − 5j − 1i + 3j = 4i − 2j N. Thus:

F = 4i − 2j = *m*a = 2a

Hence **a** = 2i − 1j m/s² and so the acceleration has a magnitude of √[2² + (−1)²] = 2.2 m/s² and is at an angle of tan⁻¹ (−1/2) = −26.6° to the **i** direction.

Vectors in space

Here we extend the consideration of components to three dimensions (Figure 2.18). A vector **r** from O to P, with coordinates (*x*, *y*, *z*), is then defined by its vector components in the three mutually perpendicular directions *x*, *y* and *z*. If **i**, **j** and **k** are the unit vectors in the directions *x*, *y* and *z*, then:

$$\mathbf{r} = x\mathbf{i} + y\mathbf{j} + z\mathbf{k} \qquad [10]$$

The magnitude of **r** is given by:

$$|\mathbf{r}| = \sqrt{x^2 + y^2 + z^2} \qquad [11]$$

The direction of a vector in three dimensions is determined by the angles it makes with the three axes, *x*, *y* and *z*, i.e. the angles *a*, *β* and *γ*. With (*x*, *y*, *z*) the coordinates of the position vector:

$$\cos a = \frac{x}{|\mathbf{r}|}, \quad \cos \beta = \frac{y}{|\mathbf{r}|} \quad \text{and} \quad \cos \gamma = \frac{z}{|\mathbf{r}|} \qquad [12]$$

These are termed the *direction cosines*. As with the two-dimensional case, the basic rule for position vectors is: *adding or subtracting position vectors is achieved by adding or subtracting their respective coordinates.*

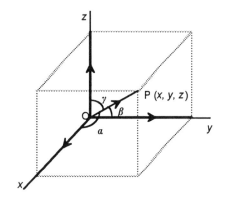

Figure 2.18   *Vector in space*

Example

Determine the magnitude and the direction cosines of the vector **r** = 2i + 3j + 6k.

Magnitude = |**r**| = $\sqrt{2^2 + 3^2 + 6^2}$ = 7

The direction cosines are:

$$l = \cos a = \frac{2}{7}, \quad m = \cos \beta = \frac{3}{7}, \quad n = \cos \gamma = \frac{6}{7}$$

Example

If **a** = 2**i** + 3**j** + 4**k** and **b** = 3**i** – 2**j** + 1**k**, determine (a) **a** + **b**, (b) **a** – **b**, (c) **a** + 2**b**.

(a) **a** + **b** = (2 + 3)**i** + (3 – 2)**j** + (4 + 1)**k** = 5**i** + 1**j** + 5**k**

(b) **a** – **b** = (2 – 3)**i** + (3 + 2)**j** + (4 – 1)**k** = –1**i** + 5**j** + 3**k**

(c) **a** + 2**b** = (2 + 6)**i** + (3 – 4)**j** + (4 + 2)**k** = 8**i** – 1**j** + 6**k**

Problems 2.1

1 If vector **a** is a velocity of 3 m/s in a north-westerly direction and **b** a velocity of 5 m/s in a westerly direction, determine: (a) **a** + **b**, (b) **a** – **b**, (c) **a** – 2**b**.

2 If vector **a** is a displacement of 5 m in a northerly direction and **b** a displacement of 12 m in an easterly direction, determine: (a) **a** + **b**, (b) **a** – **b**, (c) **b** – **a**, (d) **a** + 2**b**.

3 ABCD is a quadrilateral. Determine the single vector which is equivalent to:

(a) $\overrightarrow{AB}+\overrightarrow{BC}$, (b) $\overrightarrow{BC}+\overrightarrow{CD}$, (c) $\overrightarrow{AB}+\overrightarrow{DA}$.

4 If O, A, B, C and D are five points on a plane and $\overrightarrow{OA}$ represents the vector **a**, $\overrightarrow{OB}$ the vector **b**, $\overrightarrow{OC}$ the vector **a** + 2**b**, and $\overrightarrow{OD}$ the vector 2**a** – **b**, express (a) $\overrightarrow{AB}$, (b) $\overrightarrow{BC}$, (c) $\overrightarrow{CD}$, and (d) $\overrightarrow{AC}$ in terms of **a** and **b**.

5 ABCD is a square. A force of 6 N acts along AB, 5 N along BC, 7 N along DB and 9 N along CA. Determine the resultant force.

6 Determine the vector sums of:

(a) $\overrightarrow{AB}+\overrightarrow{BC}+\overrightarrow{CD}$, (b) $\overrightarrow{AB}-\overrightarrow{CB}+\overrightarrow{CD}+\overrightarrow{DE}$,

(c) $\overrightarrow{AB}+\overrightarrow{BC}-\overrightarrow{DC}-\overrightarrow{AD}$, (d) $\overrightarrow{AB}+\overrightarrow{BC}+\overrightarrow{CD}+\overrightarrow{DC}$

7 A point is acted on by two forces, a force of 6 N acting horizontally and a force of 4 N at 20° to the horizontal. Determine the resultant components of the forces in the vertical and horizontal directions.

8 For the following vectors determine their magnitudes and angles to the x-axis: (a) **r** = 2**i** + 3**j**, (b) **r** = 5**i** + 2**j**, (c) **r** = 3**i** + 3**j**.

9 If **a** = –2**i** + 3**j** and **b** = 6**i** + 3**j**, determine: (a) **a** + **b**, (b) **a** – **b**, (c) **a** + 2**b**.

10 If **a** = 5**i** + 2**j** and **b** = 2**i** + 3**j**, determine: (a) **a** + **b**, (b) **a** – **b**, and (c) **a** – 2**b**.

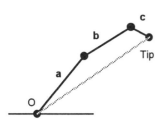

Figure 2.19   *Problem 14*

11 If **a** = 6**i** + 3**j**, **b** = –2**i** + 3**j** and **c** = 5**i** – 4**j**, determine: (a) **a** + **b** + **c**, (b) **a** – **b** – **c**, (c) **a** + 2**b** – 3**c**.

12 Determine the magnitude and direction cosines of: (a) **a** = 3**i** + 7**j** – 4**k**, (b) **a** = 2**i** + 3**j** + 5**k**, (c) **a** = –3**i** + 5**j** + 2**k**.

13 The position vectors of points P and Q are 2**i** + 3**j** – 5**k** and 4**i** – 2**j** + 2**k** respectively. Determine the length and direction cosines of the vector joining P and Q.

14 For a robot arm involving rigid links connected by flexible joints (Figure 2.19), the link vectors can be represented by **a** = 10**i** + 12**j** + 1**k**, **b** = 5**i** – 2**j** + 8**k** and **c** = 2**i** + 1**j** – 4**k**. Determine the position vector of the tip of the robot from O and the length of each link.

15 Determine the angle made by the vector **v** = –5**i** + 12**j** with the positive sense of the *x*-axis.

16 A force is specified by the vector **F** = 60**i** – 60**j** + 30**k**. Calculate the angles made by **F** with the *x*, *y* and *z* axes.

2.2 Phasors

Key point

Polar notation is when quantities such as phasors are described by their size and an angle in the form $V\angle\phi$.

Key point

Note that there is a difference between a phasor diagram and a vector diagram. A phasor diagram represents the phasors at one instant of time, a vector diagram represents the vectors without regard to time. Otherwise the mathematics of handling vectors is applicable to phasors.

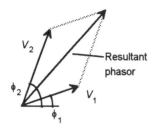

Figure 2.20   *Adding phasors*

A convenient way of specifying a phasor is, what is termed, by *polar notation*. Thus a phasor of length V and phase angle ϕ can be represented by $V\angle\phi$. Although the length of a phasor when described in the way shown in Figure 1.25 represents the maximum value of the quantity, it is more usual to specify the length as representing the root-mean-square value. The root-mean-square value is the maximum value divided by $\sqrt{2}$ and so is just a scaled version of the one drawn using the maximum value. This is because in electrical circuit work we are more usually concerned with the root-mean-square current or voltage than the maximum values.

When we are working in the time domain, i.e. the current or voltage is described by a function as time as in $v = V \sin \omega t$, and want to find, say, the sum of two voltages at some instant of time we just add the voltages. Thus, if we have a voltage across one component described by $v_1 = V_1 \sin (\omega t + \phi_1)$ and across a series component by $v_2 = V_2 \sin (\omega t + \phi_2)$, then the sum of the two voltages is:

$$v_1 + v_2 = V_1 \sin (\omega t + \phi_1) + V_2 \sin (\omega t + \phi_2)$$

This equation describes how the voltage sum varies with time.

When we are working with phasors and want to find the phasor representing the sum of two phasors we have to add the phasors in the same way that vector quantities are added. Thus if we have a voltage across one component described by $V_1\angle\phi_1$ and across a series component by $V_2\angle\phi_2$, then the phasor representing the sum of the two voltages is that indicated in Figure 2.20. While we can draw such diagrams for simple situations and obtain the resultant phasor graphically, a more useful technique is to describe a phasor by a complex number and use the techniques for manipulating complex numbers. In the next section we discuss complex numbers and consider their application to electrical circuit analysis in terms of phasors.

2.3 Complex numbers

If we square the real number +2 we obtain +4, if we square the real number –2 we obtain +4. Thus the square root of +4 is ±2. But what is the square root of –4? To give an answer we need another form of number. If we invent a number j = √–1 (mathematicians often use i rather than j but engineers and scientists generally use j to avoid confusion with i used for current in electrical circuits), then we can write √–4 = √–1 × √4 = ±j2. Thus the solution of the equation $x + 4 = 0$ is x = ±j2.

The solution of a quadratic equation of the form $ax^2 + bx + c = 0$ is given by the formula:

$$x = \frac{-b \pm \sqrt{b^2 - 4ac}}{2a}$$

Thus if we want to solve the quadratic equation $x^2 - 4x + 13 = 0$ then:

$$x = \frac{4 \pm \sqrt{16 - 52}}{2} = 2 \pm \sqrt{-9}$$

We can represent √–9 as √–1 × √+9 = j3. Thus the solution can be written as 2 ± j3, a combination of a real and either plus or minus an imaginary number. Such a pair of roots is known as a *conjugate pair* (see later in this section).

The term complex number is used for the sum of a real number and an imaginary number. Thus a complex number z can be written as $z = a + jb$, where a is the real part of the complex number and b the imaginary part.

Example

Solve the equation $x^2 - 4x + 5 = 0$.

$$x = \frac{4 \pm \sqrt{16 - 20}}{2} = 2 \pm \sqrt{-1} = 2 \pm j$$

The Argand diagram

The effect of multiplying a real number by (–1) is to move the point from one side of the origin to the other. Figure 2.21 illustrates this for (+2) being multiplied by (–1). We can think of the positive number line radiating out from the origin being rotated through 180° to its new position after being multiplied by (–1). But (–1) = j^2. Thus, multiplication by j^2 is equivalent to a 180° rotation. Multiplication by j^4 is a multiplication by (+1) and so is equivalent to a rotation through 360°. On this basis it seems reasonable to take a multiplication by j to be equivalent to a rotation through 90° and a multiplication by j^3 a rotation through 270°. This concept of multiplication by j as involving a rotation is the basis of the use of complex numbers to represent phasors in alternating current circuits.

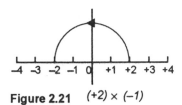

Figure 2.21   (+2) × (–1)

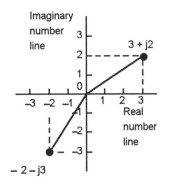

Figure 2.22 *Argand diagram*

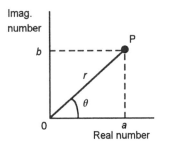

Figure 2.23 *Modulus and argument*

The above discussion leads to a diagram, called the *Argand diagram*, which we use to represent complex numbers. Since rotation by 90° from the *x*-axis on a graph gives the *y*-axis, the *y*-axis is used for imaginary numbers and the *x*-axis for real numbers (Figure 2.22). Figure 2.22 shows how we represent the complex numbers 3 + j2 and –2 – j3 on such a diagram. The line joining the number to the origin is taken as the graphical representation of the complex number.

Modulus and argument

If the complex number $z = a + jb$ is represented on an Argand diagram by the line OP, as in Figure 2.23, then the length *r* of the line OP is called the *modulus* of the complex number and its inclination θ to the real number axis is termed the *argument* of the complex number. The length of the line is denoted by $|z|$ or modulus *z* and the argument by θ or arg *z*.

Using Pythagoras' theorem:

$$|z| = \sqrt{a^2 + b^2} \qquad [13]$$

and, since $\tan \theta = b/a$:

$$\arg z = \tan^{-1}\left(\frac{b}{a}\right) \qquad [14]$$

Since $a = r \cos \theta$ and $b = r \sin \theta$, we can write a complex number *z* as:

$$z = a + jb = r \cos \theta + jr \sin \theta = r(\cos \theta + j \sin \theta) \qquad [15]$$

Thus we can specify a complex number by either stating its location on an Argand diagram in terms of its *Cartesian coordinates a* and *b* or by specifying the modulus, $|z| = r$, and the argument θ. These are termed its *polar coordinates*. The specification in polar coordinates can be written as:

$$z = |z| \angle \arg z \quad \text{or} \quad z = r\angle\theta \qquad [16]$$

Example

Determine the modulus and argument of the complex number 2 + j2.

$$|z| = \sqrt{a^2 + b^2} = \sqrt{2^2 + 2^2} = 2.8$$

$$\arg z = \tan^{-1}\left(\frac{b}{a}\right) = \tan^{-1}\left(\frac{2}{2}\right) = 45°$$

In polar form the complex number could be written as 2.8 $\angle 45°$.

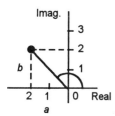

Figure 2.24 *Example*

Example

Write the complex number –2 + j2 in polar form.

$$|z| = \sqrt{a^2 + b^2} = \sqrt{(-2)^2 + 2^2} = 2.8$$

$$\arg z = \tan^{-1}\left(\frac{b}{a}\right) = \tan^{-1}\left(\frac{2}{-2}\right)$$

If we sketch an Argand diagram (Figure 2.24) for this complex number we can see that the number is in the second quadrant. The argument is thus –45° + 180° = 135°. In polar form the complex number could be written as 2.8 ∠135°.

Example

Write the complex number 10 ∠60° in Cartesian form.

$$z = r(\cos\theta + j\sin\theta) = 10(\cos 60° + j\sin 60°) = 5 + j8.7$$

2.3.1 Manipulation of complex numbers

Addition, subtraction, multiplication and division can be carried out on complex numbers in either the Cartesian form or the polar form. Addition and subtraction is easiest when they are in the Cartesian form and multiplication and division easiest when they are in the polar form.

For two complex numbers to be equal, their real parts must be equal and their imaginary parts equal. On an Argand diagram the two numbers then describe the same line. Thus 2 + j3 is *not* equal to 3 + j2 as Figure 2.25 shows.

Addition and subtraction

To add complex numbers we add the real parts and add the imaginary parts:

$$(a + jb) + (c + jd) = (a + c) + j(b + d) \qquad [17]$$

On an Argand diagram, this method of adding two complex numbers is the same as the vector addition of two vectors using the parallelogram of vectors, the line representing each complex number being treated as a vector (Figure 2.26).

To subtract complex numbers we subtract the real parts and subtract the imaginary parts:

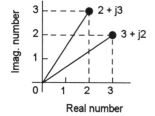

Figure 2.25 *2 + j3 and 3 + j2*

Key point

To add complex numbers, add the real parts and add the imaginary parts. To subtract complex numbers, subtract the real parts and subtract the imaginary parts.

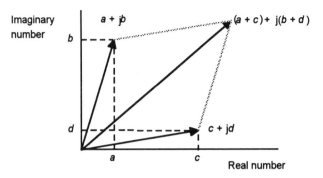

Figure 2.26 *Addition of complex numbers*

$$(a + jb) - (c + jd) = (a - c) + j(b - d) \qquad [18]$$

On an Argand diagram, this method of subtracting two complex numbers is the same as the vector subtraction of two vectors. To subtract a vector quantity you reverse its direction and then add it using the parallelogram of vectors (Figure 2.27).

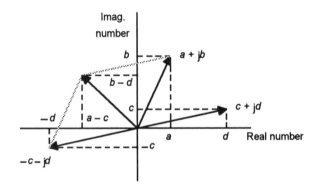

Figure 2.27 *Subtraction of complex numbers*

Example

With $z_1 = 4 + j2$ and $z_2 = 3 + j5$, determine (a) $z_1 + z_2$, (b) $z_1 - z_2$.

(a) $z_1 + z_2 = (4 + 3) + j(2 + 5) = 7 + j7$

(b) $z_1 - z_2 = (4 - 3) + j(2 - 5) = 1 - j3$

Multiplication

Consider the multiplication of the two complex numbers in Cartesian form, $z_1 = a + jb$ and $z_2 = c + jd$. The product z is given by:

$$z = (a + jb)(c + jd) = ac + j(ad + bc) + j^2bd$$

$$= ac + j(ad + bc) - bd \qquad [19]$$

Now consider the multiplication of the two complex numbers in polar form, $z_1 = |z_1|\angle\theta_1$ and $z_2 = |z_2|\angle\theta_2$. Using equation [5] we can write:

$$z_1 = |z_1|(\cos\theta_1 + j\sin\theta_1) \text{ and } z_2 = |z_2|(\cos\theta_2 + j\sin\theta_2)$$

Thus the product z is given by:

$$z = |z_1|(\cos\theta_1 + j\sin\theta_1) \times |z_2|(\cos\theta_2 + j\sin\theta_2)$$

$$= |z_1 z_2| [\cos\theta_1 \cos\theta_2 + j(\sin\theta_1 \cos\theta_2 + \cos\theta_1 \sin\theta_2) + j^2 \sin\theta_1 \sin\theta_2]$$

$$= |z_1 z_2| [(\cos\theta_1 \cos\theta_2 - \sin\theta_1 \sin\theta_2) + j(\sin\theta_1 \cos\theta_2 + \cos\theta_1 \sin\theta_2)]$$

Using the equations for $\cos(A + B)$ and $\sin(A + B)$, [28] and [29] from Chapter 1:

$$z = |z_1 z_2|[\cos(\theta_1 + \theta_2) + j\sin(\theta_1 + \theta_2)] \qquad [20]$$

Hence we can write for the complex numbers in polar form

$$z = |z_1 z_2|\angle(\theta_1 + \theta_2) \qquad [21]$$

Key point

The magnitude of the product of two complex numbers in polar form is the product of the magnitudes of the two numbers and its argument is the sum of the arguments of the two numbers.

Example

Multiply the two complex numbers $2 - j3$ and $4 + j1$.

$$(2 - j3)(4 + j1) = 8 + j2 - j12 - j^2 3$$
$$= 8 + j2 - j12 + 3 = 11 - j10$$

Example

Multiply the two complex numbers $3\angle40°$ and $2\angle70°$.

$$3\angle40° \times 2\angle70° = (3 \times 2)\angle(40° + 70°) = 6\angle110°$$

Complex conjugate

If $z = a + jb$ then the term *complex conjugate* is used for the complex number given by $z* = a - jb$. The imaginary part of the complex number changes sign to give the conjugate, conjugates being denoted as $z*$. Figure 2.28 shows an Argand diagram with a

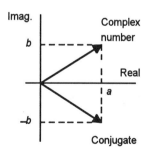

Figure 2.28 *A complex number and its conjugate*

complex number and its conjugate. The complex conjugate is the mirror image of the original complex number.

Consider now the product of a complex number and its conjugate:

$$zz* = (a + jb)(a - jb) = a^2 - j^2b = a^2 + b^2 \qquad [22]$$

The product of a complex number and its conjugate is a real number.

Example

What is the conjugate of the complex number 2 + j4?

The complex conjugate is 2 – j4.

Division

Consider the division of $z_1 = a + jb$ by $z_2 = c + jd$, i.e.

$$z = \frac{z_1}{z_2} = \frac{a + jb}{c + jd}$$

To divide one complex number by another we have to convert the denominator into a real number. This can be done by multiplying it by its conjugate. Thus:

$$z = \frac{a + jb}{c + jd} \times \frac{c - jd}{c - jd} = \frac{(a + jb)(c - jd)}{c^2 + d^2} \qquad [23]$$

Now consider the division of the two complex numbers when in polar form, $z_1 = |z_1| \angle \theta_1$ and $z_2 = |z_2| \angle \theta_2$:

$$z = \frac{|z_1|(\cos \theta_1 + j \sin \theta_1)}{|z_2|(\cos \theta_2 + j \sin \theta_2)}$$

Making the denominator into a real number by multiplying it by its conjugate:

$$z = \frac{|z_1|(\cos \theta_1 + j \sin \theta_1)}{|z_2|(\cos \theta_2 + j \sin \theta_2)} \times \frac{|z_2|(\cos \theta_2 - j \sin \theta_2)}{|z_2|(\cos \theta_2 - j \sin \theta_2)}$$

$$= \frac{|z_1|}{|z_2|} \left[\frac{(\cos \theta_1 + j \sin \theta_1)(\cos \theta_2 - j \sin \theta_2)}{\cos^2 \theta_2 + \sin^2 \theta_2} \right]$$

But $\cos^2 \theta_2 + \sin^2 \theta_2 = 1$ (chapter 3, equation [32]) and so:

$$z = \frac{|z_1|}{|z_2|} [(\cos \theta_1 \cos \theta_2 + \sin \theta_1 \sin \theta_2) + j(\sin \theta_1 \cos \theta_2 - \cos \theta_1 \sin \theta_2)]$$

Using equations cos $(A - B)$ and sin $(A - B)$, [31] and [29] from Chapter 1:

$$z = \frac{|z_1|}{|z_2|}[\cos(\theta_1 - \theta_2) + j\sin(\theta_1 - \theta_2)] \qquad [24]$$

We can express this as:

$$z = \frac{|z_1|}{|z_2|} \angle (\theta_1 - \theta_2) \qquad [25]$$

Example

Divide 1 + j2 by 1 + j1.

$$\frac{1 + j2}{1 + j1} = \frac{1 + j2}{1 + j1} \times \frac{1 - j1}{1 - j1} = \frac{1 + j1 - j^2 2}{1 - j^2}$$

$$= \frac{3 + j1}{2} = 1.5 + j0.5$$

Example

Divide 4∠40° by 2∠30°.

$$\frac{4\angle 40°}{2\angle 30°} = \frac{4}{2} \angle (40° - 30°) = 2\angle 10°$$

2.3.2 Representing phasors by complex numbers

A complex number $z = a + jb$ can be represented on an Argand diagram by a line (Figure 2.29) of length $|z|$ at an angle θ. Thus we can describe a phasor used to represent, say, a sinusoidal voltage, by a complex number in this Cartesian form as:

$$\mathbf{V} = a + jb \qquad [26]$$

An alternative way of describing a complex number, and hence a phasor, is in polar notation, i.e. the length of the phasor and its angle to some reference axis. Thus we can describe it as:

$$\mathbf{V} = V\angle\theta \qquad [27]$$

where V is the magnitude of the phasor and θ its phase angle. The magnitude $|z|$ of a complex number z and its argument θ are given by:

$$|z| = \sqrt{a^2 + b^2} \text{ and } \theta = \tan^{-1}\left(\frac{b}{a}\right) \qquad [28]$$

Since $a = |z|\cos\theta$ and $b = |z|\sin\theta$, then:

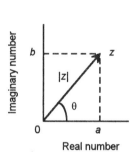

Figure 2.29 *Complex number*

$$z = |z| \cos \theta + j|z| \sin \theta = |z|(\cos \theta + j \sin \theta) \qquad [29]$$

Thus if we have the voltage across one component described by $V\angle\phi$ then we can write this as:

$$\mathbf{V} = V(\cos \phi + j \sin \phi) \qquad [30]$$

Example

Describe the signal $v = 12 \sin (314t + \pi/4)$ V by a phasor.

The phasor has a magnitude, when expressed as the maximum value, of 12 and argument $\pi/4$. Thus we can describe it as $12\angle\pi/4$ V, or by using equation [29] as $12 \cos \pi/4 + j12 \sin \pi/4 = 8.49 + j8.49$ V. If using root-mean-square values then we would have $8.49\angle\pi/4$ r.m.s.V or $6 + j6$ r.m.s.V.

Key point

Since adding or subtracting complex numbers is easier when they are in Cartesian form rather than polar form, when phasors are to be added or subtracted they should be put in Cartesian form.

Adding or subtracting phasors

If we have the voltage across one component described by $V_1\angle\phi_1$ then we can write: $\mathbf{V_1} = V_1(\cos \phi_1 + j \sin \phi_1)$. If we have the voltage across a series component described by $V_2\angle\phi_2$, then: $\mathbf{V_2} = V_2(\cos \phi_2 + j \sin \phi_2)$. The phasor for the sum of the two voltages is then obtained by adding the two complex numbers. Thus:

$$\mathbf{V} = \mathbf{V_1} + \mathbf{V_2} = V_1(\cos \phi_1 + j \sin \phi_1) + V_2(\cos \phi_2 + j \sin \phi_2)$$

$$= (V_1 \cos \phi_1 + V_2 \cos \phi_2) + j(V_1 \sin \phi_1 + V_2 \sin \phi_2) \qquad [31]$$

Subtraction is carried out in a similar manner. Since adding or subtracting complex numbers is easier when they are in Cartesian form rather than polar form, when phasors are to be added or subtracted they should be put in Cartesian form.

Example

A circuit has three components in series. If the voltages across each component are described by phasors 4 V, j2 V and 3 + j4 V, what is the voltage phasor describing the voltage across the three components?

Since the components are in series, the resultant phasor voltage is described by the phasor:

$$V = 4 + j2 + 3 + j4 = 7 + j6$$

Example

A circuit has two components in series. If the voltages across each component are described by phasors $4\angle60°$ V and $2\angle30°$ V, what is the voltage phasor describing the voltage across the two components?

For adding complex numbers it is simplest to convert the phasors into Cartesian notation. Thus:

$\mathbf{V}$ = (4 cos 60° + j4 sin 60°) + (2 cos 30° + j2 sin 30°)

  = 2 + j3.46 + 1.73 + j1 = 3.73 + j4.46 V

If we want this phasor in polar notation then:

$$V = \sqrt{3.73^2 + 4.46^2} = 5.81$$

$$\phi = \tan^{-1}\frac{4.46}{3.73} = 50°$$

Thus the phasor is $5.81\angle50°$ V.

Key point

Multiplication or division of complex numbers is easiest when they are in polar form.

Multiplication or division of phasors

Multiplication or division of complex numbers can be carried out when they are in either Cartesian form or polar form, being easiest when they are in polar form. Thus, if we have a voltage across a component described by $\mathbf{V} = V\angle\phi$ and the current by $\mathbf{I} = I\angle\theta$ then the product of the two phasors is:

$$\mathbf{VI} = VI\angle(\phi + \theta) \tag{32}$$

If the voltage and current were in Cartesian form, i.e. in the form $\mathbf{V} = a + jb$ and $\mathbf{I} = c + jd$ then the product is:

$$\mathbf{VI} = (a + jb)(c + jd) = (ac - bd) + j(bc + ad) \tag{33}$$

For division, if we have a voltage across a component described by $\mathbf{V} = V\angle\phi$ and the current by $\mathbf{I} = I\angle\theta$ then:

$$\frac{\mathbf{V}}{\mathbf{I}} = \frac{V\angle\phi}{I\angle\theta} \tag{34}$$

If the voltage and current were in Cartesian form, i.e. in the form $\mathbf{V} = a + jb$ and $\mathbf{I} = c + jd$ then:

$$\frac{\mathbf{V}}{\mathbf{I}} = \frac{a + jb}{c + jd} = \frac{a + jb}{c + jd} \times \frac{c - jd}{c - jd} = \frac{(a + jb)(c - jd)}{c^2 - d^2} \tag{35}$$

Example

If phasor **V** is represented by $10\angle 30°$ and **I** by $2\angle 45°$, determine **VI** and **V/I**.

$$\mathbf{VI} = (10 \times 2)\angle(30° + 45°) = 20\angle 75°$$

$$\frac{\mathbf{V}}{\mathbf{I}} = \frac{10\angle 30°}{2\angle 45°} = 5\angle(-15°)$$

Kirchhoff's laws and phasors

Kirchhoff's laws apply to the voltages and currents in a circuit at any instant of time. Thus the voltage law that the sum of the voltages taken round a closed loop is zero means that, with alternating voltages having values of v_1, v_2, v_3, etc. at the same instant of time:

$$v_1 + v_2 + v_3 + ... = 0$$

and so, if these voltages are sinusoidal:

$$V_1 \sin(\omega t + \phi_1) + V_2 \sin(\omega t + \phi_2) + V_3 \sin(\omega t + \phi_3) + ... = 0$$

We can consider each of these sinusoidal voltages to be the vertical projection of the phasor describing it. Thus we must have:

$$\mathbf{V_1} + \mathbf{V_2} + \mathbf{V_3} + ... = 0$$

Kirchhoff's voltage law can thus be stated as: *the sum of the phasors of all the voltages around a closed loop is zero.* Kirchhoff's current law can be stated as the sum of all the currents at a node is zero, i.e. the current entering a junction equals the current leaving it. In a similar way we can state this law for sinusoidal currents as: *the sum of the phasors of the currents at a node is zero, i.e. the sum of the phasors for currents entering a junction equals that for those leaving it.*

Example

A circuit has two components in parallel. If the currents through the components can be described by the phasors $2 + j4$ A and $4 + j1$ A, what is the phasor describing the current entering the junction?

Using Kirchhoff's current law we must have: the phasor for current entering junction = phasor sum for currents leaving the junction. Hence:

phasor for current entering = $2 + j4 + 4 + j1 = 6 + j5$ A

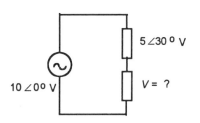

Figure 2.30 *Example*

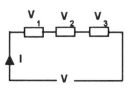

Figure 2.31 *Impedances in series*

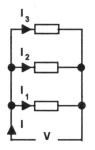

Figure 2.32 *Impedances in parallel*

Example

For the a.c. circuit shown in Figure 2.30, determine the unknown voltage.

Using Kirchhoff's voltage law, and writing the phasors in Cartesian notation:

$10 + j0 = (5 \cos 30° + j5 \sin 30°) + \mathbf{V}$

Thus:

$\mathbf{V} = 10 - 4.33 + j2.5 = 5.67 - j2.5$ V

or in polar notation:

$\mathbf{V} = \sqrt{5.67^2 + 2.5^2} \angle \tan^{-1}(-2.5/5.67 = 6.2\angle 336.2°$ V

Impedance

The term *impedance* Z is defined as the ratio of the phasor voltage across a component to the phasor current through it:

$$Z = \frac{\mathbf{V}}{\mathbf{I}} \qquad [36]$$

Thus if we have $\mathbf{V} = V\angle\theta$ and $\mathbf{I} = I\angle\phi$ then:

$$Z = \frac{V\angle\theta}{I\angle\phi} = \frac{V}{I} \angle (\theta - \phi)$$

If we have impedances connected in series (Figure 2.31), then Kirchhoff's voltage law gives:

$\mathbf{V} = \mathbf{V_1} + \mathbf{V_2} + \mathbf{V_3}$

Dividing by the phasor current, the current being the same through each:

$$\frac{\mathbf{V}}{\mathbf{I}} = \frac{\mathbf{V_1}}{\mathbf{I}} + \frac{\mathbf{V_2}}{\mathbf{I}} + \frac{\mathbf{V_3}}{\mathbf{I}}$$

Hence the total impedance Z is the sum of the impedances of the three impedances:

$$Z = Z_1 + Z_2 + Z_3 \qquad [37]$$

Consider the parallel connection of impedances (Figure 2.32). Kirchhoff's current law gives:

$\mathbf{I} = \mathbf{I_1} + \mathbf{I_2} + \mathbf{I_3}$

Dividing by the phasor voltage, the voltage being the same for each impedance:

$$\frac{I}{V} = \frac{I_1}{V} + \frac{I_2}{V} + \frac{I_3}{V}$$

Thus the total impedance Z is given by:

$$\frac{1}{Z} = \frac{1}{Z_1} + \frac{1}{Z_2} + \frac{1}{Z_3} \qquad [38]$$

Example

If the voltage across a component is 4 sin ωt V and the current through it 2 sin(ωt – 30°) A, what is its impedance?

Using equation [36] with the phasors in polar notation:

$$Z = \frac{4\angle 0°}{2\angle(-30°)} = 2\angle 30° \ \Omega$$

or Z = 2 cos 30° + 2 sin 30° = 1.73 + j1 Ω.

Example

What is the total impedance of a circuit with impedances of 2 + j5 Ω, 1 – j3 Ω and 4 + j1 Ω in series?

Z = 2 + j5 + 1 – j3 + 4 + j1 = 7 + j3 Ω

Example

What is the total impedance of impedances 4∠30° Ω in parallel with 2∠(–20°) Ω.

$$\frac{1}{Z} = \frac{1}{4\angle 30°} + \frac{1}{2\angle(-20°)} = 0.25\angle(-30°) + 0.5\angle 20°$$

$$= 0.25 \cos(-30°) + j0.25 \sin 30° + 0.5 \cos 20° + j0.5 \sin 20°$$

$$= 0.686 + j0.296$$

$$= \sqrt{0.686^2 + 0.296^2} \ \angle \ \tan^{-1}(0.296/0.686)$$

$$= 0.747\angle 23.3°$$

Hence Z = 1.339∠(–23.3°) Ω.

Circuit elements

For a *pure resistor* the current through it is in phase with the voltage across it. Thus for a voltage phasor of $V\angle 0°$ we must have a current phasor of $I\angle 0°$ and so the impedance of the circuit element is:

$$Z = \frac{\mathbf{V}}{\mathbf{I}} = \frac{V\angle 0°}{I\angle 0°} = \frac{V}{I}\angle 0°$$

The impedance is the real number V/I which is the resistance R.

For a *pure capacitance* the current leads the voltage by 90°. Thus for a voltage phasor of $V\angle 0°$ we must have a current phasor of $I\angle 90°$ and so the impedance of the circuit element is:

$$Z = \frac{\mathbf{V}}{\mathbf{I}} = \frac{V\angle 0°}{I\angle 90°} = \frac{V}{I}\angle(-90°)$$

The impedance is thus $-j(V/I)$ and is just an imaginary quantity. The term *capacitive reactance* X_C is used for the ratio of the maximum, or r.m.s., voltage and current and thus for a pure capacitance:

$$Z = -jX_C \tag{39}$$

For a *pure inductance* the current lags the voltage by 90°. Thus for a voltage phasor of $V\angle 0°$ we must have a current phasor of $I\angle(-90°)$ and so the impedance of the circuit element is:

$$Z = \frac{\mathbf{V}}{\mathbf{I}} = \frac{V\angle 0°}{I\angle(-90°)} = \frac{V}{I}\angle 90°$$

The impedance is thus $j(V/I)$ and is just an imaginary quantity. The term *inductive reactance* X_L is used for the ratio of the maximum, or r.m.s., voltage and current and thus for a pure inductance:

$$Z = jX_L \tag{40}$$

Example

Determine the impedance of a 100 Ω resistance in series with a capacitive reactance of 5 Ω.

$Z = R - jX_C = 100 - j5 \ \Omega$

Example

Express in Cartesian and polar notation, the impedance of each of the following circuits at a frequency of 50 Hz:

(a) a resistance of 20 Ω in series with an inductance of 0.1 H,
(b) a resistance of 50 Ω in series with a capacitance of 40 μF.

Also calculate the size of the current in each case and its phase relative to an applied voltage of 230 V at 50 Hz.

(a) With 50 Hz we have $\omega = 2\pi f = 2\pi \times 50 = 314.16$ rad/s. Thus:

$$Z = R + jX_L = 20 + j314.16(0.1) = 20 + j31.42 \ \Omega$$

Converting this to polar notation gives $|Z| = \sqrt{(20^2 + 31.42^2)} = 37.25 \ \Omega$. The phase is $\tan^{-1} (X_L/R) = \tan^{-1} (31.42/20) = 57.52°$ and so the impedance is $37.25\angle 57.52° \ \Omega$.

The current $I = V/Z$ and so is:

$$I = \frac{V}{Z} = \frac{230\angle 0°}{37.25\angle 57.52°} = 6.17\angle -57.52° \ A$$

(b) The capacitance reactance $X_C = 1/\omega C$ and so:

$$Z = R - jX_C = 50 - j\frac{1}{314.16 \times 40 \times 10^{-6}} = 50 - j79.58 \ \Omega$$

Converting this to polar notation $|Z| = \sqrt{(50^2 + 79.58^2)} = 93.98 \ \Omega$. The phase is $\tan^{-1} (X_C/R) = \tan^{-1} (-79.58/50) = -57.85°$ and so the impedance is $93.98\angle -57.85° \ \Omega$.

The current $I = V/Z$ and so is:

$$I = \frac{V}{Z} = \frac{230\angle 0°}{93.98\angle -57.85°} = 2.45\angle 57.85° \ A$$

Example

Calculate the resistance and the series inductance or capacitance for each of the following impedances if the frequency is 50 Hz: (a) $Z = 10 + j15 \ \Omega$, (b) $Z = -j80 \ \Omega$, (c) $Z = 50\angle 30° \ \Omega$, (d) $Z = 120\angle -60° \ \Omega$.

$\omega = 2\pi f = 2\pi \times 50 = 314$ rad/s

(a) Comparing this with $Z = R + j\omega L$, then $R = 10 \ \Omega$ and $X_L = 15 \ \Omega$. Since $X_L = \omega L =$ then $L = 15/314 = 0.048$ H.
(b) Here $R = 0$ and the capacitive reactance $X_C = 80 \ \Omega$. Since $X_C = 1/\omega C$ then $C = 1/(314 \times 80) = 39.8 \times 10^{-6}$ F or 39.8 μF.

(c) This gives in Cartesian notation (see equation [29]) Z = 50 cos 30° + j50 sin 30° = 43.3 + j25 Ω. We can compare this with $Z = R + jX_L$ and so R = 43.3 Ω and X_L = 25 Ω. Since $X_L = \omega L$ then L = 25/314 = 0.080 H.

(d) This gives in Cartesian notation (see equation [29]) Z = 120 cos –60° + j120 sin –60° = 60 – j104 Ω. We can compare this with $Z = R – jX_C$ and so R = 60 Ω and X_C = 104 Ω. Since $X_C = 1/\omega C$ then C = 1/(314 × 104) = 30.7 µF.

Problems 2.3

1 Simplify (a) j^7, (b) j^8, (c) $j^2 \times j$, (d) j^5/j^3.

2 Solve the following equations:

   (a) $x^2 + 16 = 0$, (b) $x^2 + 4x – 5 = 0$, (c) $2x^2 – 2x + 3 = 0$

3 Express the following complex numbers in polar form:

   (a) $-4 + j$, (b) $-3 – j4$, (c) 3, (d) $-j6$, (e) $1 + j$, (f) $3 – j2$

4 Express the following complex numbers in Cartesian form:

   (a) $5\angle120°$, (b) $10\angle45°$, (c) $6\angle180°$, (d) $2.8\angle76°$,

   (e) $2(\cos 30° + j \sin 30°)$, (f) $3(\cos 60° – j \sin 60°)$

5 If $z_1 = 3 + j2$ and $z_2 = -2 + j4$, determine the values of:

   (a) $z_1 + z_2$, (b) $z_1 – z_2$, (c) $z_1 z_2$, (d) $\frac{1}{z_1}$, (e) $\frac{z_1}{z_2}$

6 Evaluate the following:

   (a) $(2 + j3) + (3 – j5)$, (b) $(-4 – j6) + (2 + j5)$,

   (c) $(2 + j2) – (3 – j5)$, (d) $(2 + j4) – (1 + j4)$, (e) $4(3 + j2)$,

   (f) $j2(3 + j5)$, (g) $(1 – j2)(3 + j4)$,

   (h) $(2 + j2)(3 – j3)$, (i) $(1 + j2)(4 – j3)$,

   (j) $\frac{6 + j3}{4 – j2}$, (k) $\frac{1}{3 + j2}$, (l) $\frac{1 + j1}{1 – j1}$, (m) $\frac{3 + j2}{1 – j3}$

7 If $z_1 = 10\angle20°$, $z_2 = 2\angle40°$ and $z_3 = 5\angle60°$, evaluate the following:

   (a) $z_1 z_2$, (b) $z_1 z_3$, (c) $\frac{1}{z_1}$, (d) $\frac{1}{z_2}$, (e) $\frac{z_1}{z_2}$, (f) $\frac{z_2}{z_3}$

8 Describe the following signals by phasors written in both polar and Cartesian forms, taking the magnitude to represent the maximum value:

(a) 10 sin $(2\pi50t - \pi/6)$, (b) 10 sin $(314t + 150°)$,

(c) 22 sin $(628t + \pi/4)$

9 Determine, in both Cartesian and polar forms, the sum of the following phasors:

(a) $4\angle0°$ and $3\angle60°$, (b) $2 + j3$ and $-4 + j4$,
(c) $4\angle\pi/3$ and $2\angle\pi/6$

10 If phasors **A**, **B** and **C** are represented by **A** = $10\angle30°$, **B** = $2.5\angle60°$ and **C** = $2\angle45°$ determine:

(a) **AB**, (b) **AC**, (c) **A(B + C)**, (d) **A/B**, (e) **B/C**,
(f) **C/(A + B)**

11 If v_1 = 10 sin ωt and v_2 = 20 sin $(\omega t + 60°)$, what is (a) the phasor describing the sum of the two voltages and (b) its time-domain equation?

12 If the voltage across a component is 5 sin $(314t + \pi/6)$ V and the current through it 0.2 sin $(314t + \pi/3)$ A, what is its impedance?

13 A voltage of 100 V is applied across a circuit of impedance 40 + j30 Ω, what is, in polar notation, the current taken?

14 Determine, in Cartesian form, the total impedances of:

(a) 10 Ω in series with 2 – j5 Ω,
(b) $100\angle30°$ Ω in series with $100\angle60°$ Ω,
(c) $20\angle30°$ Ω in series with $15\angle(-10°)$ Ω,
(d) $20\angle30°$ Ω in parallel with $6\angle(-90°)$ Ω,
(e) 10 Ω in parallel with –j2 Ω,
(f) j40 Ω in parallel with j20 Ω

15 Determine, in Cartesian form, the impedance of:

(a) a resistance of 5 Ω in series with an inductive reactance of 2 Ω,
(b) a resistance of 50 Ω in series with a capacitive reactance of 10 Ω,
(c) a resistance of 2 Ω in series with an inductive reactance of 5 Ω and a capacitive reactance of 4 Ω,
(d) three elements in parallel, a resistance of 2 Ω, an inductive reactance of 10 Ω and a capacitive reactance of 5 Ω,
(e) an inductive reactance of 500 Ω in parallel with a capacitive reactance of 100 Ω

3 Mathematical models

Summary

Engineers frequently have to devise and use mathematical models for systems. Mathematical modelling is the activity by which a problem involving the real-world is translated into mathematics to form a model which can then be used to provide information about the original real problem. Such mathematical models are essential in the design-to-test phase, in particular forming the benchmark by which a computer generated simulation can be measured prior to manufacturing a prototype. This chapter is an introduction to mathematical modelling for engineering systems, later chapters involving more detailed consideration of models.

Objectives

By the end of this chapter, the reader should be able to:

- understand what is meant by a mathematical model and how such models are formulated;
- devise mathematical models for simple systems.

3.1 Modelling

Consider some real problems:

- Tall buildings are deflected by strong winds, can we devise a model which can be used to predict the amount of deflection of a building for particular wind strengths?

- Cars have suspension systems, can we devise a model which can be used to predict how a car will react when driven over a hump in the road?

- When a voltage is connected to a d.c. electrical motor, can we devise a model which will predict how the torque developed by the motor will depend on the voltage?

- Can we devise a model to enable the optimum shaft to be designed for a power transmission system connecting a motor to a load?

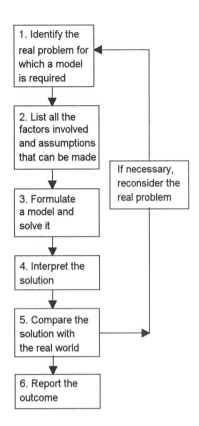

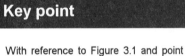

Figure 3.1 *The processes involved in devising a model*

• Can we design a model to enable an appropriate transducer to be selected as part of the monitoring/activation circuit for the safe release of an air bag in a motor vehicle under crash conditions?

• Can we design models which will enable failure to be predicted in static and dynamic systems?

• Can we design models to enable automated, robotic controlled systems to be designed?

Such questions as those above are encountered by design and manufacturing system engineers daily. Quite often it is the accuracy of a mathematical model that will determine the success or otherwise of a new design. Such models also greatly help to reduce the design-test-evaluation-manufacture process lead time.

3.1.1 How do we devise models?

The tactics adopted to devise models involve a number of stages which can be summarised by the block diagram of Figure 3.1. The first stage involves identifying what the real problem is and then identifying what factors are important and what assumptions can be made. These assumptions are used in order to simplify the model and enable an initial model to be formulated which we can check as being a reasonable approximation. Only when we are confident with the model do we build in further considerations to make the models even more accurate! For example, when modelling mechanical systems, we often initially ignore friction and the consequential heat generation; however, as the model is refined we have to consider such effects and adjust the initial 'ball park' model in order for its application to the real world to be valid. This stage will generally involve collecting data. The next stage is then to formulate a model. An essential part of this is to translate verbal statements into mathematical relationships. When solutions are then produced from the model, they need to be compared with the real world and, if necessary the entire cycle repeated.

Example of formulating a model

As an illustration, consider how we might approach the problem we started this section with:

> Tall buildings are deflected by strong winds, can we devise a model which can be used to predict the amount of deflection of a building for particular wind strengths?

The simplest form we might consider is that of a tall building which is subject to wind pressure over its entire height, the building being anchored at the ground but free to deflect at its top (Figure 3.2). If we assume that the wind pressure gives a uniform loading over the entire height of the building and does not

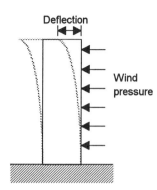

Figure 3.2 *The building deflection problem*

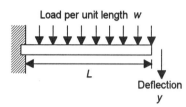

Figure 3.3 *Deflection of a uniformly loaded cantilever*

Key point

Lumped models are devised by considering each of the basic behaviour characteristics of a system and representing them by a single element.

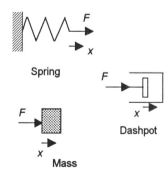

Figure 3.4 *Mechanical system building blocks*

fluctuate, then we might consider the situation is rather like the deflection of the free end of a cantilever when subject to a uniformly distributed load (Figure 3.3). For such a beam the deflection y is given by:

$$y = \frac{wL^4}{8EI} \qquad [1]$$

where L is the length of the beam, w the load per unit length, E the modulus of elasticity and I the second moment of area. The modulus of elasticity if a measure of the stiffness of the material and the second moment of area for a rectangular section is $bd^3/12$, with b being the breadth and d the depth. This would suggest that a stiff structure would deflect less and also one with a large cross-section would deflect less. Hence we might propose a model of the form:

$$\text{deflection} \propto \frac{pH^4}{\text{stiffness} \times bd^3} \qquad [2]$$

with p being the wind pressure and related to the wind velocity, H the height of the building, b its breadth and d its depth. Thus, a short, squat building will be less deflected than a tall slender one of the same building materials.

3.1.2 Lumped element modelling

Often in engineering we can devise a model for a system by considering it to be composed of a number of basic elements. We consider the characteristics of the behaviour of the system and 'lump' all the similar behaviour characteristics together and represent them by a simple element. For some elements, the relationship between their input and output is a simple proportionality, in other cases it involves a rate of change with time or even the rate of change with time of a rate of change with time.

Mechanical systems

Mechanical systems can be considered to be made up of three basic elements which represent the stiffness, damping and inertia of the system:

- *Spring element*
  The 'springiness' or 'stiffness' of a system can be represented by a spring (Figure 3.4(a)). The force F is proportional to the extension x of the spring:

  $$F = kx \qquad [3]$$

- *Damper element*
  The 'damping' of a mechanical system can be represented by a dashpot. This is a piston moving in a viscous medium in a

cylinder (Figure 3.4(b)). The damping force F is proportional to the velocity v of the damping element:

$$F = cv$$

where c is a constant. Since the velocity is equal to the rate of change of displacement x:

$$F = c\frac{dx}{dt} \qquad [4]$$

- **Mass or inertia element**

  The 'inertia' of a system, i.e. how much it resists being accelerated can be represented by mass m. Since the force F acting on a mass is related to its acceleration a by $F = ma$ and acceleration is the rate of change of velocity, with velocity being the rate of change of displacement x:

$$F = ma = m\frac{dv}{dt} = m\frac{d}{dt}\left(\frac{dx}{dt}\right) = m\frac{d^2x}{dt^2} \qquad [5]$$

To develop the equations relating inputs and outputs we use Newton's laws of motion.

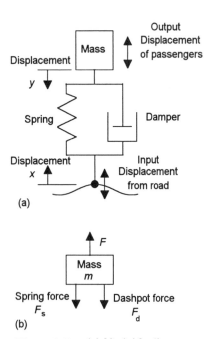

(a)

(b)

Figure 3.5 *(a) Model for the car suspension system, (b) free-body diagram*

Example

Develop a lumped-model for a car suspension system which can be used to predict how a car will react when driven over a hump in the road (the second problem listed earlier in this chapter).

Such a system can have its suspension represented by a spring, the shock absorbers by a damper and the mass of the car and its passengers by a mass. The model thus looks like Figure 3.5(a). We can then devise an equation showing how the output of the system, namely the displacement of the passengers with time depends on the input of the displacement of a car wheel as it rides over the road surface.

This is done by considering a *free-body diagram*, this being a diagram of the mass showing just the external forces acting on it (Figure 3.5(b)). We can then relate the net force acting on the mass to its acceleration by the use of Newton's law, hence obtaining an equation which relates the input to the output.

The relative extension of the spring is the difference between the displacement x of the mass and the input displacement y. Thus, the force due to the spring F_s is $k(x - y)$. The dashpot force F_d is $c(dx/dt - dy/dt)$. Hence, applying Newton's law:

$$F - F_s - F_d = ma$$

and so we can write:

$$m\frac{d^2x}{dt^2} = F - k(x - y) - c\left(\frac{dx}{dt} - \frac{dy}{dt}\right)$$

Rotational systems

For rotational systems, e.g. the drive shaft of a motor, the basic building blocks are a torsion spring, a rotary damper and the moment of inertia (Figure 3.6).

(a) Torsional spring or elastic twisting of a shaft

(b) Rotational dashpot

(c) Moment of inertia

Figure 3.6 *Rotational system elements: (a) torsional spring, (b) rotational dashpot, (c) moment of inertia*

- **Torsional spring**
  The 'springiness' or 'stiffness' of a rotational spring is represented by a torsional spring. The torque T is proportional to the angle θ rotated:

$$T = k\theta \qquad [6]$$

  where k is a constant.

- **Rotational dashpot**
  The damping inherent in rotational motion is represented by a rotational dashpot. The resistive torque T is proportional to the angular velocity ω and thus, since ω is the rate of change of angle θ with time:

$$T = c\omega = c\frac{d\theta}{dt} \qquad [7]$$

  where c is a constant.

- **Inertia**
  The inertia of a rotational system is represented by the moment of inertia I of a mass. The torque T needed to produce an acceleration a is given by $T = Ia$ and thus, since a is the rate of change of angular velocity ω with time and angular velocity is the rate of change of angle θ with time:

$$T = Ia = I\frac{d\omega}{dt} = I\frac{d}{dt}\left(\frac{d\theta}{dt}\right) = I\frac{d^2\theta}{dt^2} \qquad [8]$$

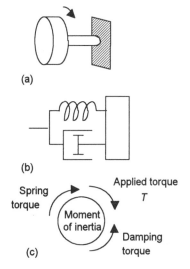

(a)

(b)

Spring torque

Applied torque T

Moment of inertia

Damping torque

(c)

Figure 3.7 *Example*

Example

Represent as a lumped-model the system shown in Figure 3.7(a) of the rotation of a disk as a result of twisting a shaft.

Figure 3.7(b) shows the lumped-model and Figure 3.7(c) the free-body diagram for the system.

The torques acting on the disk are the applied torque, the spring torque and the damping torque. The torque due to the spring is $k\theta$ and the damping torque is $c(d\theta/dt)$. Hence, since the net torque acting on the mass is:

net torque = T – spring torque – damping torque

we have:

$$I\frac{d^2\theta}{dt^2} = T - k\theta - c\frac{d\theta}{dt}$$

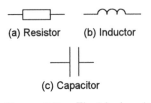

(a) Resistor (b) Inductor

(c) Capacitor

Figure 3.8 *Electrical system elements*

Electrical systems

The basic elements of electrical systems are the resistor, inductor and capacitor (Figure 3.8).

* **Resistor**
  The resistor represents the electrical resistance of the system. The potential difference v across a resistor is proportional to the current i through it:

  $$v = Ri \qquad [9]$$

* **Inductor**
  The inductor represents the electrical inductance of the system. For an inductor, the potential difference v across it depends on the rate of change of current i through it and we can write:

  $$v = L\frac{di}{dt} \qquad [10]$$

* **Capacitor**
  The capacitor represents the electrical capacitance of the system. For a capacitor, the charge q on the capacitor plates is related to the voltage v across the capacitor by $q = Cv$, where C is the capacitance. Since current i is the rate of movement of charge:

  $$i = \frac{dq}{dt}$$

  But $q = Cv$, with C being constant, so:

  $$i = \frac{d(Cv)}{dt} = C\frac{dv}{dt} \qquad [11]$$

To develop the models for systems which we describe by electrical circuits involving resistance, inductance and capacitance we use Kirchhoff's laws.

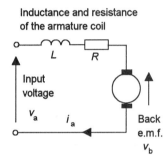

Inductance and resistance
of the armature coil

Input
voltage

v_a

i_a

Back
e.m.f.
v_b

Figure 3.9 *Lumped-model for a d.c. motor*

Example

Develop a lumped-system model for a d.c. motor relating the current through the armature to the applied voltage.

The motor consists basically of the armature coil, this being free to rotate, which is located in the magnetic field provided by either a permanent magnet or a current through field coils. When a current flows through the armature coil, forces acting on as a result of the current carrying conductors being in a magnetic field. As a result, the armature coil rotates. Since the armature is a coil rotating in a magnetic field, a voltage is induced in it in such a direction as to oppose the change producing it, i.e. there is a back e.m.f. Thus the electrical circuit we can use to describe the motor has two sources of e.m.f., that applied to produce the armature current and the back e.m.f. If we consider a motor where there is either a permanent magnet or separately excited field coils, then the lumped electrical circuit model is as shown in Figure 3.9. We can consider there are just two elements, an inductor and a resistor, to represent the armature coil. The equation is thus:

$$v_a - v_b = Ri_a + L\frac{di_a}{dt}$$

This can be considered to be the first stage in addressing the problem posed earlier in the chapter: When a voltage is connected to a d.c. electrical motor, can we devise a model which will predict how the torque developed by the motor will depend on the voltage? The torque generated will be proportional to the current through the armature.

Thermal systems

Thermal systems have two basic building blocks with thermal systems, resistance and capacitance.

- ### Thermal resistance

  The thermal resistance R is the resistance offered to the rate of flow of heat q and is defined by:

  $$q = \frac{T_1 - T_2}{R} \qquad [12]$$

  where $T_1 - T_2$ is the temperature difference through which the heat flows.

  For heat conduction through a solid we have the rate of flow of heat proportional to the cross-sectional area A and the temperature gradient. Thus, for two points at temperatures T_1 and T_2 and a distance L apart, we can write:

$$q = Ak\frac{T_1 - T_2}{L} \qquad [13]$$

with k being the thermal conductivity. With this mode of heat transfer, the thermal resistance R is L/Ak.

For heat transfer by convection between two points, Newton's law of cooling gives:

$$q = Ah(T_2 - T_1) \qquad [14]$$

where $(T_2 - T_1)$ is the temperature difference, h the coefficient of heat transfer and A the surface area across which the temperature difference is. The thermal resistance with this mode of heat transfer is thus $1/Ah$.

- *Thermal capacitance*
  The thermal capacitance is a measure of the store of internal energy in a system. If the rate of flow of heat into a system is q_1 and the rate of flow out q_2 then the rate of change of internal energy of the system is $q_1 - q_2$. An increase in internal energy can result in a change in temperature:

  change in internal energy $= mc \times$ change in temperature

  where m is the mass and c the specific heat capacity. Thus the rate of change of internal energy is equal to mc times the rate of change of temperature. Hence:

  $$q_1 - q_2 = mc\frac{\mathrm{d}T}{\mathrm{d}t} \qquad [15]$$

  This equation can be written as:

  $$q_1 - q_2 = C\frac{\mathrm{d}T}{\mathrm{d}t} \qquad [16]$$

  where the capacitance $C = mc$.

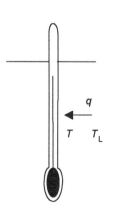

Figure 3.10 *Example*

Example

Develop a lumped-model for the simple thermal system of a thermometer at temperature T being used to measure the temperature of a liquid when it suddenly changes to the higher temperature of T_L (Figure 3.10).

When the temperature changes there is heat flow q from the liquid to the thermometer. If R is the thermal resistance to heat flow from the liquid to the thermometer then $q = (T_L - T)/R$. Since there is only a net flow of heat from the liquid to the thermometer, if the thermal capacitance of the thermometer is C, then $q = C\,\mathrm{d}T/\mathrm{d}t$.

Thus, the equation for the model is:

$$C\frac{\mathrm{d}T}{\mathrm{d}t} = \frac{T_L - T}{R}$$

Hydraulic systems

For a fluid system the three building blocks are resistance, capacitance and inertance. Hydraulic fluid systems are assumed to involve an incompressible liquid; pneumatic systems, however, involve compressible gases and consequently there will be density changes when the pressure changes. Here we will just consider the simpler case of hydraulic systems. Figure 3.11 shows the basic form of building blocks for hydraulic systems.

- **Hydraulic resistance**

  Hydraulic resistance R is the resistance to flow which occurs when a liquid flows from one diameter pipe to another (Figure 3.11(a)) and is defined as being given by the hydraulic equivalent of Ohm's law:

$$p_1 - p_2 = Rq \qquad [17]$$

- **Hydraulic capacitance**

  Hydraulic capacitance C is the term used to describe energy storage where the hydraulic liquid is stored in the form of potential energy (Figure 3.11(b)). The rate of change of volume V of liquid stored is equal to the difference between the volumetric rate at which liquid enters the container q_1 and the rate at which it leaves q_2, i.e.

$$q_1 - q_2 = \frac{\mathrm{d}V}{\mathrm{d}t}$$

But $V = Ah$ and so:

$$q_1 - q_2 = A\frac{\mathrm{d}h}{\mathrm{d}t}$$

The pressure difference between the input and output is:

$$p_1 - p_2 = p = h\rho g$$

Hence, substituting for h gives:

$$q_1 - q_2 = \frac{A}{\rho g}\frac{\mathrm{d}p}{\mathrm{d}t} \qquad [18]$$

The hydraulic capacitance C is defined as:

$$C = \frac{A}{\rho g} \qquad [19]$$

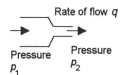

(a) Resistance

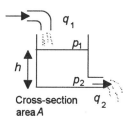

(b) Capacitance

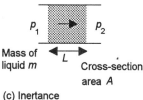

(c) Inertance

Figure 3.11 *Hydraulic system elements*

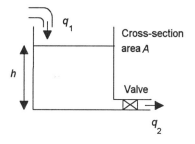

Figure 3.12 *Example*

and thus we can write:

$$q_1 - q_2 = C\frac{dp}{dt} \qquad [20]$$

- *Hydraulic inertance*

  Hydraulic inertance is the equivalent of inductance in electrical systems. To accelerate a fluid a net force is required and this is provided by the pressure difference (Figure 3.11(c)). Thus:

$$(p_1 - p_2)A = ma = m\frac{dv}{dt} \qquad [21]$$

where a is the acceleration and so the rate of change of velocity v. The mass of fluid being accelerated is $m = AL\rho$ and the rate of flow $q = Av$ and so:

$$(p_1 - p_2)A = L\rho\frac{dq}{dt}$$

$$p_1 - p_2 = I\frac{dq}{dt} \qquad [22]$$

where the inertance I is given by $I = L\rho/A$.

Example

Develop a model for the hydraulic system (Figure 3.12) where there is a liquid entering a container at one rate q_1 and leaving through a valve at another rate q_2.

We can neglect the inertance since flow rates can be assumed to change only very slowly. For the capacitance term we have:

$$q_1 - q_2 = C\frac{dp}{dt} = \frac{A}{\rho g}\frac{dp}{dt}$$

For the resistance term for the valve we have $p_1 - p_2 = Rq_1$. Thus, substituting for q_2, and recognising that the pressure difference is $h\rho g$, gives:

$$q_1 = A\frac{dh}{dt} + \frac{h\rho g}{R}$$

Problems 3.1

1 Propose a mathematical model for the oscillations of a suspension bridge when subject to wind gusts.
2 Propose a mathematical model for a machine mounted on firm ground when the machine is subject to forces when considered in terms of lumped-parameters.

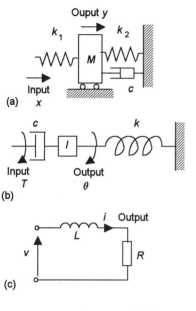

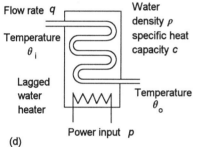

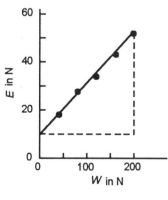

Figure 3.13 *Problem 3*

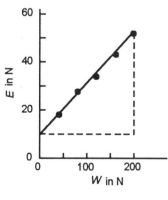

Figure 3.14 *Example*

3 Derive an equation for a mathematical model relating the input and output for each of the lumped systems shown in Figure 3.13.

3.2 Relating models and data

In testing mathematical models against real data, we often have the situation of having to check whether data fits an equation. If the relationship is linear, i.e. of the form $y = mx + c$, then it is comparatively easy to see whether the data fits the straight line and to ascertain the gradient m and intercept c. However, if the relationship is non-linear this is not so easy. A technique which can be used is to turn the non-linear equation into a linear one by changing the variables. Thus, if we have a relationship of the form $y = ax^2 + b$, instead of plotting y against x to give a non-linear graph we can plot y against x^2 to give a linear graph with gradient a and intercept b. If we have a relationship of the form $y = a/x$ we can plot a graph of y against $1/x$ to give a linear graph with a gradient of a.

Example

The following data was obtained from measurements of the load lifted by a machine and the effort expended. Determine if the relationship between the effort E and the load W is linear and if so the relationship.

E in N	18	27	32	43	51
W in N	40	80	120	160	200

Within the limits of experimental error the results appear to indicate a straight-line relationship (Figure 3.14). The gradient is 41/200 or about 0.21. The intercept with the E axis is at 10. Thus the relationship is $E = 0.21W + 10$.

Example

It is believed that the relationship between y and x for the following data is of the form $y = ax^2 + b$. Determine the values of a and b.

y	2.5	4.0	6.5	10.0	14.5
x	1	2	3	4	5

Figure 3.15 shows the graph of y against x^2. The graph has a gradient of AB/BC = 12.5/25 = 0.5 and an intercept with the y-axis of 2. Thus the relationship is $y = 0.5x^2 + 2$.

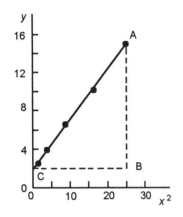

Figure 3.15   *Example*

Problems 3.2

1 Determine, assuming linear, the relationships between the following variables:

(a) The load *L* lifted by a machine for the effort *E* applied.

E in N	9.5	11.8	14.1	16.3	18.5
L in N	10	15	20	25	30

(b) The resistance *R* of a wire for different lengths *L* of that wire.

R in W	2.1	4.3	6.3	8.3	10.5
L in m	0.5	1.0	1.5	2.0	2.5

2 Determine what form the variables in the following equations should take when plotted in order to give straight-line graphs and what the values of the gradient and intercept will have.

(a) The period of oscillation *T* of a pendulum is related to the length *L* of the pendulum by the equation:

$$T = 2\pi \sqrt{\frac{L}{g}}$$

where *g* is a constant.

(b) The distance *s* travelled by a uniformly accelerating object after a time *t* is given by the equation:

$$s = ut + \tfrac{1}{2}at^2$$

where *u* and *a* are constants.

(c) The e.m.f. *e* generated by a thermocouple at a temperature θ is given by the equation;

$$e = a\theta + b\theta^2$$

where *a* and *b* are constants.

(d) The resistance *R* of a resistor at a temperature h is given by the equation:

$$R = R_0 + R_0 a\theta$$

where R_0 and α are constants.

(f) The pressure *p* of a gas and its volume *V* are related by the equation:

$$pV = k$$

where k is a constant.

(g) The deflection y of the free end of a cantilever due to it own weight of w per unit length is related to its length L by the equation:

$$y = \frac{wL^4}{8EI}$$

where w, E and I are constants.

3 The resistance R of a lamp is measured at a number of voltages V and the following data obtained. Show that the law relating the resistance to the voltage is of the form $R = (a/V) + b$ and determine the values of a and b.

R in Ω	70	62	59	56	55
V in V	60	100	140	200	240

4 The resistance R of wires of a particular material are measured for a range of wire diameters d and the following results obtained. Show that the relationship is of the form $R = (a/d^2) + b$ and determine the values of a and b.

R in Ω	0.25	0.16	0.10	0.06	0.04
d in mm	0.80	1.00	1.25	1.60	2.00

5 The volume V of a gas is measured at a number of pressures p and the following results obtained. Show that the relationship is of the form $V = ap^b$ and determine the values of a and b.

V in m³	13.3	11.4	10.0	8.9	8.0
p in 10^5 Pa	1.2	1.4	1.6	1.8	2.0

6 When a gas is compressed adiabatically the pressure p and temperature T are measured and the following results obtained. Show that the relationship is of the form $T = ap^b$ and determine the values of a and b.

p in 10^5 Pa	1.2	1.5	1.8	2.1	2.4
T in K	526	560	589	615	639

7 The cost C per hour of operating a machine depends on the number of items n produced per hour. The following data has been obtained and is anticipated to follow a relationship of the form $C = an^3 + b$. Show that this is the case and determine the values of a and b.

C in £	31	38	67	94	155
n	10	20	30	40	50

8 The following are suggested braking distances s for cars travelling at different speeds v. The relationship between s and v is thought to be of the form $s = av^2 + bv$. Show that this is so and determine the values of a and b.

s in m	5	15	30	50	75
v in m/s	5	10	15	20	25

Hint: consider s/v as one of the variables.

9 The luminosity I of a lamp depends on the voltage V applied to it. The relationship between I and V is thought to be of the form $I = aV^b$. Use the following results to show that this is the case and determine the values of a and b.

I in candela	3.6	6.4	10.0	14.4	19.6
V in volts	60	80	100	120	140

10 From a lab test, it is believed that the law relating the voltage v across an inductor and the time t is given by the relationship $v = A\,e^{t/B}$, where A and B are constant and e is the exponential function. From the lab test the results observed were:

v (volts)	908.4	394.8	171.6	32.4	14.1	6.12
t (ms)	10	20	30	50	60	70

Show that the law relating the voltage to time is, in fact, true. Then determine the values of the constants A and B.

4 Calculus

Summary

Calculus is concerned with two basic operations, differentiation and integration, and is a tool used by engineers to determine such quantities as rates of change and areas; in fact, calculus is the mathematical 'backbone' for dealing with problems where variables change with time or some other reference variable and a basic understanding of calculus is essential for further study and the development of confidence in solving practical engineering problems. This will become evident in the next chapter where physical systems will be modelled and the use of 'rates of change' equations (called differential equations) will allow the physical system to be represented, an analysis made and a solution formed under defined conditions. This chapter is an introduction to the techniques of calculus and a consideration of some of their engineering applications. The topic continues in the next chapter with a discussion of the use of differential equations to represent physical systems and their solution for various inputs.

Objectives

By the end of this chapter, the reader should be able to:

- understand the concept of a limit and its significance in rate of change relationships;
- use calculus notation for describing a rate of change (differentiation) and understand the significance of the operation;
- solve engineering problems involving rates of change;
- understand what is involved in the calculus operation of integration;
- solve engineering problems involving integration.

4.1 Differentiation

Suppose we have an equation describing how the distance covered in a straight line by a moving object varies with time. We could plot a graph of displacement against time and determine the velocity at some instant as the gradient of the tangent to the curve at that instant. By taking a number of such gradient measurements we could then determine how the velocity varied with time. However, *differentiation* is a mathematical technique which can be used to determine the rate at which functions change and hence the gradients, this thus enabling the velocity to be obtained from the equation without drawing the graph and tangents. We could

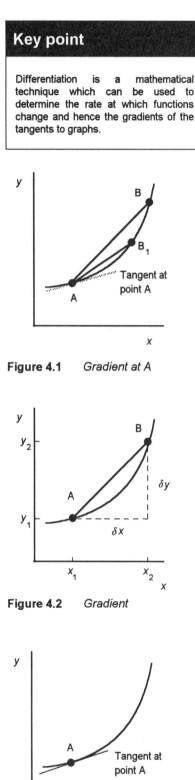

Key point

Differentiation is a mathematical technique which can be used to determine the rate at which functions change and hence the gradients of the tangents to graphs.

Figure 4.1   *Gradient at A*

Figure 4.2   *Gradient*

Figure 4.3   *Gradient at A*

also, for example, describe how the deflection of an initially horizontal beam in the y-direction alters with distance along the beam in the x-direction, or how the volume of a gas changes with temperature, or how electric charge at a point in a circuit changes with time (called current), etc. The list of uses is endless!

Limits

Consider the problem of determining the gradient of a tangent at a point on a graph. It might, for example, be a distance–time graph for a moving object to determine velocity as the rate at which distance is covered or a current–time graph for the current in an electrical circuit in order to determine the rate of change of current with time. Suppose we want to determine the gradient at point A on the curve shown in Figure 4.1. We can select another point B on the curve and join them together and then find the gradient of the line AB. The value of the gradient determined in this way will depend on where we locate the point B. If we let B slide along the curve towards A then the closer B is to A the more the line approximates to the tangent at point A. Thus the line AB_1 is a better approximation to the tangent than the line AB. The method we can use to determine the gradient of the tangent at point A is:

1   Take another point B on the same curve and determine the gradient of the line joining A and B.

2   Then move B closer and closer to A. In the limit, as the distance between A and B becomes infinitesimally small, i.e. as AB tends to zero (written as AB → 0), the gradient of the line becomes the gradient of the tangent at A.

Consider the gradient of the line AB in Figure 4.2:

$$\text{gradient} = \frac{y_2 - y_1}{x_2 - x_1}$$

The gradient is the difference in the value of y between points A and B divided by the difference in the value of x between the points. We can write this difference in the value of x as δx and this difference in the value of y as δy. The δ symbol in front of a quantity means 'a small bit of it' or 'an interval of'. Thus the equation can be written as

$$\text{gradient} = \frac{\delta y}{\delta x}$$

An alternative symbol which is often used is Δx, with the Δ symbol being used to indicate that we are referring to a small bit of the quantity x. These forms of notation do *not* mean that we have δ or Δ multiplying x. The δx or Δx should be considered as a single symbol representing a single quantity.

As we move B closer to A then the interval δx is made smaller. The gradient of the line AB then becomes closer to the tangent to the curve at point A (Figure 4.3). Eventually when the difference

Key point

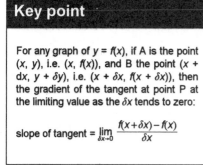

For any graph of $y = f(x)$, if A is the point (x, y), i.e. $(x, f(x))$, and B the point $(x + dx, y + \delta y)$, i.e. $(x + \delta x, f(x + \delta x))$, then the gradient of the tangent at point P at the limiting value as the δx tends to zero:

$$\text{slope of tangent} = \lim_{\delta x \to 0} \frac{f(x+\delta x)-f(x)}{\delta x}$$

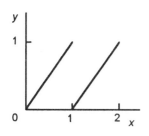

Figure 4.4   *Discontinuous function*

in x between A and B, i.e. δx, tends to zero then we have the gradient of the tangent at point A. We can write this as:

$$\lim_{\delta x \to 0} \frac{\delta y}{\delta x} = \frac{dy}{dx} \qquad [1]$$

This reads as: the limiting value of $\delta y/\delta x$ as δx tends to a zero value equals dy/dx. A *limit* is a value to which we get closer and closer as we carry out some operation. Thus dy/dx is the value of the gradient of the tangent to the curve at A. Since the tangent is the instantaneous rate of change of y with x at that point then dy/dx is the instantaneous rate of change of y with respect to x. dy/dx is called the *derivative* of y with respect to x. The process of determining the derivative for a function is called *differentiation*. The notation dy/dx should not be considered as d multiplied by y divided by d multiplied by x, but as a single symbol representing the gradient of the tangent and so the rate of change of y with x; if you like, it is a shorthand way of writing 'the rate of change of y with respect to x'.

Since we can interpret the derivative as representing the slope of the tangent to a graph of a function at a particular point, this means with a continuous function, i.e. a function which has values of y which smoothly and continuously change as x changes for all values of x, that we have derivatives for all values of x. However, with a discontinuous graph there will be some values of x for which we can have no derivative. For example, with the graph shown in Figure 4.4, there is no derivative for $x = 1$.

Example

Determine the slope of the tangent to the curve $y = f(x) = x^2$ when we have $x = 1$ and $y = 1$.

$$\text{Slope of tangent} = \lim_{\delta x \to 0} \frac{(x+\delta x)^2 - x^2}{\delta x}$$

$$= \lim_{\delta x \to 0} \frac{x^2 + 2x(\delta x) + (\delta x)^2 - x^2}{\delta x}$$

$$= \lim_{\delta x \to 0} (2x + \delta x)$$

As δx tends to zero, the slope of the tangent tends to the value $2x$. With $x = 1$ then the slope of the tangent is 2.

Maths in action

There are situations in engineering where, given an initial condition of some variable, we need to determine its rate of change with respect to some parameter.

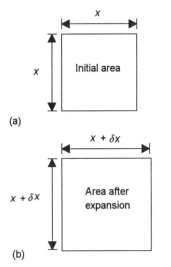

Figure 4.5   *Expansion of a plate*

As an illustration, consider a square flat metal plate of side length x (Figure 4.5(a)). The initial condition we are concerned with is the area A of the plate:

initial area A = x multiplied by x = x^2

Now suppose the plate is heated and the dimension x changes by an amount δx (which is small compared with the original dimension x but which may have an effect on such things as tolerances in assembly). The plate, which expands equally in all directions, now has sides of length $(x + \delta x)$ (Figure 4.5(b)). The new area is thus:

new area = $(x + \delta x)(x + \delta x)$ = $x^2 + 2x(\delta x) + (\delta x)^2$

If we denote the changes in area as δA, then:

δA = new area − initial area = $x^2 + 2x(\delta x) + (\delta x)^2 - x^2$

$\quad\quad = 2x(\delta x) + (\delta x)^2$

Since dx is very small, then $(\delta x)^2 \to 0$ and in this limiting condition we can write:

$dA = 2x\, dx$

$$\frac{dA}{dx} = 2x$$

We now have an expression which describes how the rate of change of the area with side length depends on the side length. We have determined the derivative.

4.1.1 Derivatives of common functions

The above examples illustrate how, given some initial condition, we can determine the derivative of a function. Rather than always work from first principles in this way, it is useful to work out some general rules we can use. The following illustrates how some commonly used functions can be differentiated.

Derivative of a constant

A graph of a constant, e.g. $y = 2$, has a gradient of 0. Thus its derivative is zero. Thus for $y = c$, where c is a constant:

$$\frac{d}{dx}(c) = 0 \tag{2}$$

Key point

The derivative of a constant is zero.

Derivative of x^n

If we differentiate from first principles $y = x$ we obtain $dy/dx = 1$. If we differentiate $y = x^2$, as in the above example and maths in action, we obtain $dy/dx = 2x$. If we differentiate $y = x^3$ we obtain $dy/dx = 3x^2$. The pattern in these differentiations is that if we have $y = x^n$, then:

$$\frac{d}{dx}(x^n) = nx^{n-1} \qquad\qquad [3]$$

This relationship applies for positive, negative and fractional values of n.

Example

Determine the derivative of the functions (a) $y = x^{3/2}$, (b) $y = x^{-4}$.

(a) $\dfrac{dy}{dx} = \dfrac{3}{2}x^{\frac{3}{2}-1} = \dfrac{3}{2}x^{\frac{1}{2}}$

(b) $\dfrac{dy}{dx} = -4x^{-4-1} = -4x^{-5}$

Derivatives of trigonometric functions

Consider the determination of how the gradients of the graph of $y = \sin x$ (Figure 4.6(a)) vary with x. Examination of the graphs shows that:

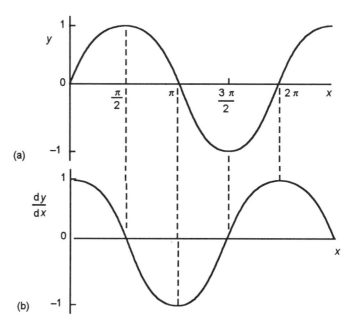

Figure 4.6 *Gradients of $y = \sin x$*

1 As x increases from 0 to $\pi/2$ then the gradient, which is positive, gradually decreases to become zero at $x = \pi/2$.

2 As x increases from $\pi/2$ to π then the gradient, which is now negative, becomes steeper and steeper to reach a maximum value at $x = \pi$.

3 As x increases from π to $3\pi/2$ then the gradient, which is negative, decreases to become zero at $x = 3\pi/2$.

4 As x increases from $3\pi/2$ to 2π the gradient, which is now positive again, increases to become a maximum at $x = 2\pi$.

Figure 4.6(b) shows the result that is obtained by plotting the gradients against x; it is a cosine curve. Thus, the derivative of $y = \sin x$ is:

$$\frac{\mathrm{d}}{\mathrm{d}x}(\sin x) = \cos x \qquad [4]$$

We can prove that the above is the case as follows. For the function $f(x) = \sin x$, we have $f(x + \delta x) - f(x) = \sin(x + \delta x) - \sin x$. Using equation [28] from Chapter 1 for the sum of two angles, $\sin(x + \delta x) = \sin x \cos \delta x + \cos x \sin \delta x$. As δx tends to 0 then $\cos \delta x$ tends to 1 and $\sin \delta x$ to δx. Thus, the derivative can be written as:

$$\frac{\mathrm{d}}{\mathrm{d}x}(\sin x) = \lim_{\delta x \to 0} \frac{\sin x + \delta x \cos x}{\delta x} = \cos x$$

If we had considered the function $\sin ax$ then we would have obtained:

$$\frac{\mathrm{d}}{\mathrm{d}x}(\sin ax) = a \cos ax \qquad [5]$$

In a similar manner we can consider $y = \cos x$ (Figure 4.7(a)) and the gradients at various points along the graph.

1 Between $x = 0$ and $x = \pi/2$ the gradient, which is negative, becomes steeper and steeper and reaches a maximum value at $x = \pi/2$.

2 Between $x = \pi/2$ and $x = \pi$ the gradient, which is negative decreases until it becomes zero at $x = \pi$.

3 Between $x = \pi$ and $x = 3\pi/2$ the gradient, which is positive, increases until it becomes a maximum at $x = 3\pi/2$.

4 Between $x = 3\pi/2$ and $x = 2\pi$ the gradient, which is positive, decreases to become zero at $x = 2\pi$.

Figure 4.7(b) shows how the gradient varies with x. The result is an inverted sine graph. Thus, for $y = \cos x$:

$$\frac{\mathrm{d}}{\mathrm{d}x}(\cos x) = -\sin x \qquad [6]$$

Key points

$\dfrac{\mathrm{d}}{\mathrm{d}x}(\sin ax) = a \cos ax$

$\dfrac{\mathrm{d}}{\mathrm{d}x}(\cos ax) = -a \sin ax$

The derivatives of tan ax, cosec ax, sec ax and cot ax can be derived using the quotient rule (see later in this chapter) as:

$\dfrac{\mathrm{d}}{\mathrm{d}x}(\tan ax) = a \sec^2 ax$

$\dfrac{\mathrm{d}}{\mathrm{d}x}(\mathrm{cosec}\, ax) = -a\, \mathrm{cosec}\, ax \cot ax$

$\dfrac{\mathrm{d}}{\mathrm{d}x}(\sec ax) = a \sec ax \tan ax$

$\dfrac{\mathrm{d}}{\mathrm{d}x}(\cot ax) = -a\, \mathrm{cosec}^2 ax$

The derivatives of sin $(ax + b)$, cos $(ax + b)$, etc. can be derived by using the chain rule (see later in this chapter) as:

$\dfrac{\mathrm{d}}{\mathrm{d}x}[\sin(ax+b)] = a\cos(ax+b)$

$\dfrac{\mathrm{d}}{\mathrm{d}x}[\cos(ax+b)] = -a\sin(ax+b)$

$\dfrac{\mathrm{d}}{\mathrm{d}x}[\tan(ax+b)] = a\sec^2(ax+b)$

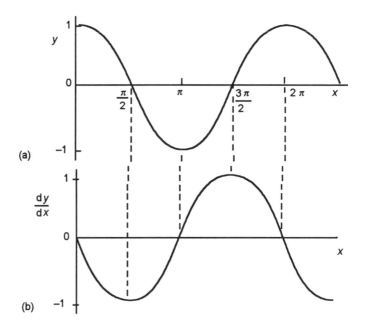

Figure 4.7 *Gradients of y = cos x*

If we had considered $y = \cos ax$, where a is a constant, then we would have obtained for the derivative:

$$\frac{d}{dx}(\cos ax) = -a\sin ax \qquad [7]$$

We can prove that the above is the case in a similar way to that used for the sine.

Example

Determine the derivatives of (a) sin 2x, (b) cos 3x.

(a) $\frac{d}{dx}(\sin 2x) = 2\cos 2x$, (b) $\frac{d}{dx}(\cos 3x) = -3\sin 3x$

Maths in action

Consider a sinusoidal current $i = I_m \sin \omega t$ passing through a *pure inductance*. A pure inductance is one which has only inductance and no resistance or capacitance. With an inductance a changing current produces a back e.m.f. of $-L \times$ the rate of change of current, i.e. $L\, di/dt$, where L is the inductance. The applied e.m.f. must overcome this back e.m.f. for a current to flow. Thus the voltage across the inductance is $L\, di/dt$. Hence:

$$v = L\frac{di}{dt} = L\frac{d}{dt}(I_m \sin \omega t) = \omega L I_m \cos \omega t$$

Since $\cos \omega t = \sin (\omega t + 90°)$, the current and the voltage are out of phase with the voltage leading the current by 90°.

Consider a circuit having just *pure capacitance* with a sinusoidal voltage $v = V_m \sin \omega t$ being applied across it. A pure capacitance is one which has only capacitance and no resistance or inductance. The charge q on the plates of a capacitor is related to the voltage v by $q = Cv$. Thus, since current is the rate of movement of charge dq/dt:

i = rate of change of q = rate of change of (Cv)

$= C \times$ (rate of change of v)

i.e. $i = C \, dv/dt$. Since current is the rate of change of charge q:

$$i = \frac{dq}{dt} = \frac{d}{dt}(Cv) = C\frac{d}{dt}(V_m \sin \omega t) = \omega C V_m \cos \omega t$$

Since $\cos \omega t = \sin (\omega t + 90°)$, the current and the voltage are out of phase, the current leading the voltage by 90°.

Derivatives of exponential functions

Consider the exponential equation $y = e^x$ and a small increase in x of δx. The corresponding increase in the value of y is δy where:

$$y + \delta y = e^{x+\delta x} = e^x e^{\delta x}$$

Thus:

$$\frac{dy}{dx} = \lim_{\delta x \to 0} \frac{\text{change in } y \text{ when } x \text{ changes by } \delta x}{\delta x}$$

$$= \lim_{\delta x \to 0} \frac{e^x e^{\delta x} - e^x}{\delta x} = \lim_{\delta x \to 0} \frac{e^x(e^{\delta x} - 1)}{\delta x}$$

If we let $\delta x = 0.01$ then $(e^{0.01} - 1)/0.01 = 1.005$. If we take yet smaller values of dx then in the limit this has the value 1. Thus:

$$\frac{d}{dx}e^x = e^x \qquad\qquad [8]$$

The derivative of e^x is e^x. Thus the gradient of the graph of $y = e^x$ at a point is equal to the value of y at that point (Figure 4.8). For example, at the point $x = 0$ on the graph the gradient is $y = e^0 = 1$. At $x = 2$ the gradient is $y = e^2 = 7.39$. At $x = -2$ the gradient is $y = e^{-2} = 0.14$.

If we had $y = e^{ax}$ then:

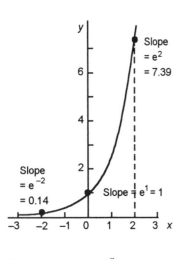

Figure 4.8 $y = e^x$

$$\frac{d}{dx}(e^{ax}) = a\,e^{ax} \tag{9}$$

Key point

$$\frac{d}{dx}(e^{ax}) = a\,e^{ax}$$

Example

Determine the derivative of $y = e^{2x}$.

$$\frac{d}{dx}e^{2x} = 2\,e^{2x}$$

Example

The variation of current i with time t in an electrical circuit is given by the equation $i = \sin 314t$. Derive an equation for the rate of change of current with time.

$$\frac{di}{dt} = \frac{d}{dt}(\sin 314t) = 314\cos 314t$$

Maths in action

The variation with time t of the displacement y of a system oscillating with simple harmonic motion is described by the equation:

$$y = A \sin \omega t$$

where A is the amplitude and ω the angular frequency. The linear velocity v is the rate of change if displacement with time, i.e. dy/dt, and so:

$$v = \frac{dy}{dt} = -A\omega \sin \omega t$$

The acceleration a is the rate of change of velocity, i.e. dv/dt, and so:

$$a = \frac{dv}{dt} = -A\omega^2 \cos \omega t = -A\omega^2 y$$

The acceleration is thus proportional to the displacement and the minus sign indicates that it is always in the opposite direction to that in which y increases, i.e. it is always directed towards the central rest position. This is the definition used for harmonic motion or cyclic motion which is referred to as *simple harmonic motion* or, for short, SHM.

4.1.2 Rules of differentiation

In this section the basic rules are developed for the differentiation of constant multiples, sums, products and quotients of functions and the chain rule for functions of functions.

Multiplication by a constant

Consider a multiple of some function, e.g. $cf(x)$ where c is a constant.

$$\frac{d}{dx}cf(x) = \lim_{\delta x \to 0}\frac{cf(x+\delta x) - cf(x)}{\delta x}$$

$$= c\lim_{\delta x \to 0}\frac{f(x+\delta x) - f(x)}{\delta x} = c\frac{d}{dx}f(x) \qquad [10]$$

The derivative of some function multiplied by a constant is the same as the constant multiplying the derivative of the function.

Example

Determine the derivatives of (a) $4x^2$, (b) $2 \sin 3x$, (c) $y = \dfrac{1}{3\sqrt{x}}$.

(a) $\dfrac{d}{dx}4x^2 = 4 \times 2x = 8x$

(b) $\dfrac{d}{dx}(2\sin 3x) = 2 \times 3\sin 3x = 6\sin 3x$

(c) $\dfrac{d}{dx}\left(\dfrac{1}{3\sqrt{x}}\right) = \dfrac{d}{dx}\left(\dfrac{1}{3}x^{-1/2}\right) = \dfrac{1}{3} \times \left(-\dfrac{1}{2}\right)x^{-3/2} = -\dfrac{1}{6}x^{-3/2}$

Sums of functions

Consider a function which can be considered to be a sum of a number of other functions, e.g. $y = f(x) + g(x)$:

$$\frac{d}{dx}[f(x) + g(x)] = \lim_{\delta x \to 0}\frac{\big[f(x+\delta x) + g(x+\delta x)\big] - \big[f(x) + g(x)\big]}{\delta x}$$

$$= \lim_{\delta x \to 0}\left[\frac{f(x+\delta x) - f(x)}{\delta x} + \frac{g(x+\delta x) - g(x)}{\delta x}\right]$$

$$= \lim_{\delta x \to 0}\left[\frac{f(x+\delta x) - f(x)}{\delta x}\right] + \lim_{\delta x \to 0}\left[\frac{g(x+\delta x) - g(x)}{\delta x}\right]$$

$$= \frac{d}{dx}f(x) + \frac{d}{dx}g(x) \qquad [11]$$

The derivative of the sum of two differentiable functions is the sum of their derivatives.

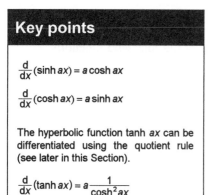

As an illustration, consider the differentiation of the hyperbolic function $y = \sinh x$. This function (see Section 1.8) can be written as $\frac{1}{2}(e^x - e^{-x})$. Thus:

$$\frac{d}{dx}(\sinh x) = \frac{d}{dx}\frac{1}{2}(e^x - e^{-x}) = \frac{1}{2}(e^x + e^x) = \cosh x \tag{12}$$

In a similar way we can differentiate sinh ax and cosh ax, obtaining:

$$\frac{d}{dx}(\sinh ax) = a \cosh ax \tag{13}$$

$$\frac{d}{dx}(\cosh ax) = a \sinh ax \tag{14}$$

Example

Determine the derivatives of:

(a) $y = 2x^3 + x^2$, (b) $y = \sin x + \cos 2x$, (c) $y = e^{4x} + x$

(a) $\dfrac{dy}{dx} = 6x^2 + 2x$ (b) $\dfrac{dy}{dx} = \cos x - 2\sin 2x$

(c) $\dfrac{dy}{dx} = 4e^{4x} + 1$

The product rule

Consider a function $y = f(x)g(x)$ which is the product of two other differentiable functions, e.g. $y = x \sin x$:

$$\frac{d}{dx}[f(x)g(x)] = \lim_{\delta x \to 0}\frac{f(x+\delta x)g(x+\delta x) - f(x)g(x)}{\delta x}$$

We can simplify this by adding and subtracting the same quantity to the numerator, namely $f(x + \delta x)g(x)$, to give:

$$\frac{d}{dx}[f(x)g(x)]$$

$$= \lim_{\delta x \to 0}\frac{f(x+\delta x)g(x+\delta x) + f(x+\delta x)g(x) - f(x+\delta x)g(x) - f(x)g(x)}{\delta x}$$

$$= \lim_{\delta x \to 0}\left[f(x+\delta x)\frac{g(x+\delta x) - g(x)}{\delta x} + g(x)\frac{f(x+\delta x) - f(x)}{\delta x}\right]$$

$$= f(x)\frac{d}{dx}g(x) + g(x)\frac{d}{dx}f(x) \tag{15}$$

This is often written in terms of u and v, where $u = f(x)$ and $v = g(x)$:

$$\frac{d}{dx}uv = u\frac{dv}{dx} + v\frac{du}{dx} \tag{16}$$

Example

Determine the derivatives of the following functions:

(a) $y = x \sin x$, (b) $y = x^2 e^{3x}$, (c) $y = (2 + x)^2$,
(d) $y = x e^x \sin x$

(a) $\dfrac{dy}{dx} = x\dfrac{d}{dx}(\sin x) + \sin x \dfrac{d}{dx}x = x\cos x + \sin x$

(b) $\dfrac{dy}{dx} = x^2 \dfrac{d}{dx}e^{3x} + e^{3x}\dfrac{d}{dx}x^2 = 3x^2\, e^{3x} + 2x\, e^{3x}$

(c) This can be written as $(2 + x)(2 + x)$ and so:

$\dfrac{dy}{dx} = (2+x)\dfrac{d}{dx}(2+x) + (2+x)\dfrac{d}{dx}(2+x) = 2(2+x)$

(d) This product has three terms and so we have to carry out the differentiation in two stages. Thus, if we first consider $x\, e^x$ as one term and the $\sin x$ as the other term:

$\dfrac{dy}{dx} = x\, e^x \dfrac{d}{dx}\sin x + \sin x \dfrac{d}{dx}(x\, e^x)$

$\quad\quad = x\, e^x \cos x + \sin x \dfrac{d}{dx}(x\, e^x)$

We can then use the product rule to evaluate the derivative of $x\, e^x$.

$\dfrac{d}{dx}(x\, e^x) = x\dfrac{d}{dx}e^x + e^x\dfrac{d}{dx}x = x\, e^x + e^x$

Hence:

$\dfrac{dy}{dx} = x\, e^x \cos x + x\, e^x + e^x$

The quotient rule

Consider obtaining the derivative of a function which is the quotient of two other functions, e.g. $f(x)/g(x)$:

$$\frac{d}{dx}\left(\frac{f(x)}{g(x)}\right) = \lim_{\delta x \to 0} \frac{\dfrac{f(x+\delta x)}{g(x+\delta x)} - \dfrac{f(x)}{g(x)}}{\delta x}$$

$$= \lim_{\delta x \to 0} \frac{g(x)f(x+\delta x) - f(x)g(x+\delta x)}{\delta x\, g(x)g(x+\delta x)}$$

Adding and subtracting $f(x)g(x)$ to the numerator enables the above equation to be simplified:

$$\frac{d}{dx}\left(\frac{f(x)}{g(x)}\right) = \lim_{\delta x \to 0} \frac{g(x)f(x+\delta x) + f(x)g(x) - f(x)g(x) - f(x)g(x+\delta x)}{\delta x g(x)g(x+\delta x)}$$

$$= \frac{\displaystyle\lim_{\delta x \to 0} \frac{g(x)\left[f(x+\delta x) - f(x)\right]}{\delta x} - \lim_{\delta x \to 0} \frac{f(x)\left[g(x+\delta x) - g(x)\right]}{\delta x}}{\displaystyle\lim_{\delta x \to 0} \left[g(x)g(x+\delta x)\right]}$$

$$= \frac{g(x)\dfrac{d}{dx}f(x) - f(x)\dfrac{d}{dx}g(x)}{\left[g(x)\right]^2} \qquad [17]$$

Key point

$$\frac{d}{dx}\left(\frac{u}{v}\right) = \frac{v\dfrac{du}{dx} - u\dfrac{dv}{dx}}{v^2}$$

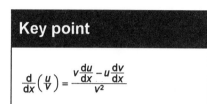

This is often written in terms of u and v, where $u = f(x)$ and $v = g(x)$:

$$\frac{d}{dx}\left(\frac{u}{v}\right) = \frac{v\dfrac{du}{dx} - u\dfrac{dv}{dx}}{v^2} \qquad [18]$$

Note that if we have just the reciprocal of some function, i.e. $1/g(x)$, then we have $f(x) = 1$ and so equation [18] gives:

$$\frac{d}{dx}\left[\frac{1}{g(x)}\right] = -\frac{1}{\left[g(x)\right]^2}\frac{d}{dx}g(x) \qquad [19]$$

Equation [18] can be used to determine the derivative of $\tan x$, since $\tan x = \sin x/\cos x$. Thus $f(x) = \sin x$ and $g(x) = \cos x$. Hence:

$$\frac{d}{dx}(\tan x) = \frac{\cos x \dfrac{d}{dx}\sin x - \sin x \dfrac{d}{dx}\cos x}{\cos^2 x}$$

$$= \frac{\cos^2 x + \sin^2 x}{\cos^2 x} = \frac{1}{\cos^2 x} \qquad [20]$$

Likewise, equation [18] can be used to determine the derivative of $\tanh x$.

$$\frac{d}{dx}(\tanh x) = \frac{d}{dx}\left(\frac{\sinh x}{\cosh x}\right)$$

$$= \frac{\cosh x \cosh x - \sinh x \sinh x}{\cosh^2 x} = \frac{1}{\cosh^2 x} \qquad [21]$$

Example

Determine the derivative of $y = (2x^2 + 5x)/(x + 3)$.

Using equation [18] with $f(x) = 2x^2 + 5x$ and $g(x) = x + 3$:

$$\frac{dy}{dx} = \frac{(x+3)(4x+5) - (2x^2+5x)(1)}{(x+3)^2} = \frac{2x^2+12x+1}{(x+3)^2}$$

Example

Determine the derivative of $y = x\,e^x/\cos x$.

This example requires the use of both the quotient and product rules for differentiation. Using equation [19] with $f(x) = x\,e^x$ and $g(x) = \cos x$:

$$\frac{dy}{dx} = \frac{\cos x \dfrac{d}{dx}(x\,e^x) - x\,e^x(-\sin x)}{\cos^2 x}$$

Now using equation [16] to obtain the derivative for the product $x\,e^x$:

$$\frac{dy}{dx} = \frac{\cos x(x\,e^x + e^x) - x\,e^x(-\sin x)}{\cos^2 x}$$

$$= \frac{e^x(x\cos x + \cos x + x\sin x)}{\cos^2 x}$$

The chain rule

Suppose we have $y = \cos x^4$ and, in order to differentiate it, write it in the form $y = \cos u$ and $u = x^4$. We can then obtain dy/du and du/dx, but how from them do we obtain dy/dx?

Consider the function $y = f(u)$ where $u = g(x)$ and the obtaining of the derivative of $y = f(g(x))$. For $u = g(x)$ a small increase of δx in the value of x causes a corresponding small increase of δu in the value of u. But $y = f(u)$ and so the small increase δu causes a correspondingly small increase of δy in the value of y. We can write, since the δu terms cancel:

$$\frac{\delta y}{\delta x} = \frac{\delta y}{\delta u} \times \frac{\delta u}{\delta x}$$

Thus:

$$\frac{dy}{dx} = \lim_{\delta x \to 0}\left(\frac{\delta y}{\delta u} \times \frac{\delta u}{\delta x}\right) = \lim_{\delta x \to 0}\left(\frac{\delta y}{\delta u}\right) \lim_{\delta x \to 0}\left(\frac{\delta u}{\delta x}\right)$$

and so:

$$\frac{dy}{dx} = \frac{dy}{du} \times \frac{du}{dx} \tag{22}$$

This is known as the *function of a function rule* or the *chain rule*.

The chain rule can be used to determine the derivative a function such as $y = \sin x^n$, for n being positive or negative or fractional. Let $u = x^n$ and so consequently $y = \sin u$. Then $du/dx = nx^{n-1}$ and $dy/du = \cos u$. Hence, using the chain rule (equation [22]) we have $dy/dx = \cos u \times nx^{n-1} = nx^{n-1}\cos x^n$.

Key point

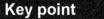

$$\frac{dy}{dx} = \frac{dy}{du} \times \frac{du}{dx}$$

Another application of the chain rule is to determine the derivatives of functions of the form $y = (ax + b)^n$, $y = e^{ax + b}$, $y = \sin(ax + b)$, etc. With such functions we let $u = ax + b$ and so then we have, for the three examples, $y = u^n$, $y = e^u$, $y = \sin u$. Then we have $du/dx = a$ and $dy/du = nx^{n-1}$, $dy/du = e^u$, $dy/du = \cos u$. Using the chain rule we then obtain $dy/dx = anu^{n-1}$, $dy/dx = a\,e^{ax+b}$ and $dy/dx = a \cos u$. Thus, for the three examples, we have:

$$\frac{dy}{dx} = an(ax + b)^{n-1}, \quad \frac{dy}{dx} = a\,e^{ax+b}, \quad \frac{dy}{dx} = a\cos(ax + b), \text{ etc.}$$

Example

Determine the derivative of $y = (2x - 5)^4$.

Let $u = 2x - 5$ and so $y = u^4$. Then $du/dx = 2$ and $dy/du = 4u^3$ and so, using equation [22]:

$$\frac{dy}{dx} = \frac{dy}{du} \times \frac{du}{dx} = 4u^3 \times 2 = 8u^3 = 8(2x - 5)^3$$

Example

Determine the derivative of $y = \sin x^3$.

Let $u = x^3$ and so $y = \sin u$. Then $dy/du = 3x^2$ and $dy/du = \cos u$ and so, using equation [22]:

$$\frac{dy}{dx} = \frac{dy}{du} \times \frac{du}{dx} = 3x^2 \times \cos u = 3x^2 \cos x^3$$

Example

Determine the derivative of $y = \sqrt{\dfrac{x^2}{x^2 + 1}}$.

Let $u = x^2/(x^2 + 1)$ and so $y = u^{1/2}$. Using the quotient rule:

$$\frac{du}{dx} = \frac{(x^2 + 1)2x - x^2(2x)}{(x^2 + 1)^2}$$

Using the chain rule:

$$\frac{dy}{dx} = \tfrac{1}{2}u^{-1/2} \times \frac{(x^2 + 1)2x - x^2(2x)}{(x^2 + 1)^2}$$

$$= \tfrac{1}{2}\left(\frac{x^2}{x^2 + 1}\right)^{-1/2} \frac{(x^2 + 1)2x - x^2(2x)}{(x^2 + 1)^2}$$

$$= \frac{x^{1/2}}{(x^2+1)^{3/2}} = \sqrt{\frac{x}{(x^2+1)^3}}$$

4.1.3 Higher-order derivatives

Consider a moving object for which we have a relationship between the displacement s of the object and time t of the form:

$$s = ut + \tfrac{1}{2}at^2$$

where u and a are constants. We can plot this equation to give a distance–time graph. If we differentiate this equation we obtain:

$$\frac{ds}{dt} = u + at$$

ds/dt is the gradient of the distance–time graph. It also happens to be the velocity. The gradient varies with time. We could thus plot a velocity–time graph, i.e. a ds/dt graph against t. Then differentiating for a second time, to obtain the gradients to this graph, we obtain the acceleration a.

$$\frac{d}{dt}\left(\frac{ds}{dt}\right) = a$$

The derivative of a derivative is called the *second derivative* and can be written as:

$$\frac{d}{dx}\left(\frac{dy}{dx}\right) \text{ or } \frac{d^2y}{dx^2}$$

The first derivative gives information about how the gradients of the tangents change. The second derivative gives information about the rate of change of the gradient of the tangents.

If the second derivative is then differentiated we obtain the third derivative.

$$\frac{d}{dx}\left(\frac{d^2y}{dx^2}\right) \text{ or } \frac{d^3y}{dx^3}$$

This, in turn, may be differentiated to give a fourth derivative, and so on.

$$\frac{d}{dx}\left(\frac{d^3y}{dx^3}\right) \text{ or } \frac{d^4y}{dx^4}$$

Example
Determine the second derivative of $y = x^3$.

The first derivative is:

$$\frac{dy}{dx} = 3x^2$$

The second derivative is given by differentiating this equation again:

$$\frac{d^2y}{dx^2} = 6x$$

Example

Determine the second derivative of $y = x^4 + 3x^2$.

The first derivative is

$$\frac{dy}{dx} = 4x^3 + 6x$$

The second derivative is

$$\frac{d^2y}{dx^2} = 12x^2 + 6$$

Maths in action

This illustrates how differential calculus may be used in the analysis of a beam which is deflected in one plane as a result of loading. Consider a beam which is bent into a circular arc and the radius R of the arc. For a segment of circular arc (Figure 4.9), the angle $\delta\theta$ subtended at the centre is related to the arc length δs by $\delta s = R\delta\theta$. Because the deflections obtained with beams are small δx is a reasonable approximation to δs and so we can write $\delta x = R\delta\theta$ and $1/R = \delta\theta/\delta x$. The slope of the straight line joining the two end points of the arc is $\delta y/\delta x$ and thus $\tan \delta\theta = \delta y/\delta x$. Since the angle will be small we can make the approximation that $\delta\theta = \delta y/\delta x$. Hence:

$$\frac{1}{R} = \frac{\delta\theta}{\delta x} = \frac{\delta}{\delta x}\left(\frac{\delta y}{\delta x}\right)$$

In the limit we can thus write:

$$\frac{1}{R} = \frac{d^2y}{dx^2}$$

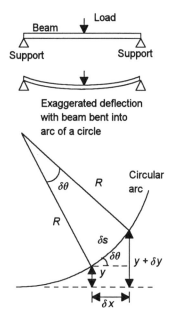

Figure 4.9 *The deflection curve of radius R*

When a beam is bent as a result of the application of a bending moment M it curves with a radius R given by the general bending equation as:

$$\frac{1}{R} = \frac{M}{EI}$$

where E is the modulus of elasticity and I the second moment of area (a property of the shape of the beam) and so we can write:

$$M = EI\frac{d^2y}{dx^2}$$

This differential equation provides the means by which the deflections of beams can be determined.

In Section 4.1.3 the determination of maximum and minimum points is discussed. We can use the criteria for a maximum in order to determine the conditions necessary for the deflection of the beam to be a maximum. In Section 4.2 we then use the above differential equation, with the condition for maximum deflection, to determine the maximum deflection of a beam.

4.1.4 Maxima and minima

There are many situations in engineering where we need to establish maximum or minimum values. For example, with a projectile we might need to determine the maximum height reached. With an electrical circuit we might need to determine the condition for maximum power to be dissipated.

Consider a graph of y against x when the values of y depend in some way on the values of x. Points on the graph at which $dy/dx = 0$ are called *turning points* and can be:

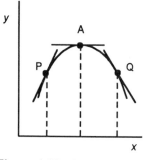

Figure 4.10 *A maximum*

- *A local maximum*

  The term *local* is used because the value of y is only necessarily a maximum for points in the locality and there could be higher values of y elsewhere on the graph. Figure 4.10 shows such a maximum. At the maximum point A we have zero gradient for the tangent, i.e. $dy/dx = 0$. Consider two points P and Q close to A, with P having a value of x less than that at A and Q having a value greater than that at A. The gradient of the tangent at P is positive, the gradient of the tangent at Q is negative. Thus for a maximum we have the gradient changing from being positive prior to the turning point to negative after it.

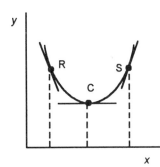

Figure 4.11   *A minimum*

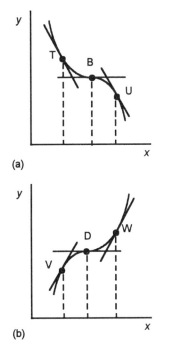

(a)

(b)

Figure 4.12   *Points of inflexion*

Key points

For the gradients in the vicinity of maxima, minima and points of inflexion, in moving from points before to after the turning point:

At a *maximum* the gradient changes from being positive to negative; the second derivative is negative.

At a *minimum* the gradient changes from being negative to positive; the second derivative is positive.

At a *point of inflexion* the sign of the gradient does not change.

• *A local minimum*

The term *local* is used because the value of y is only necessarily a minimum for points in the locality and there could be lower values of y elsewhere on the graph. Figure 4.11 shows such a minimum. At the minimum point C we have zero gradient for the tangent, i.e. $dy/dx = 0$. Consider two points R and S close to A, with R having a value of x less than that at C and S having a value greater than that at C. The gradient of the tangent at R is negative, the gradient of the tangent at S is positive. Thus for a minimum we have the gradient changing from being negative prior to the turning point to positive after it.

• *A point of inflexion*

Consider points of inflexion, as illustrated in Figure 4.12. At such points $dy/dx = 0$. However, in neither of the graphs is there a local maximum or minimum. In Figure 4.12(a), the gradient at a point T prior to the point is negative and the gradient at a point U after the point is also negative. In Figure 4.12(b), the gradient at a point V prior to the point is positive and the gradient at a point W after the point is also positive. For a point of inflexion the sign of the gradient prior to the point is the same as that after the point.

The gradient at a point on a graph is given by dy/dx. We can thus determine whether a turning point is a maximum, a minimum or a point of inflexion by considering how the value of dy/dx changes for a value of x smaller than the turning point value compared to that for a value of x greater than the turning point value.

There is an alternative method we can use to distinguish between maximum and minimum points. We need to establish how the gradient changes in going from points before to after turning points. Consider, for a maximum, a graph of the gradients plotted against x (Figure 4.13(a)). The gradients prior to the maximum are positive and decrease in value to become zero at the maximum. They then become negative and as x increases become more and more negative. The second derivative d^2y/dx^2 measures the rate of change of dy/dx with x, i.e. the gradient of the dy/dx graph. The gradient of the gradient graph is negative before, at and after the maximum point. Hence at a maximum d^2y/dx^2 is negative.

Consider a minimum (Figure 4.13(b)). The gradients prior to the minimum are negative and become less negative until they become zero at the minimum. As x increases beyond the minimum the gradients become positive, increasing in value as x increases. The second derivative d^2y/dx^2 measures the rate of change of dy/dx with x, i.e. the gradient of the dy/dx graph. The gradient of the gradient graph is positive before, at and after the minimum point. Hence at a minimum d^2y/dx^2 is positive.

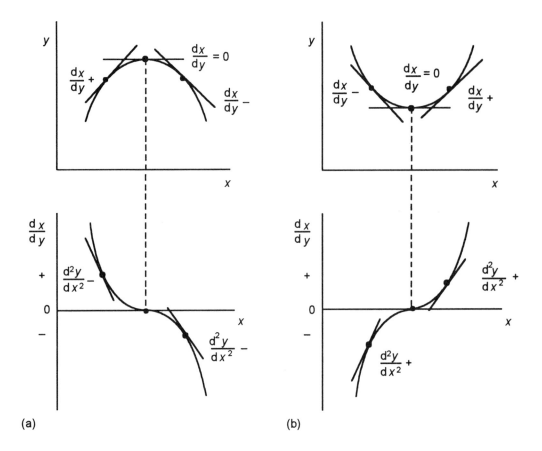

Figure 4.13 *(a) A maximum, (b) a minimum*

Example

Determine, and identify the form of, the turning points on a graph of the equation $y = 2x^3 - 3x^2 - 12x$.

Differentiating the equation gives

$$\frac{dy}{dx} = 6x^2 - 6x - 12$$

Thus the gradient of the graph is zero when $6x^2 - 6x - 12 = 0$. We can rewrite this as:

$$6(x^2 - x - 2) = 6(x + 1)(x - 2) = 0$$

The gradient is zero, and hence there are turning points, at $x = -1$ and $x = 2$.

To establish the form of these turning points consider the gradients just prior to and just after them.

Prior to the $x = -1$ turning point at $x = -2$, the gradient is $6x^2 - 6x - 12 = 6 \times (-2)^2 - 6 \times (-2) - 12 = 12$. After the point at $x = 0$ we have a gradient of -12. Thus the gradient prior to the $x = -1$ point is positive and after the point it is negative. The point is thus a maximum.

Consider the $x = 2$ turning point. Prior to the turning point at $x = 0$ the gradient is -12. After the turning point at $x = 3$ the gradient is $6 \times 3^2 - 6 \times 3 - 12 = 24$. Thus the gradient prior to the $x = 2$ point is negative and after the point it is positive. The point is thus a minimum.

Alternatively we could determine the form of the turning points by considering the sign of the second derivative at the points. The second derivative is obtained by differentiating the dy/dx equation. Thus

$$\frac{d^2y}{dx^2} = 12x - 6$$

At $x = -1$ then the second derivative is $12 \times (-1) - 6 = -18$. The negative value indicates that the point is a maximum. At $x = 2$ the second derivative is $12 \times 2 - 6 = 18$. The positive value indicates that the point is a maximum.

Figure 4.14 shows a graph of the equation $y = 2x^3 - 3x^2 - 12x$, showing the maximum at $x = -1$ and the minimum at $x = 2$.

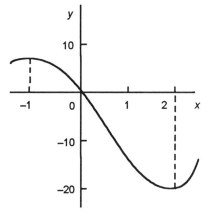

Figure 4.14   *Example*

Example

The displacement y in metres of an object is related to the time t in seconds by the equation $y = 5 + 4t - t^2$. Determine the maximum displacement.

Differentiating the equation gives:

$$\frac{dy}{dt} = 4 - 2t$$

dy/dt is 0 when $t = 2$ s. There is thus a turning point at the displacement $y = 5 + 4 \times 2 - 2^2 = 9$ m. We need to check that this is a maximum displacement. The gradient prior to the turning point at $t = 1$ has the value $4 - 2 = 2$. After the turning point at $t = 3$ it has the value $4 - 6 = -2$. The gradient changes from a positive value prior to the turning point to a negative value afterwards. It is thus a maximum.

Alternatively we could have established this by determining the second derivative. Differentiating $4 - 2t$ gives $d^2y/dx^2 = -2$. Thus, the turning point is a maximum.

Example

If the sum of two numbers is 40, determine the values which will give the minimum value for the sum of their squares.

Let the two numbers be x and y. Then we must have $x + y = 40$. We have to find the minimum value of S when we have $S = x^2 + y^2$. We need an equation which expresses the sum in terms of just one variable. Thus substituting from the previous equation gives

$$S = x^2 + (40 - x)^2 = x^2 + 1600 - 80x + x^2$$

$$= 2x^2 - 80x + 1600$$

Differentiating this equation, then

$$\frac{dS}{dx} = 4x - 80$$

The value of x to give a zero value for dS/dx is when $4x - 80 = 0$ and so when $x = 20$.

We can check that this is the value giving a minimum by considering the values of dS/dx prior to and after the point. Thus prior to the point at $x = 19$ we have $dS/dx = 4 \times 19 - 80 = -4$. After the point at $x = 21$ we have $dS/dx = 4 \times 21 - 80 = 4$. Thus dS/dx changes from being negative to positive. The turning point is thus a minimum. Alternatively we could check that this is a minimum by obtaining the second derivative. Differentiating $4x - 80$ gives $d^2S/dx^2 = 4$. Since this is positive then we have a minimum. Thus the two numbers which will give the required minimum are 20 and 20.

Example

Determine the maximum area of a rectangle with a perimeter of 32 cm.

If the width of the rectangle is w and its length L then the area $A = wL$. But the perimeter has a length of 32 cm. Thus $2w + 2L = 32$. If we eliminate w from the two equations:

$A = L(16 - L) = 16L - L^2$

Hence:

$$\frac{dA}{dL} = 16 - 2L$$

dA/dL is zero when $16 - 2L = 0$ and so when $L = 8$.

We can check that this gives a maximum area by considering values of the gradient at values of L below and above 8. At $L = 7$ the gradient is 2 and at $L = 9$ it is −2. It is thus a maximum. Alternatively we could have considered the second derivative. Since $d^2A/dL^2 = -2$ and so is negative, we have a maximum.

For a maximum area we must therefore have $L = 8$ cm, and, after substituting this value in $2w + 2L = 32$, $w = 8$ cm.

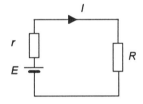

Figure 4.15   *Circuit*

Maths in action

Consider the circuit shown in Figure 4.15 where a d.c. source of e.m.f. E and internal resistance r supplies a load of resistance R. The power P supplied to the load is I^2R with the current I being $E/(R + r)$. Thus:

$$P = \frac{E^2R}{(R+r)^2}$$

Differentiating with respect to x by using the quotient rule gives:

$$\frac{dP}{dR} = \frac{(R+r)^2 - 2R(R+r)}{(R+r)^4} = \frac{E^2}{(R+r)^2} - \frac{2E^2R}{(R+r)^3}$$

$dP/dR = 0$ when:

$$\frac{2E^2R}{(R+r)^3} = \frac{E^2}{(R+r)^2}$$

and so $R = r$. We can check that this is the condition for maximum power transfer by considering the second derivative. We have the sum of two terms and so for the $E^2/(R + r)^2$ term, let $u = R + r$ and $y = E^2/u^2$. Then, $du/dR = 1$ and $dy/du = -2E^2/u^3$ and $dy/dR = -2E^2/(R + r)^3$. For $y = 2E^2R/(R + r)^3$ we can use the quotient rule to give $dy/dR = [(R + r)^3 2E^2 - 2E^2R3(R + r)^2]/(R + r)^6$ which can be simplified to $2E^2/(R + r)^3 - 6E^2R/(R + r)^4$. Hence:

$$\frac{d^2P}{dR^2} = -\frac{4E^2}{(R+r)^3} + \frac{6E^2R}{(R+r)^4}$$

With $R = r$ the second derivative is negative and so we have maximum power transfer.

4.1.5 Inverse functions

If we have a function y which is a continuous function of x then the derivative, i.e. the slope of the tangent to a graph of y plotted against x, is dy/dx. However, if we have x as a continuous function of y then the derivative, i.e. the slope of the tangent to a graph of x plotted against y, is dx/dy. How are these derivatives related? We might, for example, have $y = x^2$ and so $dy/dx = 2x$. For the inverse function $x = \sqrt{y}$ and $dx/dy = \frac{1}{2}y^{-1/2}$.

If we have a function $y = f(x)$ then we can write for the inverse $x = g(y)$. Thus $x = g\{f(x)\}$. Differentiating both sides of this equation with respect to x, using the chain rule for the right-hand side, gives:

$$1 = \frac{dx}{dy} \times \frac{dy}{dx}$$

Hence, with $y = f(x)$, the derivatives of the inverse function can be derived by using:

$$\frac{dy}{dx} = \frac{1}{\dfrac{dx}{dy}} \tag{23}$$

For example, for $y = x^2$ we have $dy/dx = 2x$; the inverse function is $x = \sqrt{y}$ and $dx/dy = \frac{1}{2}y^{-1/2}$. Then $dy/dx = 1/(\frac{1}{2}y^{-1/2}) = 2\sqrt{y} = 2x$.

Example

Determine dy/dx for the function described by the equation $x = y^2 + 2y$.

It is easier to obtain dx/dy from the equation and thus the problem is tackled by doing that operation first. Thus $dx/dy = 2y + 2$. Then, using equation [23]:

$$\frac{dy}{dx} = \frac{1}{2y+2}$$

Logarithmic functions

Consider the function $y = \ln x$. We can write this as $x = e^y$. Differentiating x with respect to y gives:

$$\frac{dx}{dy} = e^y = x$$

Hence, using equation [23]:

$$\frac{d}{dx}(\ln x) = \frac{1}{x} \qquad\qquad\qquad [24]$$

Note that since x must be positive for $\ln x$ to have any meaning, equation [24] only applies for positive values of x.

Example

Determine the derivative of $y = e^{-x} \ln x$.

Using the product rule then:

$$\frac{dy}{dx} = e^{-x}\left(\frac{1}{x}\right) - \ln x\ (e^{-x}) = e^{-x}\left(\frac{1}{x} - \ln x\right)$$

Problems 4.1

1 Determine the derivatives of the following functions:

(a) $y = x^5$, (b) $y = 2x^{-4}$, (c) $y = -3x^2$, (d) $y = \frac{1}{2}x$,

(e) $y = 2\pi x^2$, (f) $y = \tan 3x$, (g) $y = 5\cos 2x$,

(h) $y = 4\ e^{x/2}$, (i) $y = 2\ e^{-2x}$, (j) $y = 3\ e^{3x}$, (k) $y = \dfrac{5}{3\sqrt{x}}$,

(l) $y = \dfrac{6}{5x^2}$, (m) $y = \dfrac{7}{\sqrt{3x}}$, (n) $y = \dfrac{5}{(2x)^3}$, (o) $y = \sqrt{3x}$,

(p) $y = (3x - 2x^2)(5 + 4x)$, (q) $y = 5x\sin x$, (r) $x\ e^{x/2}$,

(s) $(x^2 + 1)\sin x$, (t) $y = \dfrac{2x+1}{x-6}$, (u) $y = \dfrac{x+1}{\sqrt{x}}$,

(v) $y = \dfrac{\sin x}{x}$, (w) $y = \dfrac{e^{2x}}{x^2+1}$, (x) $y = \dfrac{\sinh 2x}{\cosh 3x}$,

(y) $y = \dfrac{1}{2-7x}$, (z) $y = \dfrac{1}{\sqrt{3-x}}$

2 Determine the second derivatives of the following functions:

(a) $y = x^2 + 2x$, (b) $y = \sin 2x$, (c) $y = \dfrac{1}{x^2}$,

(d) $y = 3x^4 - x^2 - \dfrac{1}{x}$, (e) $y = x^4 + 2x^3 - 8x + 5$,

(f) $y = \dfrac{x+1}{\sqrt{x}}$

3 Determine the velocity and acceleration after a time of 2 s for an object which has a displacement x which is a function of time t and given by $x = 12 + 15t - 2t^2$, with t being in seconds.

4 Determine the velocity and acceleration at a time t for an object which has a displacement x in metres given by $x = 3 \sin 2t + 3 \cos 3t$, t being in seconds.

5 The voltage v, in volts, across a capacitor of capacitance 2 μF varies with time t, in seconds, according to the equation $v = 3 \sin 5t$. Determine how the current varies with time.

6 The current i, in amps, through an inductor of inductance 0.05 H varies with time t, in seconds, according to the equation $i = 10(1 - e^{-100t})$. Determine how the potential difference across the inductor varies with time.

7 The volume of a cone is one-third the product of the base area and the height. For a cone with a height equal to the base radius, determine the rate of change of cone volume with respect to the base radius.

8 The volume of a sphere of radius r is $\frac{4}{3}\pi r^3$. Determine the rate of change of the volume with respect to the radius.

9 With the Doppler effect, the frequency f_o heard by an observer when a sound source of frequency f_s is moving away from the observer with a velocity v is given by $f_o = f_s/(1 + v/c)$, where c is the velocity of sound. Determine the rate of change of the observed frequency with respect to the velocity.

10 The length L of a metal rod is a function of temperature T and is given by the equation $L = L_0(1 + aT + bT^2)$. Determine an equation for the rate of change of length with temperature.

11 Determine and identify the form of the turning points on graphs of the following functions:

(a) $y = x^2 - 4x + 3$, (b) $y = x^3 - 6x^2 + 9x + 3$,

(c) $y = x^5 - 5x$, (d) $y = \sin x$ for x between 0 and 2π,

(e) $y = 2x^3 + 3x^2 - 12x + 3$

12 A cylindrical container, open at one end, has a height of h m and a base radius of r m. The total surface area of the container is to be 3π m^2. Determine the values of h and r which will make the volume a maximum.

13 A cylindrical metal container, open at one end, has a height of h cm and a base radius of r cm. It is to have an internal volume of 64π cm^3. Determine the dimensions of the container which will require the minimum area of metal sheet in its construction.

14 The bending moment M of a uniform beam of length L at a distance x from one end is given by $M = \frac{1}{2}wLx - \frac{1}{2}wx^2$, where w is the weight per unit length of beam. Determine the value of x at which the bending moment is a maximum.

15 The deflection y of a beam of length L at a distance x from one end is found to be given by $y = 2x^4 - 5Lx^3 + 2L^2x^2$. Determine the values of x at which the deflection is a maximum.

16 Determine the maximum value of the alternating voltage described by the equation $v = 40 \cos 1000t + 15 \sin 1000t$ V.

17 The intensity of illumination from a point light source of intensity I at a distance d from it is I/d^2. Determine the point along the line between two sources 10 m apart at which the intensity of illumination is a minimum if one of the sources has eight times the intensity of the other.

18 Determine the maximum rate of change with time of the alternating current $i = 10 \sin 1000t$ mA, the time t being in seconds.

19 The deflection y of a propped cantilever of length L at a distance x from the fixed end is given by:

$$y = \frac{1}{EI}\left(\frac{5wLx^3}{48} - \frac{wL^2x^2}{16} - \frac{wx^4}{24} \right)$$

where w is the weight per unit length and E and I are constants. Determine the value of x at which the deflection is a maximum.

20 The e.m.f. E produced by a thermocouple depends on the temperature T and is given by $E = aT + bT^2$. Determine the temperature at which the e.m.f. is a maximum.

21 The horizontal range R of a projectile projected with a velocity v at an angle θ to the horizontal is given by $R = (v^2/g) \sin 2\theta$. Determine the angle at which the range is a maximum for a particular velocity.

22 A 100 cm length of wire is to be bent to form two squares, one with side x and the other with side y. Determine the values of x and y which give the minimum area enclosed by the squares.

23 The rate r at which a chemical reaction proceeds depends on the quantity x of a chemical and is given by $r = k(a - x)(b + x)$. Determine the maximum rate.

24 A cylinder has a radius r and height h with the sum of the radius and height being 2 m. Determine the radius giving the maximum volume.

25 A rectangle is to have an area of 36 cm². Determine the lengths of the sides which will give a minimum value for the perimeter.

26 An open tank is to be constructed with a square base and vertical sides and to be able to hold, when full to the brim, 32 m³ of water. Determine the dimensions of the tank if the area of sheet metal used is to be a minimum.

4.2 Integration

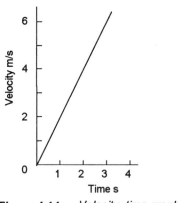

Figure 4.14 *Velocity–time graph*

Key point

Differentiation is the determination of the relationship for the gradient of a graph. We can define integration as the mathematical process which reverses differentiation, i.e. given the gradient relationship then finding the equation which was responsible for it.

Integration can be considered to be the mathematical process which is the reverse of the process of differentiation. It also turns out to be a process for finding areas under graphs.

As an illustration of the application of integration in engineering as the reverse of differentiation, consider the situation where the velocity v of an object varies with time t, say $v = 2t$ (Figure 4.14). Since velocity is the rate of change of distance x with time we can write this as:

$$\frac{dx}{dt} = 2t$$

Thus we know how the gradient of the distance–time graph varies with time. Integration is the method we can use to determine from this how the distance varies with time. We thus start out with the gradients and find the distance–time graph responsible for them, the reverse of the process used with differentiation.

4.2.1 Integration as the reverse of differentiation

Suppose we have an equation $y = x^2$. When this equation is differentiated we obtain the derivative of $dy/dx = 2x$. Thus, in this case, when given the gradient as $2x$ we need to find the equation which on being differentiated gave $2x$. Thus, integrating $2x$ should give us x^2. However, the derivative of $x^2 + 1$ is also $2x$, likewise the derivative of $x^2 + 2$, the derivative of $x^2 + 3$, and so on. Figure 4.15 shows part of the family of graphs which all have the gradients given by $2x$.

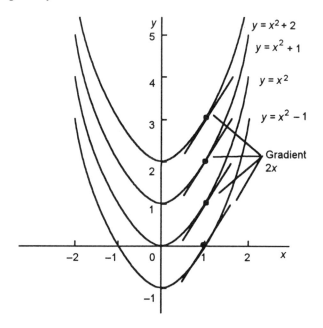

Figure 4.15 *dy/dx = 2x. All the above graphs have gradients which are 2x.*

Thus, for each of the graphs, at a particular value of x, such as $x = 1$, they all give the same gradient of 2. Thus in the integration of $2x$ we are not sure whether there is a constant term or not, or what value it might have. Hence a constant C has to be added to the result. Thus the outcome of the integration of $2x$ has to be written as being $x^2 + C$. The integral, which has to have a constant added to it, is referred to as an *indefinite integral*.

To indicate the process of integration a special symbol:

$$\int f(x)\, dx \qquad\qquad [25]$$

is used. This sign indicates that integration is to be carried out and the dx that x is the variable we are integrating with respect to. Thus the integration referred to above can be written as

$$\int 2x\, dx = x^2 + C$$

Integrals of common functions

The integrals of functions can be determined by considering what equation will give the function when differentiated. For example, consider:

$$\int x^n\, dx$$

Considering integration as the inverse of differentiation, the question becomes as to what function gives x^n when differentiated. The derivative of x^{n+1} is $(n + 1)x^n$. Thus, we have the derivative of $x^{n+1}/(n + 1)$ as x^n. Hence:

$$\int x^n\, dx = \frac{x^{n+1}}{n+1} + C \qquad\qquad [26]$$

This is true for positive, negative and fractional values of n other than $n = -1$, i.e. the integral of x^{-1}. For the integral of x^{-1}, i.e.

$$\int \frac{1}{x}\, dx$$

then since the derivative of $\ln x$ is $1/x$:

$$\int \frac{1}{x}\, dx = \ln x + C \qquad\qquad [27]$$

This only applies if x is positive, i.e. $x > 0$. If x is negative, i.e. $x < 0$, then the integral of $1/x$ in such a situation is *not* $\ln x$. This is because we cannot have the logarithm of a negative number as a real quantity. To show that only positive values of a quantity are to be considered, we write it as $|x|$.

Consider the integral of the exponential function e^x, i.e.

$$\int e^x\, dx$$

The derivative of e^x is e^x. Thus:

Key points

Function being integrated	Outcome of the integration
$\int ax^n\, dx$	$\frac{ax^n}{n+1} + C$, n not -1
$\int a\, dx$	$ax + C$
$\int \frac{1}{x}\, dx$	$\ln x + C$
$\int e^{ax}\, dx$	$\frac{1}{a} e^{ax} + C$
$\int \sin ax\, dx$	$-\frac{1}{a}\cos ax + C$
$\int \cos ax\, dx$	$\frac{1}{a}\sin ax + C$

$$\int e^x\, dx = e^x + C \qquad\qquad [28]$$

The key points shows some functions and their integrals.

Example

Evaluate the integrals:

(a) $\int x^4\, dx$, (b) $\int x^{1/2}\, dx$, (c) $\int x^{-4}\, dx$, (d) $\int x^{-1}\, dx$,

(e) $\int \cos 4x\, dx$, (f) $\int e^{2x}\, dx$, (g) $\int 5\, dx$.

(a) Using $\int x^n\, dx = \frac{x^{n+1}}{n+1} + C$:

$$\int x^4\, dx = \frac{x^{4+1}}{4+1} + C = \frac{x^5}{5} + C$$

(b) Using $\int x^n\, dx = \frac{x^{n+1}}{n+1} + C$:

$$\int x^{1/2}\, dx = \frac{x^{1/2+1}}{1/2+1} + C = \frac{x^{3/2}}{3/2} + C = \frac{2}{3} x^{3/2} + C$$

(c) Using $\int x^n\, dx = \frac{x^{n+1}}{n+1} + C$:

$$\int x^{-4}\, dx = \frac{x^{-4+1}}{-4+1} + C = \frac{x^{-3}}{-3} + C = -\frac{1}{3} x^{-3} + C$$

(d) Using the relationship giving in the table:

$$\int x^{-1}\, dx = \ln x + C$$

(e) Using $\int \cos ax\, dx = \frac{1}{a}\sin x + C$:

$$\int \cos 4x\, dx = \frac{1}{4}\sin 4x + C$$

(f) Using $\int e^{ax}\, dx = \frac{1}{a} e^x + C$:

$$\int e^{2x}\, dx = \frac{1}{2} e^{2x} + C$$

(g) Using $\int a\, dx = ax + C$:

$$\int 5\, dx = \int 5x^0\, dx = 5x + C$$

The above is just a particular version of the standard integral:

$$\int x^n\, dx = \frac{x^{n+1}}{n+1} + C$$

with $n = 0$ since $x^0 = 1$.

Integral of a sum

The derivative of, for example, $x^2 + x$ is the derivative of x^2 plus the derivative of x, i.e. it is $2x + 1$. The integral of $2x + 1$ is thus $x^2 + x + C$. Thus, the integral of the sum of a number of functions is the sum of their separate integrals.

Example

Determine the integral $\int (x^3 + 2x + 4)\,dx$.

We can write this as:

$$\int (x^3 + 2x + 4)\,dx = \int x^3\,dx + \int 2x\,dx + \int 4\,dx$$

Hence the integral is:

$$\frac{x^4}{4} + P + x^2 + Q + 4x + R$$

where P, Q and R are constants. We can combine these constants into a single constant C. Hence the integral is:

$$\frac{x^4}{4} + x^2 + 4x + C$$

Finding the constant of integration

The solution given by the above integration is a general solution and includes a constant. As was indicated earlier in Figure 4.15 the integration of $2x$ gives $y = x^2 + C$. This solution indicates a family of possible equations which could give $dy/dx = 2x$. We can, however, find a *particular solution* if we are supplied with information giving specific coordinate values which have to fall on the graph curve. Thus, in this case, we might be given the condition that when $y = 1$ we have $x = 1$. This must fit the equation $y = x^2 + C$ and can only be the case when $C = 0$. Hence the solution is $y = x^2$.

Example

Determine the equation of a graph if it has to have y = 0 when x = 2 and has a gradient given by:

$$\frac{dy}{dx} = 3x + 2$$

To obtain the general solution, i.e. the family of curves which fit the above gradient equation, we integrate. Thus:

$$y = \int (3x + 2)\,dx = \tfrac{3}{2}x^2 + 2x + C$$

The particular curve we require must have $y = 0$ when $x = 2$. Putting this data into the equation gives $2 = 0 + 0 + C$. Hence $C = 2$ and so the particular solution is:

$$y = \tfrac{3}{2}x^2 + 2x + 2$$

Example

A curve is such that its gradient is described by the equation $dy/d\theta = \cos\theta$ and $y = 1$ when $\theta = \pi/2$ radians. Find the equation of the curve.

Here we have the relationship $dy/d\theta = \cos\theta$, and so integration gives:

$$y = \int \cos\theta \, d\theta = \sin\theta + C$$

This is the general solution. To find the specific equation we need to evaluate C by substituting the known conditions, namely that $y = 1$ when $\theta = \pi/2$ radians. Thus:

$1 = \sin(\pi/2) + C$

$1 = 1 + C$

Therefore $C = 0$ and the required equation is:

$y = \sin\theta$

Example

At any point on a curve we have a gradient of $dy/dt = 3\sin t$. Find the equation of the curve given that $y = 2$ when t has the value of 25°.

Given the relationship $dy/dt = 3\sin t$, integration gives:

$$y = \int 3\sin t \, dt = 3\int \sin t \, dt = 3(-\cos t) + C$$

The general equation is thus $y = -3\cos t + C$. But $y = 2$ when $t = 25°$ and so:

$2 = -3\cos 25° + C = -3(0.91) + C$

Thus $C = 4.73$ and the specific equation, which is the required equation, is:

$y = 4.73 - 3\cos t.$

Maths in action

The bending of beams

See the Maths in action in Section 4.1.3 for a preliminary discussion of the bending of beams.

The deflection y of a beam can be obtained by integrating the differential equation:

$$\frac{d^2y}{dx^2} = -\frac{M}{EI}$$

with respect to x to give:

$$\frac{dy}{dx} = -\frac{1}{EI}\int M\,dx + A$$

with A being the constant of integration and then carrying out a further integration with respect to x to give:

$$y = -\frac{1}{EI}\int\left[\int M\,dx + A\right] + B = -\frac{1}{EI}\int\int M\,dx + Ax + B$$

with B being a constant of integration.

At the point where maximum deflection occurs, the slope of the deflection curve will be zero and thus the point of maximum deflection can be determined by equating dy/dx to zero.

As an illustration, consider a horizontal cantilever supporting a load at its free end (Figure 4.16). The bending moment a distance x from the fixed end is given by $M = -F(L - x)$ and so the differential equation becomes:

$$EI\frac{d^2y}{dx^2} = -M = F(L - x)$$

Integrating with respect to x gives:

$$EI\frac{dy}{dx} = FLx - \frac{Fx^2}{2} + A$$

Since the slope of the beam $dy/dx = 0$ at the fixed end where $x = 0$, then we must have $A = 0$ and so:

$$EI\frac{dy}{dx} = FLx - \frac{Fx^2}{2}$$

Integrating again gives:

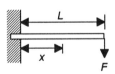

Figure 4.16   *Example*

$$Ely = \frac{FLx^2}{2} - \frac{Fx^3}{6} + B$$

Since, at the fixed end we have zero deflection, i.e. we have $y = 0$ at $x = 0$, then we must have $B = 0$ and so:

$$Ely = \frac{FLx^2}{2} - \frac{Fx^3}{6} = \frac{Fx^2}{6}(3L - x)$$

When $x = L$:

$$y = \frac{FL^3}{3EI}$$

For beams with a number of concentrated loads, there will be discontinuities in the bending moment diagram and so we cannot write a single bending moment equation to cover the entire beam but have to write separate equations for each part of the beam between adjacent loads. Integration of each expression then gives the deflections relationship for each part of the beam. There is an alternative and that involves writing a single equation using, what are termed, Macaulay's brackets. For a discussion and examples of this method, see the companion book: *Mechanical Engineering Systems* by R. Gentle, P. Edwards and W. Bolton.

4.2.2 Integration as the area under a graph

Consider a moving object and its graph of velocity v against time t (Figure 4.17). The distance travelled between times of t_1 and t_2 is the area under the graph between those times. If we divide the area into a number of equal width strips then we can represent this area under the velocity–time graph as being the sum of the areas of these equal width strip areas, as illustrated in Figure 4.17. If t is the value of the time at the centre of a strip of width δt and v the velocity at this time, then a strip has an area of $v\,\delta t$. Thus the area under the graph between the times t_1 and t_2 is equal to the sum of the areas of all such strips between the times t_1 and t_2,

distance travelled = sum of the areas of all the strips between t_1 and t_2

We can write this summation as:

$$x = \sum_{t=t_1}^{t=t_2} v\,\delta t$$

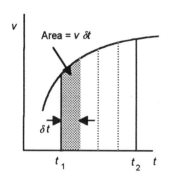

Figure 4.17   *Velocity-time graph*

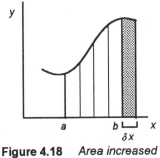

Figure 4.18 *Area increased by one strip*

The $\sum$ sign is used to indicate that we are carrying out a summation of a number of terms. The limits between which this summation is to be carried out are indicated by the information given below and above the sign. If we make δt very small, i.e. let δt tend to 0, then we denote it by dt. The sum is then the sum of a series of very narrow strips and is written as:

$$x = \lim_{\delta t \to 0} \sum_{t=t_1}^{t=t_2} v\,\delta t = \int_{t_1}^{t_2} v\,dt \qquad [29]$$

The integral sign is an "S" for summation and the t_1 and t_2 are said to be the limits of the range of the variable t. Here x is the *integral* of the v with time t between the limits t_1 and t_2. The process of obtaining x in this way is termed *integration*. Because the integration is between specific limits it is referred to as a *definite integral*.

Integration as reverse of differentiation and area under a graph

The definitions of integration in terms of the reverse of differentiation and as the area under a graph describe the same concept. Suppose we increase the area A under a graph of y plotted against x by one strip (Figure 4.18). Then the increase in the area δA is the area of this strip. Thus:

increase in area $\delta A = y\,\delta x$

So we can write:

$$\frac{\delta A}{\delta x} = y$$

In the limit as δx tends to 0 then we can write dA/dx and so

$$\frac{dA}{dx} = y$$

With integration defined as the inverse of differentiation then the integration of the above equation gives the area A, i.e.

$$A = \int y\,dx \qquad [30]$$

This is an indefinite integral, which is the same as that given by the definition for integration as the area under a graph when limits are imposed. An *indefinite integral* has no limits and the result has a constant of integration. Integration between specific limits gives a *definite integral*.

Areas under graphs

Consider the integration of y with respect to x when we have $y = 2x$. This has no specified limits and so is an indefinite integral,

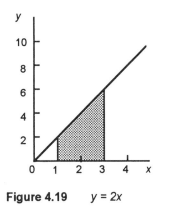

Figure 4.19 $y = 2x$

with the solution as the function which differentiated would give $2x$:

$$\int 2x \, dx = x^2 + C$$

Now consider the area under the graph of $y = 2x$ between the limits of $x = 1$ and $x = 3$ (Figure 4.19). We can write this as the definite integral:

$$\int_1^3 2x \, dx$$

$$\int_1^3 2x \, dx = [x^2 + C]_1^3$$

The square brackets round the $x^2 + C$ are used to indicate that we have to impose the limits of 3 and 1 on it. Thus the integral is the value of $x^2 + C$ when $x = 3$ minus the value of $x^2 + C$ when $x = 1$.

$$\int_1^3 2x \, dx = (9 + C) - (1 + C) = 8 \text{ square units}$$

The constant term C vanishes when we have a definite integral.

Note that an area below the x-axis is negative. If the area is required when part of it is below the x-axis then the parts below and above the x-axis must be found separately and then added, disregarding the sign of the area.

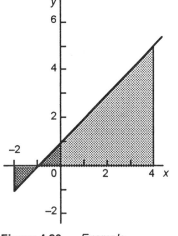

Figure 4.20   *Example*

Example

Determine the area between a graph of $y = x + 1$ and the x-axis between $x = -2$ and $x = 4$.

Figure 4.20 shows the graph. The area required is that between the values of x of -2 and 4. We can break this area down into a number of elements. The area under the graph between $x = 0$ and $x = 4$ is that of a rectangle 4×1 plus a triangle $\frac{1}{2}(4 \times 4)$ and so is +12 square units. The area between $x = -1$ and 0 is that of a triangle $\frac{1}{2}(1 \times 1) = 0.5$ square units and the area between $x = -1$ and $x = -2$ is a triangular area below the axis and so is negative and given by $\frac{1}{2}(1 \times 1) = -0.5$ square units. Hence the total area under the graph is $+12 + 0.5 - 0.5 = 12$ square units.

Alternatively we can consider this area as the integral:

$$\text{area} = \int_{-2}^4 (x + 1) \, dx = \left[\frac{x^2}{2} + x + C \right]_{-2}^4$$

and so:

$$\text{area} = (8 + 4 + C) - (2 - 2 + C) = 12 \text{ square units}$$

Example

Determine the value of the integral $\int_{-2}^{4} e^{2x}\,dx$.

We can consider that this integral represents the area under the graph between $x = -2$ and $x = 4$ of e^{2x} plotted against x.

$$\int_{-2}^{4} e^{2x}\,dx = \left[\tfrac{1}{2} e^{2x} + C \right]_{-2}^{4}$$

$$= \tfrac{1}{2} e^{8} - \tfrac{1}{2} e^{-4} = 1490.479 - 0.009$$

The value of the integral is thus 1490.470.

Example

Determine the value of the integral $\int_{0}^{\pi/3} \cos 2x\,dx$.

We can consider that this integral represents the area under the graph between $x = 0$ and $x = \pi/3$ of $\cos 2x$ plotted against x.

$$\int_{0}^{\pi/3} \cos 2x\,dx = \left[\tfrac{1}{2} \sin 2x + C \right]_{0}^{\pi/3}$$

$$= \tfrac{1}{2} \sin 2\pi/3 - \tfrac{1}{2} \sin 0 = 0.433$$

The value of the integral is thus 0.433.

Example

Find the areas under the curve $y = x^3$ between (a) $x = 0$ and $x = 1$, (b) $x = -1$ and $x = 1$.

(a) Figure 4.21 shows the graph. The area is:

$$\int_{0}^{1} x^3 = \left[\frac{x^4}{4} \right]_{0}^{1} = \frac{1}{4} - 0 = \frac{1}{4}$$

(b) The area taking into account the sign of y, is:

$$\int_{-1}^{1} x^3 = \left[\frac{x^4}{4} \right]_{-1}^{1} = \frac{1}{4} - \frac{1}{4} = 0$$

The area is zero because the area between $x = -1$ and $x = 0$ is negative. What we have is the sum of two areas:

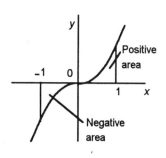

Figure 4.21   *Example*

$$\int_{-1}^{1} x^3 = \int_{-1}^{0} x^3 \, dx + \int_{0}^{1} x^3 \, d$$

$$\int_{-1}^{0} x^3 \, dx = \left[\frac{x^4}{4} \right]_{-1}^{0} = 0 - \frac{1}{4} = -\frac{1}{4}$$

$$\int_{0}^{1} x^3 \, dx = \left[\frac{x^4}{4} \right]_{0}^{1} = \frac{1}{4} - 0 = \frac{1}{4}$$

The sum of the two areas is thus zero. If we want the total area, regardless of sign, between the axis and the curve for any function then we have to determine the positive and negative elements separately and then, ignoring the sign, add them. For this curve this gives ½.

Maths in action

Work

With a constant force F acting on a body and a displacement x in the direction of the force, then the work done W on a body, i.e. the energy transferred to it, is given by $W = Fx$.

Consider a variable force described by the graph shown in Figure 4.22 for the force applied to an object and how it varies with the displacement of that object. For a small displacement δx we can consider the force to be effectively constant at F. Thus the work done for that displacement is $F \delta x$. This is the area of the strip under the force–distance graph. If we want the work done in changing the displacement from x_1 to x_2 then we need to determine the sum of all such strips between these displacements, i.e. the total area under the graph between the ordinates for x_1 and x_2. Thus:

$$\text{work done} = \sum_{x=x_1}^{x=x_2} F \delta x$$

If we make the strips tend towards zero thickness then the above summation becomes the integral, i.e.

$$\text{work done} = \int_{x_1}^{x_2} F \, dx$$

Consider the work done in stretching a spring when a force F is applied and causes a displacement change in its point of application, i.e. an extension, from 0 to x if $F = kx$, where k is a constant. Figure 4.23 shows the force–distance graph.

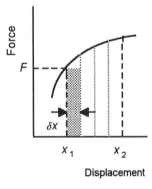

Figure 4.22 *Force-displacement graph*

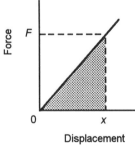

Figure 4.23 *Stretching a spring*

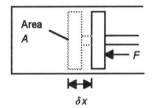

Figure 4.24 *Compressing a gas*

The work done is the area under the graph between 0 and x. This is the area of a triangle and so the work done is ½Fx. Since $F = kx$ we can write this as ½kx^2.

We could have solved this problem by integration. Thus:

$$\text{work done} = \int_0^x F\,dx = \int_0^x kx\,dx = \left[\tfrac{1}{2}kx^2 + C\right]_0^x = \tfrac{1}{2}kx^2$$

As a further illustration, consider the work done as a result of a piston in Figure 4.24 being moved to reduce the volume of a gas. The work done in moving the piston through a small distance δx when the force is F is $F\,\delta x$. Since pressure is force per unit area, then if the force acts over an area A the pressure $p = F/A$. Thus:

$$\text{work done} = F\,\delta x = pA\,\delta x$$

But $A\,\delta x$ is the change in volume δV of the gas. Hence, the work done $= p\,\delta V$. The total work done in changing the volume of a gas from V_1 to V_2 is thus:

$$\text{work done} = \sum_{V_1}^{V_2} p\delta V$$

If we consider δV tending to zero then we can write

$$\text{work done} = \int_{V_1}^{V_2} p\,dV$$

For a gas that obeys Boyle's law, i.e. pV = a constant k, the work done in compressing a gas from a volume V_1 to V_2 is thus:

$$\text{work done} = \int_{V_1}^{V_2} p\,dV = \int_{V_1}^{V_2} \frac{k}{V}\,dV = [\ln V + C]_{V_1}^{V_2}$$

Hence the work done is $\ln V_2 - \ln V_1$.

Maths in action

Centre of gravity and centroid
The weight of a body is made up of the weights of each constituent particle, each such particle having its weight acting at a different point. However, it is possible to replace all the weight forces of an object by a single weight force acting at a particular point, this point being termed the *centre of gravity*.

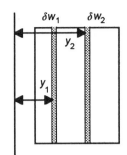

Axis about which moments
are taken

Figure 4.25 *Moments of elements*

If we consider a sheet to be made of a large number of small strip elements of mass at different distances from an axis (Figure 4.25) then the weight of each element will give rise to a moment about that axis. Thus the total moment due to all the weight elements is $\delta w_1 x_1 + \delta w_2 x_2 + \delta w_3 x_3 + \dots$. If a single weight W at a distance $\bar{x}$ is to give the same moment, then:

$$W\bar{x} = \sum \delta w\, x \text{ for all the strips in the sheet}$$

Thus the distance of the centre of gravity from the chosen axis is:

$$\bar{x} = \frac{\sum_1^n \delta w_i x_i}{W}$$

For a thin flat plate of uniform density, the weight of an element is proportional to its area. We then refer to the *centroid* since it is purely geometric. The distance of the centroid from the chosen axis is thus:

$$\bar{x} = \frac{\sum_1^n \delta a_i x_i}{A}$$

where δa represents the area of an elemental strip. The product of an area and its distance from an axis is known as the *first moment of area* of that area about the axis. Thus the centroid distance from an axis is the sum of the first moments of all the area elements divided by the sum of all the areas of the elements.

If we consider infinitesimally small elements, i.e. $\delta a \to 0$, then we can write:

$$\bar{x} = \frac{\int x\, \mathrm{d}a}{A}$$

Consider the determination of the centroid of a triangular area (Figure 4.26). Consider a small strip of area $\delta A = x\, \delta y$. By similar triangles $x/(h - y) = b/h$ and so $x\, \delta y = [b(h - y)/h]\, \delta y$. The total area $A = \frac{1}{2}bh$. Hence, the y coordinate of the centroid is:

$$\bar{y} = \frac{2}{bh} \int_0^h y\, \frac{b(h - y)}{h}\, \mathrm{d}y = \tfrac{1}{3}h$$

The centroid is located at one-third the altitude of the triangle. The same result is obtained if we consider the location with respect to the other sides. The centroid is at one-third the altitude along of the lines drawn from each apex to the opposite side.

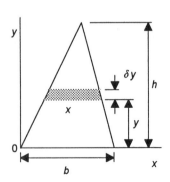

Figure 4.26 *Triangular area*

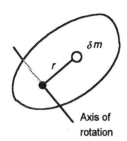

Figure 4.27 *Rotation of a rigid body*

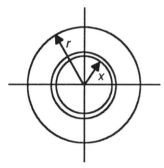

Figure 4.28 *Disc*

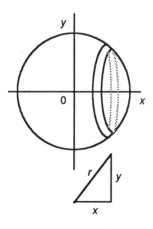

Figure 4.29 *Sphere*

Maths in action

Moment of inertia

Consider a rigid body rotating with a constant angular acceleration a about some axis (Figure 4.27). We can consider the body to be made up of small elements of mass δm. For such an element a distance r from the axis of rotation we have a linear acceleration of $a = ra$. Thus the force acting on the element is $\delta m \times ra$. The moment of this force is thus $Fr = r^2a\,\delta m$. The total moment, i.e. torque T, due to all the elements of mass in the body is thus:

$T = \sum r^2 a\,\delta m$ for all the elements

Thus if we have elements of mass at radial distance from 0 to R, in the limit as $\delta m \to 0$:

$T = \int_0^R r^2 a\,\mathrm{d}m$

Since a is a constant we can write the above equation as:

$T = \left(\int_0^R r^2\,\mathrm{d}m\right)a = Ia$

where I *is the moment of inertia*

As an illustration consider the determination of the moment of inertia of a uniform disc about an axis through its centre and at right angles to its plane. Figure 4.28 shows the disc with an element of mass being chosen as a disc with a radius x and width δx. The element is a strip of length $2\pi x$ and so an area of $2\pi x\,\delta x$. If the mass of the disc is m per unit area, then the mass of the element is $\delta m = 2\pi m x\,\delta x$. The moment of inertia of the element is $x^2\,\delta m = 2\pi m x^3\,\delta x$. Thus the moment of inertia of the disc is:

$I = \int_0^r 2\pi m x^3\,\mathrm{d}x = 2\pi m\left[\dfrac{x^4}{4}\right]_0^r = \tfrac{1}{2}\pi m r^4$

As another illustration, consider a sphere of radius r and mass per unit volume m. If we take a thin slice of thickness δx of the sphere perpendicular to the diameter about which the moment of inertia is to be determined and a distance x from the sphere centre (Figure 4.29), then with the slice radius y we have an element of volume $\pi y^2\,\delta x$ and hence mass $\pi m y^2\,\delta x$. The moment of inertia of a disc is ½ mass × radius² (see the previous example) and thus the moment of inertia of the slice is $\tfrac{1}{2}(\pi m y^2\,\delta x)y^2$ and the moment of inertia of the sphere as the sum of all the slices as $\delta x \to 0$ is:

$$I = \int_{-r}^{r} \tfrac{1}{2}\pi m y^4 \, dx$$

Since $r^2 = y^2 + x^2$:

$$I = \tfrac{1}{2}\pi m \int_{-r}^{r} (r^2 - x^2)^2 \, dx = \tfrac{1}{2}\pi m \int_{-r}^{r} (r^4 - 2r^2 x^2 + x^4) \, dx$$

$$= \tfrac{1}{2}\pi m \left[r^4 x - \tfrac{2}{3}r^2 x^3 + \tfrac{1}{5}x^5 \right]_{-r}^{r} = \tfrac{8}{15}\pi m r^5$$

The total mass M of the sphere is $\tfrac{4}{3}\pi m r^3$ so $I = \tfrac{2}{5}Mr^2$.

Upper surface stretched

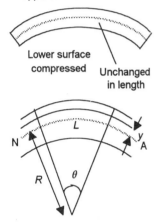

Lower surface compressed Unchanged in length

Figure 4.30 *Bending stretches the upper surface and contracts the lower surface, in-between there is an unchanged in length surface*

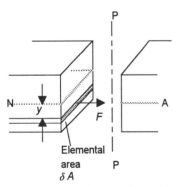

Elemental area δA

Figure 4.31 *Elemental area in a section PP*

Maths in action

Second moment of area

Consider a beam that has been bent into the arc of a circle so that the uppermost surface is in tension and the lower surface in compression (Figure 4.30). The upper surface has increased in length and the lower surface decreased in length; between the two there is a plane which is unchanged in length; this is called the neutral plane and the line where the plane cuts the cross-section of the beam is the neutral axis.

An initially horizontal plane through the beam which is a distance y from the neutral axis changes in length as a consequence of the beam being bent and the strain it experiences is the change in length ΔL divided by its initial unstrained length L. For circular arcs, the arc length is the radius of the arc multiplied by the angle it subtends, and thus, $L + \Delta L = (R + y)\theta$. The neutral axis NA will, by definition, be unstrained and so for it we have $L = R\theta$. Hence, the strain on aa is:

$$\text{strain} = \frac{\Delta L}{L} = \frac{(R+y)\theta - R\theta}{R\theta} = \frac{y}{R}$$

Provided we can use Hooke's law, the stress due to bending which is acting this plane is:

$$\text{stress } \sigma = E \times \text{strain} = \frac{Ey}{R}$$

Looking at a cross-sectional slice of the beam cut by PP we have Figure 4.31. The moment M of the elemental force F about the neutral axis is Fy and the stress σ acting on the elemental area is $F/\delta A$. Therefore the moment is $(\sigma \, \delta A)y$. Hence, using the equation we derived above for the stress, the moment of this element about the neutral axis is:

$$\text{moment} = \frac{Ey}{R}\delta A \times y = \frac{E}{R}y^2\delta A$$

The total moment M produced over the entire cross-section is the sum of all the moments produced by all the elements of area in the cross-section. Thus, if we consider each such element of area to be infinitesimally small, we can write:

$$M = \int \frac{E}{R}y^2 \, dA = \frac{E}{R} \int y^2 dA$$

The integral is termed the *second moment of area I* of the section:

$$I = \int y^2 \, dA$$

Thus we can write:

$$M = \frac{EI}{R}$$

For a rectangular cross-section of breadth b and depth d (Figure 4.32) with a segment of thickness δy a distance y from the neutral axis, the second moment of area for the segment is:

second moment of area of strip $= y^2 \delta A = y^2 b \delta y$

The total second moment of area for the section about the neutral axis is thus:

$$\text{second moment of area} = \int_{-d/2}^{d/2} y^2 b \, dy = \frac{bd^3}{12}$$

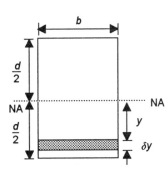

Figure 4.32 *Second moment of area*

4.2.3 Techniques for integration

There are a number of techniques which can aid in the integration of functions. In this section we look at integration by substitution, integration by parts and partial fractions.

Integration by substitution

This involves simplifying integrals by making a *substitution*. The term *integration by change of variable* is often used since the variable has to be changed as a result of the substitution. The aim of making a substitution is to put the integral into a simpler form for integration. As an illustration, consider the integral:

$$\int e^{5x} \, dx$$

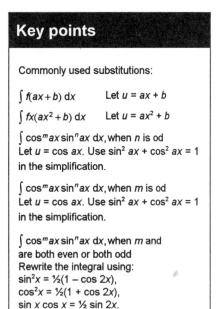

Key points

Commonly used substitutions:

$\int f(ax+b)\,dx$ Let $u = ax + b$

$\int fx(ax^2+b)\,dx$ Let $u = ax^2 + b$

$\int \cos^m ax \sin^n ax\,dx$, when n is od
Let $u = \cos ax$. Use $\sin^2 ax + \cos^2 ax = 1$
in the simplification.

$\int \cos^m ax \sin^n ax\,dx$, when m is od
Let $u = \cos ax$. Use $\sin^2 ax + \cos^2 ax = 1$
in the simplification.

$\int \cos^m ax \sin^n ax\,dx$, when m and
are both even or both odd
Rewrite the integral using:
$\sin^2 x = \frac{1}{2}(1 - \cos 2x)$,
$\cos^2 x = \frac{1}{2}(1 + \cos 2x)$,
$\sin x \cos x = \frac{1}{2}\sin 2x$.

The substitution $u = 5x$ reduces e^{5x} to e^u. However, we also need to change dx in the variable to du for the integration. Since $du/dx = 5$, we can write the integral as:

$$\int e^u \frac{du}{5} = \frac{1}{5}\int e^u\,du = \frac{1}{5}e^u + C = \frac{1}{5}e^{5x} + C$$

In the above case the substitution of u for $5x$ seemed a sensible way to simplify the integral. However, there are no general rules for finding suitable substitutions and the key points show some of the more commonly used substitutions.

Example

Determine the indefinite integral $\int(4x+1)^3\,dx$.

If we let $u = 4x + 1$ then $du/dx = 4$ and $dx = \frac{1}{4}\,du$:

$$\int(4x+1)^3\,dx = \int u^3 \frac{1}{4}\,du = \frac{1}{4}\frac{u^4}{4} + C = \frac{1}{16}(4x+1)^4 + C$$

Example

Determine the indefinite integral $\int \frac{x}{3x^2+4}\,dx$.

If we let $u = 3x^2 + 4$, then $du/dx = 6x$ and so $x\,dx = (1/6)\,du$. Hence:

$$\int \frac{x}{3x^2+4}\,dx = \int \frac{1}{u}\frac{1}{6}\,du = \frac{1}{6}\int \frac{1}{u}\,du$$

$$= \frac{1}{6}\ln|u| + C = \frac{1}{6}\ln(3x^2+4) + C$$

The modulus sign is used with the integration of $1/u$ because no assumption is made at that stage as to whether u is positive or negative. The sign is dropped when the substitution is made because $3x^2 + 4$ is always positive.

Example

Determine the indefinite integral $\int \cos^2 x \sin^3\,dx$.

If we let $u = \cos x$, then $du/dx = -\sin x$ and so $\sin x\,dx = -du$. The integral then can be written as:

$$\int \cos^2 x \sin^3 x\, dx = \int \cos^2 x \sin^2 x \sin x\, dx$$

$$= \int \cos^2 x (1 - \cos^2 x) \sin x\, dx$$

$$= \int u^2 (1 - u^2)\, du = \int (u^2 - u^4)\, du$$

$$= \frac{u^3}{3} - \frac{u^5}{5} + C = \tfrac{1}{3} \cos^3 x - \tfrac{1}{5} \cos^5 x + C$$

Example

Determine the indefinite integral $\int \cos x \sin^2 x\, dx$.

Let $u = \sin x$. Then $du/dx = \sin x$ and so $\sin x\, dx = du$. The integral can then be written as:

$$\int \cos x \sin^2 x\, dx = \int \sin^2 x \cos x\, dx$$

$$= \int u^2\, du = \frac{u^3}{3} + C = \tfrac{1}{3} \sin^3 x + C$$

Trigonometric substitutions

A useful group of substitutions is to use trigonometric functions. For example, for integrals involving $\sqrt{(a^2 - x^2)}$ terms, we can use the substitution $x = a \sin \theta$. Then $\sqrt{(a^2 - x^2)} = \sqrt{(a^2 - a^2 \sin^2 \theta)} = a \cos \theta$, since we have $1 - \sin^2 \theta = \cos^2 \theta$. Since $dx/d\theta = a \cos \theta$ then $a \cos \theta\, d\theta$. The key points give other such substitutions.

Key points

Useful trigonometric substitutions:

$\int \sqrt{(a^2 - x^2)}\, dx$ Let $x = a \sin \theta$

$\int \sqrt{(a^2 + x^2)}\, dx$ Let $x = a \tan \theta$

$\int \sqrt{(x^2 - a^2)}\, dx$ Let $x = a \sec \theta$

Example

Determine the indefinite integral $\int \sqrt{(1 - x^2)}\, dx$.

Let $x = \sin \theta$. Then $\sqrt{(1 - x^2)} = \sqrt{(1 - \sin^2 \theta)} = \cos \theta$. Since $dx/d\theta = \cos \theta$ then $dx = \cos \theta\, d\theta$. Thus the integral becomes:

$$\int \sqrt{(1 - x^2)}\, dx = \int \cos \theta \cos \theta\, d\theta = \int \cos^2 \theta\, d\theta$$

Since $\cos 2\theta = 2 \cos^2 \theta - 1$, we have:

$$\int \tfrac{1}{2}(1 + \cos 2\theta)\, d\theta = \tfrac{1}{2}\theta + \tfrac{1}{4} \sin 2\theta + C$$

Back substitution using $\theta = \sin^{-1} x$ gives:

$$\tfrac{1}{2} \sin^{-1} x + \tfrac{1}{4} \sin(2 \sin^{-1} x) + C$$

However, a simpler expression is obtained if we first replace the sin 2θ using sin 2θ = 2 sin θ cos θ = 2sinθ $\sqrt{(1 - \sin^2 \theta)}$.

$$\tfrac{1}{2}\theta + \tfrac{1}{4}\sin 2\theta + C = \tfrac{1}{2}\sin^{-1}x + \tfrac{1}{2}x\sqrt{1-x^2} + C$$

Example

Determine the indefinite integral $\int \dfrac{1}{x^2+4}\,dx$.

Let $x = 2\tan\theta$. Then $dx/d\theta = 2\sec^2\theta$ and so:

$$\int \frac{1}{x^2+4}\,dx = \int \frac{1}{4\tan^2\theta+4}2\sec^2\theta\,d\theta$$

$$= \int \frac{1}{4\sec^2\theta}2\sec^2\theta\,d\theta$$

$$= \int \tfrac{1}{2}\,d\theta = \tfrac{1}{2}\theta + C = \tfrac{1}{2}\tan^{-1}\frac{x}{2} + C$$

Another form of useful substitution, when we have integrals involving sin x, cos x, tan x terms, is to let $u = \tan \frac{1}{2}x$. Then $du/dx = \frac{1}{2}\sec^2 \frac{1}{2}x$. But $\sec^2 x = 1 + \tan^2 x$, thus $du/dx = \frac{1}{2}(1 + \tan^2 x) = \frac{1}{2}(1 + u^2)$. Thus $dx = 2\,du/(1 + u^2)$. The trigonometric functions can all be expressed in terms of u. Thus:

$$\sin x = 2\sin\frac{x}{2}\cos\frac{x}{2} = 2\sin\frac{x}{2}\cos\frac{x}{2}\frac{\cos\frac{x}{2}}{\cos\frac{x}{2}} = \frac{2\tan\frac{x}{2}}{\sec^2\frac{x}{2}}$$

$$= \frac{2\tan\frac{x}{2}}{1+\tan^2\frac{x}{2}} = \frac{2u}{1+u^2}$$

Figure 4.33 shows the right-angled triangle with such an angle. Hence:

$$\cos x = \frac{1-u^2}{1+u^2} \quad \text{and} \quad \tan x = \frac{2u}{1-u^2}$$

Note that integration of the squares of trigonometric functions can be obtained by using trigonometric identities to put the functions in non-squared form. Thus:

$$\int \sin^2 x\,dx = \int \tfrac{1}{2}(1-\cos 2x)\,dx = \tfrac{1}{2}(x - \tfrac{1}{2}\sin 2x) + C$$

$$\int \cos^2 x\,dx = \int \tfrac{1}{2}(1+\cos 2x)\,dx = \tfrac{1}{2}(x + \tfrac{1}{2}\sin 2x) + C$$

$$\int \tan^2 x\,dx = \int (\sec^2 x - 1)\,dx = \tan x - x + C$$

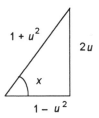

Figure 4.33 *Angle x*

Example

Determine the indefinite integral $\int \frac{1}{\sin x}\, dx$.

Let $u = \tan \frac{1}{2}x$, then $du/dx = \frac{1}{2} \sec^2 \frac{1}{2}x = \frac{1}{2}(1 + \tan^2 \frac{1}{2}x)$ $= \frac{1}{2}(1 + u^2)$ and replacing $\sin x$ by $2u/(1 + u^2)$:

$$\int \frac{1}{\sin x}\, dx = \int \frac{1+u^2}{2u}\, \frac{2}{1+u^2}\, du = \int \frac{1}{u}\, du$$

$$= \ln|u| + C = \ln\left| \tan \frac{x}{2} \right| + C$$

Substitution with definite integrals

The above has discussed the substitution procedure with indefinite integrals where the variable was changed from x to u. When we have definite integrals we can do the same procedure and take account of the limits of integration at the end *after* reversing the substitution. The limits are in terms of values of x. However, it is often simpler to express the limits in terms of u and take account of the limits *before* reversing the substitution. To illustrate this, consider the integration of $\cos^3 x$ between the limits 0 and $\frac{1}{2}\pi$. If we let $u = \sin x$ then $du/dx = \cos x$ and so $\cos x\, dx = du$. When $x = 0$ then $u = 0$ and when $x = \frac{1}{2}\pi$ then $u = 1$. Thus the integral can be written as:

$$\int_0^{\pi/2} \cos^3 x\, dx = \int_0^{\pi/2} \cos^2 x \cos x\, dx = \int_0^{\pi/2} (1 - \sin^2 x) \cos x\, dx$$

$$= \int_0^1 (1 - u^2)\, du = \left[u - \frac{u^3}{3} \right]_0^1 = \frac{2}{3}$$

Integration by parts

The product rule for differentiation gives:

$$\frac{d}{dx}[f(x)g(x)] = f(x)\frac{d}{dx}g(x) + g(x)\frac{d}{dx}f(x)$$

Integrating both sides of this equation with respect to x gives:

$$\int \frac{d}{dx}[f(x)g(x)]\, dx = \int f(x)\frac{d}{dx}g(x)\, dx + \int g(x)\frac{d}{dx}f(x)\, dx$$

Hence:

$$\int f(x)\frac{d}{dx}g(x)\, dx = \int \frac{d}{dx}[f(x)g(x)]\, dx - \int g(x)\frac{d}{dx}f(x)\, dx$$

$$= f(x)g(x) - \int g(x)\frac{d}{dx}f(x)\, dx \qquad\qquad [31]$$

This is the formula for *integration by parts*. This is often written in terms of $u = f(x)$ and $v = g(x)$ as:

Key point

Integration by parts:

$\int u\frac{dv}{dx}\,dx = uv - \int v\frac{du}{dx}\,dx$

$$\int u\frac{dv}{dx}\,dx = uv - \int v\frac{du}{dx}\,dx \qquad [32]$$

With a definite integral the equation becomes:

$$\int_a^b u\frac{dv}{dx}\,dx = [uv]_a^b - \int_a^b v\frac{du}{dx}\,dx \qquad [33]$$

Example

Determine the indefinite integral $\int x\,e^x\,dx$.

The integral consists of the product of two factors. If we let $u = x$ and $dv/dx = e^x$, then $v = \int e^x\,dx$ and equation [32] gives:

$$\int x\,e^x\,dx = x\,e^x - \int (e^x)(1)\,dx = x\,e^x - e^x + C$$

Example

Determine the indefinite integral $\int e^x \sin x\,dx$.

Let $u = e^x$ and $dv/dx = \sin x$. Then:

$$v = \int \sin x\,dx = -\cos$$

Hence, using equation [32] gives:

$$\int e^x \sin x\,dx = e^x(-\cos x) - \int (-\cos x)(e^x)\,dx$$

$$= -e^x \cos x + \int e^x \cos x\,dx + C$$

Applying integration by parts again, with $u = e^x$ and $dv/dx = \cos x$. Then $v = \int \cos x\,dx = \sin x$. Hence, using equation [32] gives:

$$\int e^x \sin x\,dx = -e^x \cos x + e^x \sin x - \int e^x \sin x\,dx + C$$

Thus:

$$\int e^x \sin x\,dx = \tfrac{1}{2}(-e^x \cos x + e^x \sin x + C)$$

Example

Determine the definite integral $\int_0^1 x^2\,e^x\,dx$.

Let $u = x^2$ and $dv/dx = e^x$. Then $v = \int e^x\, dx = e^x$. Thus, using equation [9]:

$$\int_0^1 x^2\, e^x\, dx = [x^2\, e^x]_0^1 - \int_0^1 e^x (2x)\, dx$$

Applying integration by parts again, with $u = x$ and $dv/dx = e^x$. Then $v = \int e^x\, dx = e^x$. Thus, using equation [33]:

$$\int_0^1 x^2\, e^x\, dx = [x^2\, e^x]_0^1 - 2[x e^x]_0^1 + 2 \int_0^1 e^x\, dx$$

$$= [x^2\, e^x]_0^1 - 2[x e^x]_0^1 + 2[e^x]_0^1$$

$$= e^1 - 2\, e^1 + 2\, e^1 - 2\, e^0 = e - 2$$

Key points

The procedure for obtaining partial fractions can be summarised as:

1. If the degree of the denominator is equal to, or less than, that of the numerator, divide the denominator into the numerator to obtain the sum of a polynomial plus a fraction which has the degree of the denominator greater than that of the numerator.

2. Write the denominator in the form of linear factors, i.e. of the form $(ax + b)$, or irreducible quadratic factors, i.e. of the form $(ax^2 + bx + c)$.

3. Write the fraction as a sum of partial fractions involving constants A, B, etc.

4. Determine the unknown constants which occur with the partial fractions by equating the fraction with the partial fractions and either solving the equation for specific values of x or equating the coefficients of equal powers of x.

5. Replace the constants in the partial fractions with their values.

Integration by partial fractions

Integrals involving fractions can often by simplified by expressing the integral as the sum or difference of two or more partial fractions which then lend themselves to easier integration. For example:

$$\frac{3x + 4}{x^2 + 3x + 2} = \frac{3x + 4}{(x + 1)(x + 2)}$$

can be expressed as the partial fractions:

$$\frac{3x + 4}{(x + 1)(x + 2)} = \frac{1}{x + 1} + \frac{1}{x + 2}$$

When the degree of the denominator is greater than that of the numerator then an expression can be directly resolved into partial fractions. The form taken by the partial fractions depends on the type of denominator concerned.

- If the denominator contains a *linear factor*, i.e. a factor of the form $(x + a)$, then for each such factor there will be a partial fraction of the form:

$$\frac{A}{(x + a)}$$

where A is some constant.

- If the denominator contains *repeated linear factors*, i.e. a factor of the form $(x + a)^n$, then there will be partial fractions:

$$\frac{A}{(x + a)} + \frac{B}{(x + a)^2} + \ldots + \frac{C}{(x + a)^n}$$

with one partial fraction for each power of $(x + a)$.

- If the denominator contains an *irreducible quadratic factor*, i.e. a factor of the form $ax^2 + bx + c$, then there will be a partial fraction of the form:

$$\frac{Ax+B}{ax^2+bx+c}$$

for each such factor.

- If the denominator contains *repeated quadratic factors*, i.e. a factor of the form $(ax^2 + bx + c)^n$, there will be partial fractions of the form:

$$\frac{Ax+B}{ax^2+bx+c} + \frac{Cx+D}{(ax^2+bx+c)^2} + \dots + \frac{Ex+F}{(ax^2+bx+c)^n}$$

with one for each power of the quadratic.

The values of the constants A, B, C, etc. can be found by either making use of the fact that the equality between the fraction and its partial fractions must be true for all values of the variable x or that the coefficients of x^n in the fraction must equal those of x^n when the partial fractions are multiplied out.

When the degree of the denominator, i.e. the power of its highest term, is equal to or less than that of the numerator, the denominator must be divided into the numerator until the result is the sum of terms with the remainder fraction term having a denominator which is of higher degree than its numerator. Consider, for example, the fraction:

$$\frac{x^3-x^2-3x+1}{x^2-3x+2}$$

The numerator has a degree of 3 and the denominator a degree of 2. Thus, dividing has to be used. Thus

$$\frac{x^3-x^2-3x+1}{x^2-3x+2} = x+2+\frac{x-3}{x^2-3x+2}$$

The fractional term can then be simplified using partial fractions.

$$\frac{x-3}{x^2-3x+2} = \frac{x-3}{(x-1)(x-2)} = \frac{A}{x-1} + \frac{B}{x-2}$$

to give:

$$\frac{x^3-x^2-3x+1}{x^2-3x+2} = x+2+\frac{2}{x-1} - \frac{1}{x-2}$$

Example

Simplify into its partial fraction form: $\dfrac{3x+4}{(x+1)(x+2)}$.

This has two linear factors in the denominator and so the partial fractions are of the form:

$$\frac{A}{x+1} + \frac{B}{x+2}$$

with one partial fraction for each linear term. Thus for the expressions to be equal we must have:

$$\frac{3x+4}{(x+1)(x+2)} = \frac{A}{x+1} + \frac{B}{x+2} = \frac{A(x+2)+B(x+1)}{(x+1)(x+2)}$$

Thus

$$3x + 4 = A(x + 2) + B(x + 1)$$

Consider the requirement that this relationship is true for all values of x. Then, when $x = -1$ we must have:

$$-3 + 4 = A(-1 + 2) + B(-1 + 1)$$

Hence $A = 1$. When $x = -2$ we must have:

$$-6 + 4 = A(-2 + 2) + B(-2 + 1)$$

Hence $B = 2$.

Alternatively, we could have determined these constants by multiplying out the expression and considering the coefficients, i.e.

$$3x + 4 = A(x + 2) + B(x + 1) = Ax + 2A + Bx + B$$

Thus, for the coefficients of x to be equal we must have $3 = A + B$ and for the constants to be equal $4 = 2A + B$. These two simultaneous equations can be solved to give A and B. The partial fractions are thus:

$$\frac{3x+4}{(x+1)(x+2)} = \frac{1}{x+1} + \frac{2}{x+2}$$

Example

Determine the indefinite integral $\int \frac{1}{x^2 - 1}\, dx$.

The fraction $1/(x^2 - 1)$ can be written as:

$$\frac{1}{(x-1)(x+1)} = \frac{A}{x-1} + \frac{B}{x+1} = \frac{A(x+1)+B(x-1)}{(x-1)(x+1)}$$

Hence, equating coefficients of x gives $A + B = 0$ and equating integers gives $A - B = 1$. Thus $A = \frac{1}{2}$ and $B = -\frac{1}{2}$. Hence the integral can be expressed as:

$$\int \frac{1}{x^2 - 1}\, dx = \frac{1}{2} \int \frac{1}{x - 1}\, dx - \frac{1}{2} \int \frac{1}{x + 1}\, dx$$

We can determine these integrals by substitution. Thus if we let $u = x - 1$ then $du/dx = 1$ and so:

$$\int \frac{1}{x - 1}\, dx = \int \frac{1}{u}\, du = \ln|u| = \ln|x - 1| + A$$

Likewise the integral of $1/(x + 1)$ is $\ln|x + 1| + B$. Hence:

$$\int \frac{1}{x^2 - 1}\, dx = \frac{1}{2} \ln|x - 1| + \frac{1}{2} \ln|x + 1| + C$$

Example

Determine the indefinite integral $\int \frac{x^3}{x - 2}\, dx$.

This fraction has a numerator of higher degree than the denominator and so the numerator must be divided by the denominator until the remainder is of lower degree than the denominator. Thus:

$$
\begin{array}{r}
x^2 + 2x + 4 \\
x - 2 \overline{\smash{\big)}\, x^3 } \\
\underline{x^3 - 2x^2} \\
2x^2 \\
\underline{2x^2 - 4x} \\
4x \\
\underline{4x - 8} \\
8
\end{array}
$$

Hence the integral becomes:

$$\int \frac{x^3}{x - 2}\, dx = \int \left(x^2 + 2x + 4 + \frac{8}{x - 2} \right) dx$$

$$= \frac{x^3}{3} + x^2 + 4x + 8 \ln|x - 2| + C$$

Example

Determine the indefinite integral $\int \frac{1}{x(x^2 + 1)}\, dx$.

Expressed as partial fractions:

$$\frac{1}{x(x^2+1)} = \frac{A}{x} + \frac{Bx+C}{x^2+1} = \frac{A(x^2+1)+(Bx+C)x}{x(x^2+1)}$$

Equating the constant terms gives $A = 1$. Equating the coefficients of x gives $C = 0$. Equating the coefficients of x^2 gives $A + B = 0$, and so $B = -1$. Thus the integral becomes:

$$\int \frac{1}{x(x^2+1)}\, dx = \int \left(\frac{1}{x} - \frac{x}{x^2+1}\right) dx$$

The integration of $1/(x^2 + 1)$ can be carried out by using a substitution. Let $u = x^2 + 1$ and so $du/dx = 2x$. Thus:

$$\int \frac{x}{x^2+1}\, dx = \int \frac{1}{2u}\, du = \tfrac{1}{2} \ln|u| + C = \tfrac{1}{2} \ln|x^2+1| +$$

and so:

$$\int \frac{1}{x(x^2+1)}\, dx = \ln|x| - \tfrac{1}{2}\ln|x^2+1| + C$$

Maths in action

The technique of using partial fractions to simplify expressions has many uses. In Chapter 6, we shall see how partial fractions can help in the solution of differential equations using the Laplace transform. As an illustration, consider a differential equation relating rotational displacement θ to time t for a rotating power transmission shaft:

$$\frac{d^2\theta}{dt^2} - 6\frac{d\theta}{dt} - 10\theta = 20 - e^{2t}$$

Given that when $t = 0$ we have $\theta = 4$ and $d\theta/dt = 25/2$, the Laplace transform enables the differential equation to be written in the form:

$$\mathcal{L}\{\theta\} = \frac{4s^3 - \frac{39}{2}s^2 + 42s - 40}{s(s-2)(s^2-6s+10)}$$

We can use the method of partial fractions to simplify the expression. Let the fraction be replaced by:

$$\frac{A}{s} + \frac{B}{s-2} + \frac{Cs+D}{s^2-6s+10}$$

Then we must have:

$$4s^3 - \frac{39}{2}s^2 + 42s - 40 = A(s-2)(s^2 - 6s + 10)$$
$$+ B(s)(s^2 - 6s + 10) + (Cs + D)(s)(s-2)$$

If we let $s = 2$, then $32 - 78 + 84 - 40 = B(2)(4 - 12 + 10)$ and so $B = -1/2$. If we let $s = 0$, then $-40 = A(-2)(10)$ and so $A = 2$. Comparing coefficients of s gives $42 = 22A + 10B - 2D$ and so $D = -3/2$. Comparing coefficients of s^3 gives $4 = A + B + C$ and so $C = 5/2$. Putting these values into the partial fraction equation gives:

$$\frac{2}{s} + \frac{\left(-\frac{1}{2}\right)}{s-2} + \frac{\left(\frac{5}{2}\right)s + \left(-\frac{3}{2}\right)}{s^2 - 6s + 10}$$

$$= \frac{2}{s} - \frac{1}{2(s-2)} + \frac{5s - 3}{2(s^2 - 6s + 10)}$$

This is a lot easier to handle than the original equation.

4.2.4 Means

The *mean* of a set of numbers is their sum divided by the number of numbers summed. The *mean value of a function* between $x = a$ and $x = b$ is the mean value of all the ordinates between these limits. Suppose we divide the area into n equal width strips (Figure 4.34), then if the values of the mid-ordinates of the strips are y_1, y_2, ... y_n the mean value is:

mean value of $y = \dfrac{y_1 + y_2 + ... + y_n}{n}$

If δx is the width of the strips, then $n \, \delta x = b - a$. Thus:

mean value of $y = \dfrac{\left(y_1 + y_2 + ... + y_n\right)\delta x}{b - a}$

Hence, as $\delta x \to 0$:

mean value of $y = \dfrac{1}{b - a} \int_a^b y \, \mathrm{d}x$ [34]

Since the sum of all the $y \, \delta x$ terms is the area under the graph between $x = a$ and $x = b$:

mean value of $y = \dfrac{\text{area under graph}}{b - a}$

But the product of the mean value and $(b - a)$ is the area of a rectangle of height equal to the mean value and width $(b - a)$. Figure 4.35 shows this mean value rectangle.

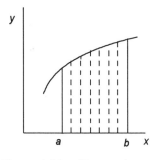

Figure 4.34 *Mean value*

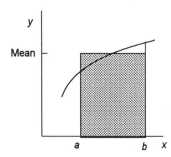

Figure 4.35 *Mean value rectangle*

Example

Determine the mean value of the function $y = \sin x$ between $x = 0$ and $x = \pi$.

The mean value of function is:

$$\frac{1}{b-a}\int_a^b y\, dx = \frac{1}{\pi-0}\int_0^\pi \sin x\, dx$$

$$= \frac{1}{\pi}[-\cos x]_0^\pi = \frac{2}{\pi} = 0.637$$

Root-mean-square values

The power dissipated by an alternating current i when passing through a resistance R is i^2R. The mean power dissipated over a time interval from $t = 0$ to $t = T$ will thus be:

$$\text{mean power} = \frac{1}{T-0}\int_0^T i^2R\, dt = \frac{R}{T}\int_0^T i^2\, dt$$

If we had a direct current I generating the same power then we would have:

$$I^2R = \frac{R}{T}\int_0^T i^2\, dt$$

and:

$$I = \sqrt{\frac{1}{T}\int_0^T i^2\, dt} \qquad [35]$$

This current I is known as the *root-mean-square* current. There are other situations in engineering and science where we are concerned with determining root-mean-square quantities. The procedure is thus to determine the mean value of the squared function over the required interval and then take the square root.

Example

Determine the root-mean-square current value of the alternating current $i = I \sin \omega t$ over the time interval $t = 0$ to $t = 2\pi/\omega$.

The root-mean-square value is:

$$I = \sqrt{\frac{1}{T}\int_0^T i^2\, dt} = \sqrt{\frac{\omega}{2\pi}\int_0^{2\pi/\omega} I^2 \sin^2\omega t\, dt}$$

$$= \sqrt{\frac{I^2\omega}{2\pi}\int_0^{2\pi/\omega} \tfrac{1}{2}(1-\cos 2\omega t)\, dt}$$

$$= \sqrt{\frac{I^2\omega}{4\pi}\left[t - \frac{1}{2\omega}\sin 2\omega t\right]_0^{2\pi/\omega}} = \frac{I}{\sqrt{2}}$$

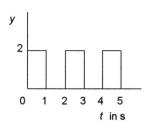

Figure 4.36 *Example*

Example

Determine the root-mean-square value of the waveform shown in Figure 4.36 over a period of 0 to 2 s.

From $t = 0$ to $t = 1$ s the waveform is described by $y = 2$. From $t = 1$ s to $t = 2$ s the waveform is described by $y = 0$. Thus the root-mean-square value is given by:

$$y_{rms} = \sqrt{\frac{1}{2}\left(\int_0^1 4\,dt + \int_1^2 0\,dt\right)} = \sqrt{\frac{1}{2}[4t]_0^1} = \sqrt{2}$$

Example

An alternating current is defined by the equation:

$i = 25 \sin 100\pi t$ mA

Determine its mean value over half-a-cycle and the root-mean-square value over a cycle.

We have $100\pi = 2\pi f = 2\pi/T$, where f is the frequency and T the periodic time. Hence the periodic time is 0.02 s and the time for half-a-cycle is 0.01 s.

Using equation [34], the mean value over half-a-cycle is:

$$\frac{1}{0.01 - 0} \int_{t=0}^{t=0.01} 25 \sin 100\pi t\,dt$$

We can use the standard form for the integral of sin ax to give the mean value as:

$$\frac{25}{0.01}\left[-\frac{1}{100\pi}\cos 100\pi t + C\right]_0^{0.01}$$

$$= \frac{1}{0.04\pi}\left((-\cos\pi) - (-\cos 0)\right) = \frac{1}{0.04\pi}(1 + 1)$$

Thus the mean value over half-a-cycle is 15.92 mA. Note that for a sinusoidal signal the mean value over a full cycle is zero.

The root-mean-square value over a cycle is given by equation [35] as:

$$\sqrt{\frac{1}{T}\int_0^T i^2\,dt} = \sqrt{\frac{1}{0.02}\int_0^{0.02} 25^2 \sin^2 100\pi t\,dt}$$

Since $\sin^2\theta = \frac{1}{2}(1 - \cos 2\theta)$, we can write:

$$\sqrt{\frac{625}{0.02}\int_0^{0.02} \tfrac{1}{2}(1 - \cos 200\pi t\,dt}$$

$$\sqrt{\frac{625}{0.04}\left[t - \frac{1}{200\pi}\sin 200\pi t\right]_0^{0.02}} = \sqrt{\frac{625}{0.04}\times 0.02}$$

and so the r.m.s. current $25/\sqrt{2} = 17.68$ mA. Note, that in general, the root-mean-square value of a sinusoidal signal is always the maximum value divided by $\sqrt{2}$.

Problems 4.2

1 Determine the integrals of the following:

(a) 4, (b) $2x^3$, (c) $2x^3 + 5x$, (d) $x^{2/3} - 3x^{1/2}$, (e) $4 + \cos 5x$,

(f) $2\,e^{-3x}$, (g) $4\,e^{x/2} + x^2 + 2$, (h) $4/x$

2 Determine the areas under the following curves between the specified limits and the x-axis:

(a) $y = 4x^3$ between $x = 1$ and $x = 2$,
(b) $y = x$ between $x = 0$ and $x = 4$,
(c) $y = 1/x$ between $x = 1$ and $x = 3$,
(d) $y = x^3 - 3x^2 - 2x + 2$ between $x = -1$ and $x = 2$,
(e) $y = x^2 - x - 2$ between $x = -1$ and $x = 2$,
(f) $y = x^2 - 1$ between $x = -1$ and $x = 2$,
(g) the area between $x = 0$ and $x = 2$ for the curve defined by $y = x^2$ between $x = 0$ and $x = 1$ and by $y = 2 - x$ between $x = 1$ and $x = 2$.

3 Determine the areas bounded by graphs of the following functions and between the specified ordinates:

(a) $y = 9 - x^2$, $y = -2$, $x = -2$ and $x = 2$,
(b) $y = 4$, $y = x^2$, $x = 0$ and $x = 1$

4 Determine the geometrical area enclosed between the graph of the function $y = x(x - 1)(x - 2)$ and the x-axis.
5 Determine the area bounded by graphs of $y = x^3$ and $y = x^2$.
6 Determine the area bounded by the graph of $y = \sin x$, the x-axis and the line $x = \pi/2$.
7 Determine the area bounded by graphs of $y = x^2 - 2x + 2$ and $y = 4 - x$.

8 Determine the values, if they exist, of the following definite integrals:

(a) $\int_{-\infty}^{1} x\,dx$, (b) $\int_{1}^{\infty} \frac{1}{x^3}\,dx$, (c) $\int_{0}^{\infty} e^{-3x}\,dx$, (d) $\int_{-\infty}^{-1} x^4\,dx$

9 Determine the following indefinite integral by using the given substitutions:

(a) $\int (x^2+1)x^3\,dx$ using $u = x^2+1$,

(b) $\int 2\,e^{4x-1}\,dx$ using $u = 4x-1$,

(c) $\int \frac{x+1}{\sqrt{2x+1}}\,dx$ using $u = \sqrt{2x+1}$,

(d) $\int x\sin x^2\,dx$ using $u = x^2$,

(e) $\int \sqrt{x^2+4}\,dx$ using $x = 2\sinh u$,

(f) $\int x\sqrt{x-1}\,dx$ using $u = \sqrt{x-1}$,

(g) $\int \sec x\,dx$ using $u = \tan\frac{1}{2}x$,

(h) $\int \frac{1}{\sqrt{9-x^2}}\,dx$ using $x = 3\sin\theta$,

(i) $\int \sin^2 2x\cos^3 2x\,dx$ using $u = \sin x$

10 Determine the following indefinite integrals by making appropriate substitutions:

(a) $\int x\sqrt{x+2}\,dx$, (b) $\int \frac{1}{(x^2+1)^{3/2}}\,dx$, (c) $\int \sin^3 x\,dx$,

(d) $\int \frac{1}{\sqrt{1-4x^2}}\,dx$, (e) $\int x\sqrt{x^2+2}\,dx$, (f) $\int \frac{1}{4+25x^2}\,dx$,

(g) $\int \cos^3 x\sin^4 x\,dx$, (h) $\int \tan^3 x\sec^2 x\,dx$,

(i) $\int \frac{1}{5+4\cos x}\,dx$

11 By making appropriate substitutions, evaluate the following definite integrals:

(a) $\int_{0}^{1} \frac{1}{2-x}\,dx$, (b) $\int_{0}^{1} \frac{3x^2}{(x^3+9)^2}\,dx$, (c) $\int_{0}^{1} \frac{x^2}{\sqrt{1-x^2}}\,dx$,

(d) $\int_{-1}^{1} x^2\sqrt{2-x^2}\,dx$, (e) $\int_{0}^{2} \frac{1}{4+x^2}\,dx$

12 Using the method of integration by parts, determine the following indefinite integrals:

(a) $\int x^2 \ln x \, dx$, (b) $\int x \, e^{2x} \, dx$, (c) $\int x^3 \cos x \, dx$,

(d) $\int x \sin 5x \, dx$ (e) $\int x \ln 3x \, dx$, (f) $\int \sin^2 x \, dx$

13 Using the method of integration by parts, evaluate the following definite integrals:

(a) $\int_0^{\pi/2} x \cos x \, dx$, (b) $\int_0^{\pi/2} x \cos^2 x \, dx$,

(c) $\int_0^{\pi} (\pi - x) \cos x \, dx$

14 Determine the following indefinite integrals:

(a) $\int \frac{x^2}{2x-3} \, dx$, (b) $\int \frac{x}{1-2x} \, dx$, (c) $\int \frac{x^2}{2x^2+x-3} \, dx$,

(d) $\int \frac{x^2}{(x^2-1)(2x+1)} \, dx$, (e) $\int \frac{x+1}{x(x-2)(x+2)} \, dx$,

(f) $\int \frac{3x-1}{(2x+1)(x-1)} \, dx$, (g) $\int \frac{2x^3+3x^2-3}{2x^2-x-1} \, d$

(h) $\int \frac{1}{(x-2)(x-3)} \, dx$, (i) $\int \frac{5x^2+20x+6}{x(x+1)^2} \, dx$,

(j) $\int \frac{2x^3-4x-8}{x(x-1)(x^2+4)} \, dx$, (k) $\int \frac{1}{x^2(x^2+1)} \, dx$

15 Determine the moment of inertia for a uniform triangular sheet of mass M, base b and height h about (a) an axis through the centroid and parallel to the base and (b) about the base. The centroid is at one-third the height.

16 Determine the moment of inertia of a flat circular ring with an inner radius r, outer radius $2r$ and mass M about an axis through its centre and at right angles to its plane.

17 Determine the moment of inertia of a uniform square sheet of mass M and side L about (a) an axis through its centre and in its plane, (b) an axis in its plane a distance d from its centre.

18 Determine the mean values of the following functions between the specified limits:

(a) $y = 2x$ between $x = 0$ and $x = 1$,
(b) $y = x^2$ between $x = 1$ and $x = 4$,
(c) $y = 3x^2 - 2x$ between $x = 1$ and $x = 4$,
(d) $y = \cos^2 x$ between $x = 0$ and $x = 2\pi$.

19 With simple harmonic motion, the displacement x of an object is related to the time t by $x = A \cos \omega t$. Determine the mean value of the displacement during one-quarter of an oscillation, i.e. between when $\omega t = 0$ and $\omega t = 0$.

20 The number N of radioactive atoms in a sample is a function of time t, being given by $N = N_0 \, e^{-\lambda t}$. Determine

the mean number of radioactive atoms in the sample between $t = 0$ and $t = 1/\lambda$.

21 Determine the root-mean-square values of the following functions between the specified limits:

(a) $y = x^2$ from $x = 1$ to $x = 3$,
(b) $y = x$ from $x = 0$ to $x = 2$,
(c) $y = \sin x + 1$ from $x = 0$ to $x = 2\pi$,
(d) $y = \sin 2x$ from $x = 0$ to $x = \pi$,
(e) $y = e^x$ from $x = -1$ to $x = +1$

22 Determine the root-mean-square value of a half-wave rectified sinusoidal voltage. Between the times $t = 0$ and $t = \pi/\omega$ the equation is $v = V \sin \omega t$ and between $t = \pi/\omega$ and $t = 2\pi/\omega$ we have $v = 0$.

5 Differential equations

Summary

This chapter introduces ordinary differential equations, shows how they can be used to model the behaviour of systems in engineering and looks at their solution for different inputs to the systems. Differential equations arise from such situations as the lumped models designed to represent systems (see Chapter 3), the motion of projectiles, the cooling of a solid or liquid, transient currents and voltages in electrical circuits, oscillations with mechanical or electrical systems and the rate of decay of radioactive substances.

Objectives

By the end of this chapter, the reader should be able to:

- represent engineering systems by differential equations;
- solve first- and second-order differential equations;
- solve the differential equations representing models of engineering systems for step and ramp inputs.

5.1 Differential equations

A *differential equation* is an equation involving derivatives of a function. Thus examples of differential equations are:

$$\frac{dy}{dx} + 2y = 5 \text{ and } \frac{d^2y}{dx^2} + 3\frac{dy}{dx} + 2y = 5$$

The term *ordinary differential equation* is used when there is only one independent variable, the above examples having only y as a function of x and so being ordinary differential equations.

Chapter 3 showed how differential equations can be evolved for the mathematical models of lumped engineering systems. The following extends that analysis to illustrate how ordinary differential equations can be evolved for some simple systems.

Mechanical systems

Consider a freely falling body of mass m in air (Figure 5.1). The gravitational force acting on the body is mg, where g is the

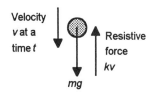

Figure 5.1 *Body falling in air*

Key point

The order of a differential equation is equal to the order of the highest derivative that appears in the equation.

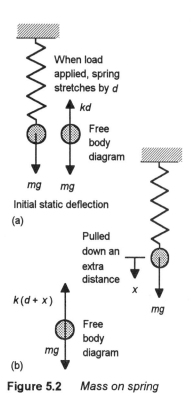

Figure 5.2 *Mass on spring*

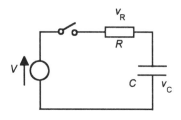

Figure 5.3 *Series RC circuit*

acceleration due to gravity. Opposing the movement of the body through the air is air resistance. Assuming that the air resistance force is proportional to the velocity v, the net force F acting on the body is $mg - kv$, where k is a constant. But Newton's second law gives the net force F acting on a body as the product of its mass m and acceleration a, i.e. $F = ma$. But acceleration is the rate of change of velocity v with time t. Thus we can write:

$$F = m\frac{dv}{dt} = mg - kv$$

and so the differential equation describing this system is:

$$m\frac{dv}{dt} + kv = mg \qquad [1]$$

This is a *first-order differential equation* because the highest derivative is dv/dt. It describes how the velocity varies with time.

Consider another mechanical system, an object of mass m suspended from a support by a spring (Figure 5.2). When the mass is placed on the spring it stretches by d, called the static displacement (Figure 5.2(a)). Assuming Hooke's law, and so the displacement proportional to the force exerted by the spring, we can write $F = kd$. At equilibrium, considering the vertical forces and applying $F = ma$, we have $mg - kd = ma = 0$ as there is zero acceleration. Now if we pull the body down a distance x from this equilibrium position (Figure 5.2(b)) and again apply $F = ma$ to the system when released, the net restoring force acting on the body is $mg - k(d + x) = ma$ and so $-kx = ma$. Since acceleration is the rate of change of velocity with time, with velocity being the rate of change of displacement with time:

$$ma = m\frac{dv}{dt} = m\frac{d}{dt}\left(\frac{dx}{dt}\right) = m\frac{d^2x}{dt^2} = -kx$$

and so:

$$m\frac{d^2x}{dt^2} + kx = 0 \qquad [2]$$

This is a *second-order differential equation* because the highest derivative is d^2x/dt^2. It describes the resulting oscillations of the body after it has been released.

Electrical systems

For an electrical circuit with a resistor in series with a capacitor (Figure 5.3), the supply voltage V equals the sum of the voltages across the resistor and capacitor:

$$V = v_R + v_C = Ri + v_C$$

When a pure capacitor has a potential difference v applied across it, the charge q on the plates is given by $q = Cv$, where C is the capacitance. Current i is the rate of movement of charge and so:

$$i = \frac{dq}{dt} = C\frac{dv}{dt} \qquad [3]$$

and thus we have:

$$RC\frac{dv_C}{dt} + v_C = V \qquad [4]$$

This first-order differential equation describes how the capacitor voltage changes with time from when the switch is closed.

When a charged capacitor discharges through a resistance (Figure 5.4) then $v_R + v_C = 0$ and so:

$$RC\frac{dv_C}{dt} + v_C = 0 \qquad [5]$$

This first-order differential equation describes how the capacitor voltage changes with time from when the switch is closed.

When a pure inductor has a current i flowing through it, then the induced e.m.f. produced in the component is proportional to the rate of change of current, the induced e.m.f. being $-L\,di/dt$ where L is the inductance. If the component has only inductance and no resistance, then there can be no potential drop across the component due to the current through the resistance and thus to maintain the current through the inductor the voltage source must supply a potential difference v which just cancels out the induced e.m.f. Thus the potential difference across an inductor is:

$$v = L\frac{di}{dt} \qquad [6]$$

If we have an electrical circuit containing an inductor in series with a resistor (Figure 5.5) then, when the supply voltage V is applied, we have $V = v_L + v_R$. Thus, using equation [6]:

$$L\frac{di}{dt} + Ri = V$$

The steady state current I will be attained when the current ceases to change with time. We then have $RI = V$ and so the first-order differential equation can be written as:

$$\frac{L}{R}\frac{di}{dt} + i = I \qquad [7]$$

Consider now a circuit including a resistor, a capacitor and an inductor in series (Figure 5.6). When the switch is closed the supply voltage v is applied across the three components and $V = v_R + v_L + v_C$. Thus, using equation [6]:

$$Ri + L\frac{di}{dt} + v_C = V$$

Since $i = C\,dv_C/dt$ (equation [3]), then:

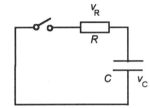

Figure 5.4　*RC discharge circuit*

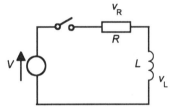

Figure 5.5　*Series RL circuit*

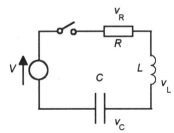

Figure 5.6　*Series RLC circuit*

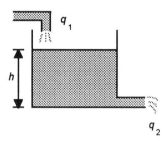

Figure 5.7 *Liquid level in a tank*

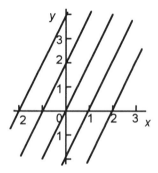

Figure 5.8 *General solution*

$$LC\frac{\mathrm{d}^2 v_C}{\mathrm{d}t^2} + RC\frac{\mathrm{d}v_C}{\mathrm{d}t} + v_C = V \qquad [8]$$

This second-order differential equation describes how the voltage across the capacitor varies with time.

Hydraulic systems

Consider an open tank into which liquid can enter at the top through one pipe and leave at the base through another (Figure 5.7). If the liquid enters at the rate of a volume of q_1 per second and leaves at the rate of q_2 per second, then the rate at which the volume V of liquid in the tank changes with time is:

$$\frac{\mathrm{d}V}{\mathrm{d}t} = q_1 - q_2$$

But $V = Ah$, where A is the cross-sectional area of the tank and h the height of the liquid in the tank. Thus:

$$\frac{\mathrm{d}(Ah)}{\mathrm{d}t} = A\frac{\mathrm{d}h}{\mathrm{d}t} = q_1 - q_2$$

The rate at which liquid leaves the tank, when flowing from the base of the tank into the atmosphere, is given by Torricelli's theorem as $q_2 = \sqrt{(2gh)}$. Thus we have:

$$A\frac{\mathrm{d}h}{\mathrm{d}t} + \sqrt{2gh} = q_1 \qquad [9]$$

Solving differential equations

The differential equation $\mathrm{d}y/\mathrm{d}x = 2$ describes a straight line with a constant gradient of 2 (Figure 5.8). There are, however, many possible graphs which fit this specification, the family of such lines having equations of the form $y = 2x + A$, where A is a constant. These are all solutions for the differential equation.

Thus the differential equation $\mathrm{d}y/\mathrm{d}x = 2$ has many solutions given by $y = 2x + A$, this being termed the *general solution*. Only if constraints are specified which enable constants like A to be evaluated will there be just one solution, this being then termed a *particular solution*. The term *initial conditions* are used for the constraints if specified at $y = 0$ and *boundary conditions* if specified at some other value of y. Thus if, for a general solution $y = 2x + A$, we have the initial condition that $y = 0$ when $x = 0$ then A is 0 and so the particular solution is $y = 2x$.

Example

Verify that $y = e^x$ is a particular solution of the differential equation $\mathrm{d}y/\mathrm{d}x = y$.

If $y = e^x$ then $dy/dx = e^x$. Thus for all values of y we have $dy/dx = y$ and so $y = e^x$ is a solution.

Example

$y = A e^x + B e^{2x}$ is a general solution of the differential equation:

$$\frac{d^2y}{dx^2} - 3\frac{dy}{dx} + 2y = 0$$

Determine the particular solution for the boundary conditions $y = 3$ when $x = 0$ and $dy/dx = 5$ when $x = 0$.

For $y = A e^x + B e^{2x}$ with $y = 3$ when $x = 0$ we have $3 = A + B$. With $y = A e^x + B e^{2x}$ we have $dy/dx = A e^x + 2B e^{2x}$ and thus with $dy/dx = 5$ when $x = 0$ we have $5 = A + 2B$. This pair of simultaneous equations gives $A = 1$ and $B = 2$. Thus the particular solution is:

$$y = e^x + 2 e^{2x}$$

We can check that this is a valid solution by substituting it in the differential equation:

$$e^x + 8 e^{2x} - 3(e^x + 4 e^{2x}) + 2(e^x + 2 e^{2x}) = 0$$

Problems 5.1

1 Derive differential equations to represent the following situations:

(a) The velocity v of an object of mass m in terms of time t when thrown vertically upwards against air resistance proportional to the square of its velocity.

(b) The displacement x of a mass m on a spring when the mass is pulled down from its equilibrium position and released when there is a damping force proportional to the velocity.

(c) The velocity v of a boat of mass m on still water in terms of time t after the engines are switched off if the drag forces acting on the boat are proportional to the velocity.

(d) The velocity v of an object falling from rest in air if the drag forces are proportional to the square of the velocity.

(e) The intensity I of a beam of light emerging from a block of glass in terms of the thickness x of the glass if the intensity decreases at a rate proportional to the block thickness.

(f) The rate at which the pressure p at the base of a tank changes with time if liquid of density q enters the tank at the volume rate of q_1 and leaves at the rate of q_2.

(g) The height h of a liquid in a tank open to the atmosphere as a function of time t after a leak from the base of the tank occurs.

2 Verify that the following are solutions of the given differential equations:

(a) $y = \cos 2x$ for $\dfrac{d^2y}{dx^2} + 4y = 0,$

(b) $y = 2\sqrt{x} - \sqrt{x}\ln x$ for $4x^2\dfrac{d^2y}{dx^2} + y = 0,$

(c) $y = e^x \cos x$ for $\dfrac{d^2y}{dx^2} - 2\dfrac{dy}{dx} + 2y = 0,$

(d) $y = 2 e^x + 3x\, e^x$ for $\dfrac{d^2y}{dx^2} + 3\dfrac{dy}{dx} + 2y = 0$

3 For the following general solutions of differential equations, verify that they are solutions and determine the particular solution for the given boundary conditions:

(a) $y = A\, e^x + Bx\, e^x$ for:

$$\dfrac{d^2y}{dx^2} - 2\dfrac{dy}{dx} + y = 0,\ y = 0 \text{ and } \dfrac{dy}{dx} = 1 \text{ at } x = 0,$$

(b) $y = A \sin \omega t + B \cos \omega t$ for:

$$\dfrac{d^2y}{dt^2} + \omega^2 y = 0,\ y = 2 \text{ and } \dfrac{dy}{dt} = 1 \text{ at } t = 0,$$

(c) $y = (A + x^2)\, e^{-x}$ for:

$$\dfrac{dy}{dx} + y = 2x\, e^{-x},\ y = 2 \text{ at } x = 0$$

4 The differential equation relating the deflection y with distance x from the fixed end of a cantilever with a uniformly distributed load is:

$$\dfrac{d^2y}{dx^2} = -\dfrac{w}{2EI}(L^2 - 2Lx + x^2)$$

The general solution is given as:

$$y = -\dfrac{w}{2EI}\left(\tfrac{1}{2}L^2x^2 - \tfrac{1}{3}Lx^3 + \tfrac{1}{12}x^4\right) + Ax + B$$

Verify that this is the general solution and determine the particular solution for $y = 0$ and $dy/dx = 0$ at $x = 0$.

5.2 First-order differential equations

First-order differential equations are often used to model the behaviour of engineering systems. For example, the exponential growth system where the rate of change dN/dt of some quantity is proportional to the quantity N present can be represented by:

$$\frac{dN}{dt} = kN$$

or exponential decay, e.g. radioactivity, where the rate at which a quantity decreases is proportional to the quantity present:

$$\frac{dN}{dt} = -kN$$

Such differential equations are of the form:

$$\frac{dy}{dx} = f(y) \tag{10}$$

Another form of differential equation is illustrated by the growth of the voltage v_C across a capacitor in an electrical circuit having a capacitor C in series with a resistor R and connected to a step voltage input of V:

$$RC\frac{dv_C}{dt} + v_C = V$$

Such equations are of the form:

$$\frac{dy}{dx} + Py = Q \tag{11}$$

where P and Q are constants or functions of x.

This section looks at probably the most common method that is used for the solution of such differential equations, the separation of variables, and how it can be used to determine the output of systems which are modelled by first-order differential equations.

Separation of variables

A first-order equation is said to be *separable* if the variables x and y can be separated. To solve such equations we simply separate the variables and then integrate both sides of the equation with respect to x. The following shows solutions of the various forms taken by separable equations:

- **Equations of the form $\dfrac{dy}{dx} = f(x)$**

   If we integrate both sides of the equation with respect to x:

$$\int \frac{dy}{dx}\, dx = \int f(x)\, dx$$

This is equivalent to separating the variables and writing:

$$\int dy = \int f(x)\, dx \qquad\qquad [12]$$

Example

Solve the differential equation $dy/dx = 2x$.

Separating the variables gives:

$$\int dy = \int 2x\, dx$$

and thus $y = x^2 + A$.

Example

If $dp/dt = (3 - t)^2$, find p in terms of t given the condition that $p = 3$ when $t = 2$.

Separating the variables gives:

$$dp = (3 - t)^2\, dt$$

and so:

$$\int p = \int (3 - t^2)\, dt = \int (9 - 6t + t^2)\, dt$$

$$p = 9t - \frac{6t^2}{2} + \frac{t^3}{3} + C$$

This is the general solution. Using the given conditions that $p = 3$ when $t = 2$ gives:

$$3 = 9(2) - 3(2)^2 + \frac{(2)^3}{3} + C$$

Hence $C = -5.67$ and so the specific solution is:

$$p = 9t - 3t^2 + \frac{t^3}{3} - 5.67$$

- **Equations of the form $\dfrac{dy}{dx} = f(y)$**

  This can be rearranged to give:

  $$\frac{1}{f(y)}\frac{dy}{dx} = 1$$

  Integrating both sides with respect to x:

  $$\int \frac{1}{f(y)}\frac{dy}{dx}\, dx = \int 1\, dx$$

This is equivalent to separating the variables:

$$\int \frac{1}{f(y)}\, dy = \int 1\, dx \qquad [13]$$

Example

Solve the differential equation dy/dx = 2y.

Separating the variables gives:

$$\int \frac{1}{y}\, dy = \int 2\, dx$$

Thus ln y = 2x + A. We can write this as y = e$^{2x + A}$ = e^{2x} e^{A} = B e^{2x}, where B is a constant.

- **Equations of the form $g(y)\dfrac{dy}{dx} = f(x)$**

Integrating both sides of the equation with respect to x gives:

$$\int g(y)\frac{dy}{dx}\, dx = \int f(x)\, dx$$

This is equivalent to:

$$\int g(y)\, dy = \int f(x)\, dx \qquad [14]$$

Example

Solve the differential equation dy/dx = 2x/y.

Separating the variables give:

$$\int y\, dy = \int 2x\, dx$$

Thus ½y^2 = x^2 + A.

- **Equations of the form $\dfrac{dy}{dx} = f(x)g(y)$**

This can be rearranged and integrated with respect to x to give:

$$\int \frac{1}{g(y)}\frac{dy}{dx}\, dx = \int f(x)\, dx$$

This is equivalent to:

$$\int \frac{1}{g(y)}\, dy = \int f(x)\, dx \qquad [15]$$

Key point

To solve first-order differential equations by separation of variables:

1. Write the differential equation in the form f(y) dy = g(x) dx.

2. Solve by integrating both sides of the equation.

Example

Solve the differential equation dy/dx = 2yx.

Separating the variables gives:

$\int \frac{1}{y}\,dy = \int 2x\,dx$

Thus ln $y = x^2 + A$.

Equations which are not of any of the above forms may often be put into one of the forms by a *change of variable*. As an illustration, consider the differential equation dy/dx = $y/(y + x)$. This can be written as:

$$\frac{dy}{dx} = \frac{\frac{y}{x}}{\frac{y}{x} + 1}$$

If we let $v = y/x$ then $y = vx$ and dy/dx = $v + x$ dv/dx. Thus the above equation can be written as:

$$v + x\frac{dv}{dx} = \frac{v}{v+1}$$

$$x\frac{dv}{dx} = \frac{v}{v+1} - v = -\frac{v^2}{v+1}$$

$$\frac{v+1}{v^2}\frac{dv}{dx} = -\frac{1}{x}$$

Integrating with respect to x:

$$\int\left(\frac{1}{v} + \frac{1}{v^2}\right)\frac{dv}{dx}\,dx = -\int \frac{1}{x}\,dx$$

This is equivalent to:

$$\int\left(\frac{1}{v} + \frac{1}{v^2}\right)dv = -\int \frac{1}{x}\,dx$$

Hence ln $v - (1/v) = -\ln x + A$ and so $\ln(y/x) - (x/y) = -\ln x + A$.

Example

Solve the differential equation dy/dx = cos^2 y if $y = \pi/4$ when $x = 0$.

We can write the equation as: $\sec^2 y\dfrac{dy}{dx} = 1$

Hence, separating the variables gives:

$$\int \sec^2 y \, dy = \int 1 \, dx$$

and so we have $\tan y = x + A$. Since $y = \pi/4$ when $x = 0$ then $\tan \pi/4 = A$ and so $A = 1$. Thus $\tan y = x + 1$ or $y = \tan^{-1}(x + 1)$.

5.2.1 The responses of first-order systems

This section looks at how, when differential equations are involved in modelling a system, the dynamic responses of systems can be predicted. For example, if the input signal to a measurement system suddenly changes, the output will not instantaneously change to the new value but some time will elapse before it reaches a steady-state value. If the voltage applied to an electrical circuit suddenly changes to a new value, the current in the circuit will not change instantly to the new value but some time will elapse before it reaches the steady new value. If a continually changing signal is applied to a system, the response of the system may lag behind the input. The way in which a system reacts to input changes is termed its *dynamic characteristic*.

First-order systems and step inputs

Consider a thermometer (Figure 5.9) at temperature T_0 inserted into a liquid at a temperature T_1. We can thus think of the thermometer being subject to a step input, i.e. the input abruptly changes from T_0 to T_1. The thermometer will then, over a period of time, change its temperature until it becomes T_1. Thus we have a measurement system, the thermometer, which has a step input and an output which changes from T_0 to T_1 over some time. How does the output, i.e. the reading of the thermometer T, vary with time.

The rate at which energy enters the thermometer from the liquid is proportional to the difference in temperature between the liquid and the thermometer. Thus, at some instant of time when the temperature of the thermometer is T, we can write:

$$\frac{dQ}{dt} = h(T_1 - T)$$

where h is a constant called the *heat transfer coefficient*. For a thermometer with a specific heat capacity c and a mass m, the relationship between heat input Q and the consequential temperature change is:

$$Q = mc \text{ (temperature change)}$$

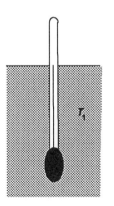

Figure 5.9   *Thermometer inserted in liquid*

When the rate at which heat enters the thermometer is dQ/dt, we can write for the rate at which the temperature changes:

$$\frac{dQ}{dt} = mc\frac{dT}{dt}$$

Thus:

$$mc\frac{dT}{dt} = h(T_1 - T)$$

We can rewrite this with all the output terms on one side of the equals sign and the input on the other, thus:

$$mc\frac{dT}{dt} + hT = hT_1 \tag{16}$$

We no longer have a simple relationship between the input and output but a relationship which involves time. The form of this equation is typical of first-order systems.

We can solve this equation by separation of the variables:

$$\int \frac{1}{T_1 - T}\, dT = \int \frac{h}{mc}\, dt$$

$$-\ln(T_1 - T) = (h/mc)t + A$$

where A is a constant. This can be rewritten as:

$$T_1 - T = e^A\, e^{t/\tau} = C\, e^{t/\tau}$$

where $\tau = mc/h$ and is termed the *time constant*. The time constant can be defined as the value of the time which makes the exponential term become e^{-1}. $T = T_0$ at $t = 0$ and so $C = T_1 - T_0$. Thus:

$$T = T_1 + (T_0 - T_1)\, e^{-t/\tau} \tag{17}$$

The first term is the *steady-state value*, i.e. the value that will occur after sufficient time has elapsed for all transients to die away, and the second term a transient one which changes with time, eventually becoming zero. Figure 5.10 shows graphically how the temperature T indicated by the thermometer changes with time.

After a time equal to one time constant the output has reached about 63% of the way to the steady-state temperature, after a time equal to two time constants the output has reached about 86% of the way, after three time constants about 95% and after about four time constants it is virtually equal to the steady-state value. The error at any instant is the difference between what the thermometer is indicating and what the temperature actually is. Thus:

$$\text{error} = T - T_1 = (T_0 - T_1)\, e^{-t/\tau} \tag{18}$$

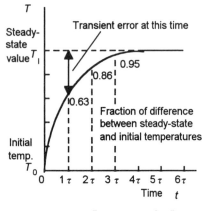

Figure 5.10   *Response of a first-order system to a step input*

This error changes with time and eventually will become zero. Thus it is a transient error.

If a thermometer is required to be fast reacting and quickly attain the temperature being measured, it needs to have a small time constant. Since $\tau = mc/h$, this means a thermometer with a small mass, a small thermal capacity and a large heat transfer coefficient. If we compare a mercury-in-glass thermometer with a thermocouple, then the smaller mass and specific heat capacity of the thermocouple will give it a smaller time constant and hence a faster response to temperature changes.

Example

A thermometer indicates a temperature of 20°C when it is suddenly immersed in a liquid at a temperature of 60°C. If the thermometer behaves as a first-order system and has a time constant of 5 s what will its readings be after (a) 5 s, (b) 10 s, (c) 15 s.

The temperature T of the thermometer varies with time according to equation [17]:

$$T = T_i + (T_0 - T_i)\, e^{-t/\tau} = 60 - 40\, e^{-t/5}$$

After 5 s the thermometer reading will have reached about 63% of the way to the steady-state value, after 10 s about 86%, after 15 s about 95% and after 20 s it is virtually at the steady-state value. Thus after 5 s the reading is 45.3°C, after 10 s it is 54.6°C, after 15 s it is 58.0°C.

Example

A thermometer which behaves as a first-order element has a time constant of 15 s. Initially it reads 20°C. What will be the time taken for the temperature to rise to 90% of the steady-state value when it is immersed in a liquid of temperature 100°C, i.e. a temperature of 92°C?

Equation [17], $T = T_i + (T_0 - T_i)\, e^{-t/\tau}$, can be arranged as:

$$\frac{T - T_i}{T_i - T_0} = e^{-t/\tau}$$

With $T - T_0$ as 90% of $T_i - T_0$, then we have $T - T_i$ as 10% of $T_i - T_0$ and thus:

$$0.10 = e^{-t/15}$$

Taking logarithms gives $-2.30 = -t/15$ and so $t = 34.5$ s.

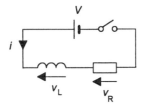

Figure 5.11 *Circuit with series inductance and resistance*

Maths in action

Transients in electrical circuits

Consider the growth of current in a circuit possessing inductance and resistance (Figure 5.11). At some time t after the switch is closed and the current is i, we have:

$$V = v_L + v_R$$

Since $v_R = Ri$ and $v_L = L\,di/dt$ (equation [6]):

$$V = L\frac{di}{dt} + Ri$$

This is a first-order differential equation. It can be solved by separating the variables:

$$\frac{di}{V - Ri} = \frac{dt}{L}$$

If the switch is closed at time $t = 0$ then $i = 0$ when $t = 0$. For the other limit of integration we look for the current to be i at time t. Thus:

$$\int_0^i \frac{1}{V - Ri}\,di = \frac{1}{L}\int_0^t dt$$

$$\left[-\frac{1}{R}\ln(V - Ri)\right]_0^i = \frac{1}{L}[t]_0^t$$

$$-\frac{1}{R}\ln(V - Ri) + \frac{1}{R}\ln V = \frac{t}{L}$$

$$\frac{1}{R}\ln\left(\frac{V}{V - Ri}\right) = \frac{t}{L}$$

$$\ln\left(\frac{V}{V - Ri}\right) = \frac{Rt}{L}$$

$$\frac{V}{V - Ri} = e^{Rt/L} \quad \text{or} \quad \frac{V - Ri}{V} = e^{-Rt/L}$$

$$V - Ri = V\,e^{-Rt/L}$$

$$i = \frac{V}{R}(1 - e^{-Rt/L})$$

The maximum circuit current I is V/R and so:

$$i = I(1 - e^{-Rt/L})$$

When $t = L/R$ then $i = I(1 - e^{-1}) = 0.63I$. This is the same as in Figure 5.10 and so L/R is the time constant of the circuit.

First-order systems in general

In general, a first-order system has a differential equation which can be written in the form:

$$a_1 \frac{dx}{dt} + a_0 x = by \qquad [19]$$

where x is the output, t the time and y the input; a_0, a_1 and b are constants for the system represented by the equation. The left-hand side of the equals sign contains the output related terms and the right-hand side the input related terms. This equation can be rearranged as:

$$\frac{a_1}{a_0} \frac{dx}{dt} + \frac{a_0}{a_0} x = \frac{b}{a_0} y$$

and, if we let $\tau = a_1/a_0$ and $k = b/a_0$, then we have:

$$\tau \frac{dx}{dt} + x = ky \qquad [20]$$

τ defines the *time constant* of the system and k the *static system sensitivity*.

The steady-state value of the output occurs when $dx/dt = 0$ and so $x = ky$ and thus:

steady-state output value $= ky$ $\qquad [21]$

The solution of the differential equation for a step input from some initial value to final value at time $t = 0$ is of the form:

$x =$ steady-state value + (initial value − steady-state value) $e^{-t/\tau}$ [22]

Table 5.1 shows the percentage of the response, i.e. $(x -$ initial value)/ (steady − initial values) × 100%, that will have been achieved after various multiples of the time constant. The percentage dynamic error is (steady-state value − x)/(steady − initial value) × 100%. With a step input, the time constant can be defined as the time taken for the output to reach 63.2% of the steady-state value (see Figure 5.10).

There is an alternative way of defining the time constant. At the instant the input starts and we have $t = 0$, then $x = 0$ and so equation [20] gives $\tau \, dx_0/dt = ky$, where dx_0/dt is the initial gradient of the graph of output with time. Thus, since ky is the steady-state value:

initial gradient of graph = (steady-state value)/τ $\qquad [23]$

Thus, on a graph of output plotted against time for a step input (Figure 5.12), if we draw the tangent to the curve at time $t = 0$, equation [23] gives the initial gradient and so the time constant can be considered to be the time taken for the output to reach the steady-state value if the initial rate of change of output with time were maintained.

Table 5.1 *First-order system response*

Time	% response	% dynamic error
0	0.0	100.0
1τ	63.2	36.8
2τ	86.5	13.5
3τ	95.0	5.0
4τ	98.2	1.8
5τ	99.3	0.7
∞	100.0	0.0

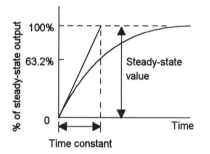

Figure 5.12 *Step response of a first-order system*

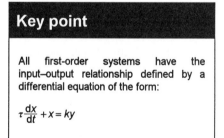

Key point

All first-order systems have the input–output relationship defined by a differential equation of the form:

$$\tau \frac{dx}{dt} + x = ky$$

and all give a response to a step input of the form shown in Figure 5.10.

The time constant can be defined as the time taken, when there is a step input, as:

• the output to reach 63.2% of the steady-state value;

• the output to reach the steady-state value if the initial rate of change of output with time were maintained.

Key point

A way of looking at differential equations which you might come across is in terms of the D-operator. The term *operator* is used for a function which transforms one function into another function. The *D-operator* is such a function which is sometimes used with differential equations. With such an operator we regard dy/dx as the result of an operator applied to the function y and write this as Dy.

$$\frac{dy}{dx} = \frac{d}{dx}(y) = D(y)$$

The differential equation:

$$\tau\frac{dx}{dt} + x = ky$$

thus becomes written as:

$$\tau Dx + x = ky$$

D behaves like an ordinary algebraic quantity and so we can write:

$$\frac{\text{output } x}{\text{input } y} = \frac{k}{\tau D + 1}$$

Thus $k/(\tau D + 1)$ is the quantity we operate on the input by in order to give the output and is called a *transfer function*.

Likewise, we have:

$$\frac{d^2y}{dx^2} = \frac{d}{dy}\left(\frac{dy}{dx}\right) = D(Dy) = D^2y$$

and can write second-order differential equations in terms of the D-operator and obtain a transfer function.

Example

An electrical circuit consisting of resistance R in series with an initially uncharged capacitor of capacitance C has an input of a step voltage V at time $t = 0$. Determine (a) how the potential difference across the capacitor will change with time and (b) with $R = 1$ MΩ, $C = 4$ μF and a step voltage of 12 V, the potential difference across the capacitor after 2 s.

(a) The differential equation, equation [5], is:

$$RC\frac{dv_C}{dt} + v_C = V$$

This equation is the same form as equation [19] so we can recognise that the solution must be of the form given by equation [21] with the time constant being RC:

x = steady-state value + (initial value
$\qquad$ − steady-state value) $e^{-t/\tau}$

$$v_C = V + (0 - V)\,e^{-t/\tau} = V(1 - e^{-t/\tau})$$

(b) The time constant is $RC = 1 \times 10^6 \times 4 \times 10^{-6} = 4$ s. Thus after 2 s, $v_C = 12(1 - e^{-2/4}) = 4.72$ V.

Example

A thermocouple in a protective sheath has an output voltage θ_0 in volts related to the input temperature θ_i in °C by the equation:

$$30\frac{d\theta_0}{dt} + 3\theta_0 = 1.5 \times 10^{-5}\theta_i$$

Determine the time constant τ and the static system sensitivity k.

To put the equation in the standard form of equation [20] we divide by 3:

$$10\frac{d\theta_0}{dt} + \theta_0 = 0.5 \times 10^{-5}\theta_i$$

The time constant is thus 10 s and the static system sensitivity 0.5×10^{-5} V/°C. The thermocouple thus takes 10 s to reach 63.2% of its steady-state output and we need to wait at least three times this time for the output to be close to the voltage corresponding to the temperature being measured.

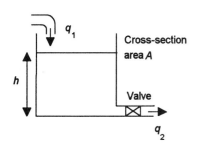

Figure 5.13 *Example*

Example

Determine the time constant and the static system sensitivity for the hydraulic system shown in Figure 5.13 in which a liquid flows into a container at a constant rate and liquid also flows out of the container through a valve at a constant rate.

The differential equation for this system was developed in Chapter 3. The rate of change of liquid volume in the container with time is $A\,\mathrm{d}h/\mathrm{d}t$ and so:

$$q_1 - q_2 = A\frac{\mathrm{d}h}{\mathrm{d}t}$$

For the resistance term for the valve we have $p_1 - p_2 = Rq_1$ and so, since the pressure difference is $h\rho g$:

$$h\rho g = Rq_2$$

Thus, substituting for q_2 gives:

$$q_1 - \frac{h\rho g}{R} = A\frac{\mathrm{d}h}{\mathrm{d}t}$$

and so we can write:

$$A\frac{\mathrm{d}h}{\mathrm{d}t} + \frac{\rho g}{R}h = q_1$$

We can put this equation in the standard form by dividing by $\rho g/R$:

$$\frac{AR}{\rho g}\frac{\mathrm{d}h}{\mathrm{d}t} + h = \frac{R}{\rho g}q_1$$

Comparison with equation [20] thus gives:

time constant $\tau = \dfrac{AR}{\rho g}$

and:

static system sensitivity $k = \dfrac{R}{\rho g}$

Note, that in terms of the D-operator, we can write the differential equation as:

$$\tau Dh + h = kq_1$$

and so a transfer function of:

$$\frac{h}{q_1} = \frac{k}{\tau D + 1}$$

Problems 5.2

1 Solve, by separation of the variables, the following differential equations:

(a) $\dfrac{dy}{dx} = \dfrac{1}{x}$, (b) $\dfrac{dy}{dx} = \cos\tfrac{1}{2}x$, (c) $\dfrac{dy}{dx} = y^2$, (d) $\dfrac{dy}{dx} = -2y$,

(e) $\dfrac{dy}{dx} = 2x(y^2 + 1)$, (f) $\dfrac{dy}{dx} = 3x^2\,e^{-y}$, (g) $\dfrac{dy}{dx} = 4 + 3x^2$,

(h) $\dfrac{dy}{dx} = 2y^2$, (i) $\dfrac{dy}{dx} = 3x^2\,e^{-y}$, (j) $\dfrac{dy}{dx} = \dfrac{1}{2y}$,

(k) $x^2\dfrac{dy}{dx} - y + 1 = 0$, (l) $\dfrac{dy}{dx} = x^2 y$

2 Determine the solution of the differential equation $dy/dx = 2xy^2$ if $y = \tfrac{1}{2}$ when $x = 0$.

3 A capacitor of capacitance C which has been charged to a voltage V_0 is discharged through a resistance R. Determine how the voltage v_C across the capacitor changes with time t if $dv_C/dt = -V/RC$.

4 The rate at which radioactivity decays with time t is given by the differential equation $dN/dt = -kN$, where N is the number of radioactive atoms present at time t. If at time $t = 0$ the number of radioactive atoms is N_0, solve the differential equation and show how the number of radioactive atoms varies with time.

5 When a steady voltage V is applied to a circuit consisting of a resistance R in series with inductance L, determine how the current i changes with time t if $L\,di/dt + Ri = V$ and $i = 0$ when $t = 0$.

6 Determine the solution of the differential equation $dy/dx = 2 - y$ if $y = 1$ when $x = 0$.

7 A stone freely falls from rest and is subject to air resistance which is proportional to its velocity. Derive the differential equation describing its motion and hence determine how its velocity v varies with time t if $v = 0$ at $t = 0$. Take the acceleration due to gravity as 10 m/s^2.

8 For a belt drive, the difference in tension T between the slack and tight sides of the belt over a pulley is related to the angle of lap θ on the pulley by $dT/d\theta = \mu T$, where μ is the coefficient of friction. Solve the differential equation if $T = T_0$ when $\theta = 0°$.

9 A rectangular tank is initially full of water. The water, however, leaks out through a small hole in the base at a rate proportional to the square root of the depth of the water. If the tank is half empty after one hour, how long must elapse before it is completely empty?

10 For a circuit containing resistance R in series with capacitance C, the potential difference v_C across the capacitor varies with time, being given by $v_C = V - V\,e^{-t/RC}$. What is the time constant for the circuit?

11 A hot object cools at a rate proportional to the difference between its temperature and that of its surroundings. If it initially is at 75°C and cooling at a rate of 2° per minute, what will be its temperature after 15 minutes if the surroundings are at a temperature of 15°C?

12 A sphere of ice melts so that its volume V changes at the rate given by $dV/dt = -4\pi kr^2$, where k is a constant and r is the radius at time t after it began to melt. Show that, if R is the initial radius, $r = R e^{-kt}$.

13 A 1000 μF capacitor has been charged to a potential difference of 12 V. At time $t = 0$ it is discharged through a 20 kΩ resistor. What will be the potential difference across the capacitor after 2 s?

14 Determine how the circuit current varies with time when there is a step voltage V input to a circuit having an inductance L in series with resistance R.

15 A sensor behaves as a capacitance of 2 μF in series with a 1 MΩ resistance. As such the relationship between its input y and output x is given by $2(dx/dy) + x = y$. How will the output vary with time when the input is a unit step input at time $t = 0$?

16 A system is specified as being first order with a differential equation relating output x to input y by:

$$a_1 \frac{dx}{dt} + a_0 x = y$$

If it has a time constant of 10 s and a steady-state value of 5. How will the output of the system vary with time when subject to a step input?

17 A sensor is first order with a differential equation relating its output x for input y by:

$$a_1 \frac{dx}{dt} + a_0 x = y$$

If it has a time constant of 1 s, what will be the percentage dynamic error after (a) 1 s, (b) 2 s, from a unit step input signal to the sensor?

18 How long must elapse for the dynamic error of a sensor with a differential equation of the form:

$$a_1 \frac{dx}{dt} + a_0 x = y$$

and subject to a step input to drop below 5% if the sensor is first order with a time constant of 4 s?

19 A thermometer originally indicates a temperature of 20°C and is then suddenly inserted into a liquid at 45°C. The thermometer has a time constant of 2 s. (a) Derive a differential equation showing how the thermometer reading is related to the temperature input and (b) give its solution showing how the thermometer reading varies with time.

5.3 Second-order differential equations

As an example of a second-order ordinary differential equation, consider the displacement y of a freely falling object in a vacuum as a function of time t. It falls with the acceleration due to gravity g and is described by the second-order differential equation:

$$\text{acceleration} = \frac{d^2y}{dt^2} = g \qquad [24]$$

Another example is the displacement y of an object when freely oscillating with simple harmonic motion when there is damping, this being described by the second-order differential equation:

$$m\frac{d^2y}{dt^2} + c\frac{dy}{dt} + ky = 0 \qquad [25]$$

If the oscillating object is not left freely to oscillate when some external force is applied, say $F \sin \omega t$, then we have:

$$m\frac{d^2y}{dt^2} + c\frac{dy}{dt} + ky = F\sin \omega t \qquad [26]$$

With a series electrical circuit containing resistance R, capacitance C and inductance L, the potential difference v_C across the capacitor when it is allowed to discharge is described by the second-order differential equation:

$$LC\frac{d^2v_C}{dt^2} + RC\frac{dv_C}{dt} + v_C = 0 \qquad [27]$$

If such a circuit has a voltage V applied to it we have:

$$LC\frac{d^2v_C}{dt^2} + RC\frac{dv_C}{dt} + v_C = V \qquad [28]$$

Arbitrary constants

Consider an object falling freely with the acceleration due to gravity g. If we take g to be 10 m/s^2 then equation [1] becomes:

$$\frac{d^2y}{dt^2} = 10$$

If we integrate both sides of the equation with respect to t we have:

$$\int \frac{d^2y}{dt^2}\, dt = \int 10\, dt$$

$$\frac{dy}{dt} = 10t + A$$

where A is the constant of integration. If we now integrate this equation with respect to t:

Key point

In general, a second-order differential equation has the form:

$$a_2\frac{d^2y}{dx^2} + a_1\frac{dy}{dx} + a_0y = b$$

where a_2, a_1, a_0 and b are functions of x, b often being termed the forcing function.

$$\int \frac{dy}{dt} \, dt = \int (10t + A) \, dt$$

$$y = 5t^2 + At + B$$

where B is the constant arising from this integration. Thus the above general solution for the second-order differential equation has two arbitrary constants. With all second-order differential equations there will be two arbitrary constants because two integrations are needed to obtain the solution.

Because there are two arbitrary constants with a second-order differential equation, two sets of values are needed to determine them. This is generally done by specifying two initial conditions: the value of the solution and the value of the derivative at a single point. Thus we might have the initial conditions that $y = 20$ at $t = 0$ and $dy/dt = 0$ at $x = 0$.

Example

If the general solution to the differential equation:

$$\frac{d^2 y}{dx^2} - 2\frac{dy}{dx} + y = 0$$

is $y = A\, e^x + Bx\, e^x$, determine the solution if $y = 1$ at $x = 0$ and $dy/dx = -1$ at $x = 0$.

From the initial condition $y = 1$ at $x = 0$ we have, when substituting these values in the general solution, $1 = A + 0$. Thus $A = 1$. If we differentiate the general solution to give $dy/dx = A\, e^x + Bx\, e^x + B\, e^x$ and substitute the initial condition $dy/dx = 0$ at $x = 0$, then $-1 = A + 0 + B$ and so $B = -2$. Thus the solution is $y = e^x - 2x\, e^x$.

5.3.1 Second-order homogeneous differential equations

Consider a second-order differential equation of the basic form:

$$a_2 \frac{d^2 y}{dx^2} + a_1 \frac{dy}{dx} + a_0 y = 0 \tag{29}$$

where a_2, a_1 and a_0 are constants. Such a differential equation is said to be *homogeneous* since, when all the dependent variables are moved to the left of the equal sign, there is just a zero on the right. In the case of a homogeneous linear first-order differential equation with constant coefficients:

$$a_1 \frac{dy}{dx} + a_0 y = 0$$

we have $dy/dx = -(a_0/a_1)y$ and thus, by separation of the variables, the solution is $\ln y = -(a_0/a_1)x + A$ or $y = C\,e^{kx}$, where $k = -(a_0/a_1)$. To solve the constant coefficient second-order differential equation it seems reasonable to consider that it might have a solution of the form $y = A\,e^{sx}$, where A and s are constants. Thus, trying this as a solution, the second-order differential equation [6] becomes:

$$a_2As^2\,e^{sx} + a_1As\,e^{sx} + a_0A\,e^{sx} = 0$$

Since the exponential function is never zero we must have, if $y = A\,e^{sx}$ is to be a solution:

$$a_2s^2 + a_1s + a_0 = 0 \tag{30}$$

Equation [30] is called the *auxiliary equation* or *characteristic equation* associated with the differential equation [29]. This quadratic equation has the roots:

$$s = \frac{-a_1 \pm \sqrt{a_1^2 - 4a_2a_0}}{2a_2} \tag{31}$$

The roots of the auxiliary equation, as given by equation [31], can be:

- **Two distinct real roots if $a_1^2 > 4a_2a_0$**
  The general solution to the differential equation is then:

  $$y = A\,e^{s_1x} + B\,e^{s_2x} \tag{32}$$

Example

Determine the general solution of the differential equation:

$$\frac{d^2y}{dx^2} - \frac{dy}{dx} - 6y = 0$$

Trying $y = A\,e^{sx}$ as a solution gives the auxiliary equation:

$$s^2 - s + 6 = 0$$

which factors as $(s - 3)(s + 2) = 0$ and so $s_1 = 3$ and $s_2 = -2$. Thus the general solution is:

$$y = A\,e^{3x} + B\,e^{-2x}$$

- **Two equal real roots if $a_1^2 = 4a_2a_1$**
  This gives $s_1 = s_2 = -a_1/2a_2$. In order to have a solution with two arbitrary constants we *cannot* have a general solution of:

$$y = A \ e^{s_1 x} + B \ e^{s_2 x} = (A + B) \ e^{sx} = C \ e^{sx}$$

since this can be reorganised to imply only one constant. Thus we try a second solution of the form $y = Bx \ e^{sx}$. Then, since $dy/dx = B \ e^{sx} + Bsx \ e^{sx}$ and $d^2y/dx^2 = 2Bs \ e^{sx} + Bs^2x \ e^{sx}$, substituting into equation [29] gives:

$$a_2(2s + s^2x) + a_1(1 + sx) + a_0x = 0$$

$$(a_2s^2 + a_1s + a_0)x + (2a_2s + a_1) = 0$$

But $a_2s^2 + a_1s + a_0 = 0$ is the auxiliary equation and so the first term is zero. Also $s = -a_1/2a_2$ and so the second term is zero. Thus $y = Bx \ e^{sx}$ is a solution. The general solution is thus:

$$y = A \ e^{s_1 x} + Bx \ e^{s_2 x} \qquad \qquad [33]$$

Example

Determine the general solution of the differential equation:

$$\frac{d^2y}{dx^2} + 8\frac{dy}{dx} + 16y = 0$$

Trying $y = A \ e^{sx}$ as a solution gives the auxiliary equation:

$$s^2 + 8s + 16 = 0$$

This factors as $(s + 4)(s + 4) = 0$ and so we have two roots of $s = -4$. The solution is thus of the form given in equation [33]:

$$y = A \ e^{-4x} + Bx \ e^{-4x}$$

- **Two distinct complex roots if $a_1^2 < 4a_2a_0$**

  With this condition, equation [8] can be written as:

$$s = \frac{-a_1 \pm j\sqrt{4a_2a_0 - a_1^2}}{2a_2} = a \pm j\beta$$

where $a = -(a_1/2a_2)$ and $\beta = \sqrt{(a_0/a_2) - (a_1/a_2)^2/4}$. Thus the general solution is:

$$y = A \ e^{(a+j\beta)x} + B \ e^{(a-j\beta)x}$$

This can be written as:

$$y = A \ e^{ax} \ e^{j\beta x} + B \ e^{ax} \ e^{-j\beta x} = e^{ax}(A \ e^{j\beta x} + B \ e^{-j\beta x})$$

Euler's equation
A complex number z can be expressed as:

$z = |z|(\cos \theta + j \sin \theta)$

Hence:

$\frac{dz}{d\theta} = |z|(-\sin \theta + j\cos \theta) = j(\cos \theta + j\sin \theta)$

But this means that the derivative is just j times z. A function with this property that the derivative is proportional to itself is the exponential. Thus we can write:

$e^{j\theta} = \cos \theta + j\sin \theta$

This is known as Euler's equation.

An alternative way of arriving at this equation is to consider sines and cosines expressed as series, and so write:

$z = |z|\left(1 - \frac{\theta^2}{2!} + \frac{\theta^2}{4!} - ...\right) +$
$\quad |z|j\left(\theta - \frac{\theta^3}{3!} + \frac{\theta^5}{5!} - ..\right)$

Since $j^2 = -1$, $j^3 = -j$, $j^4 = 1$, $j^5 = j$, etc. we can write the equation as:

$z = |z|\left(1 + j\theta + \frac{j^2\theta^2}{2!} + \frac{j^3\theta^3}{3!} + \frac{j^4\theta^4}{4!} + ...\right)$

But this is the form of the series for e^x. Thus we can write:

$e^{j\theta} = \cos \theta + j\sin \theta$

If the roots of the auxiliary equation are both real, i.e. $a_1^2 > 4a_2a_0$, then:

$y = A e^{s_1 x} + B e^{s_2 x}$

If the roots are real and equal, i.e. $a_1^2 = 4a_2a_0$

$y = A e^{s_1 x} + Bx e^{s_2 x}$

If the roots are imaginary, i.e. $a_1^2 < 4a_2a_0$, then:

$y = e^{\alpha x}\left[C \cos \beta x + D \sin \beta x\right]$

There is a relationship called Euler's formula which enables the above equation to be written as:

$y = e^{\alpha x}\left[A(\cos \beta x + j \sin \beta x) + B(\cos \beta x - j \sin \beta x)\right]$

$\quad = e^{\alpha x}\left[(A + B)\cos \beta x + j(A - B)\sin \beta x\right]$

$\quad = e^{\alpha x}\left[C \cos \beta x + D \sin \beta x\right]$ [34]

Example

Determine the general solution of the differential equation:

$$\frac{d^2y}{dx^2} - 2\frac{dy}{dx} + 5y = 0$$

and the particular solution if $y = 1$ and $dy/dx = 2$ at $x = 0$.

Trying $y = A e^{sx}$ as a solution gives, since $dy/dx = sA e^{sx}$ and $d^2y/dx^2 = s^2A e^{sx}$, the auxiliary equation:

$s^2 - 2s + 5 = 0$

This has roots:

$$s = \frac{2 \pm \sqrt{4 - 20}}{2} = 1 \pm j2$$

The general solution will thus be of the form given for equation [34]:

$y = e^x(C \cos 2x + D \sin 2x)$

With $y = 1$ when $x = 0$ we have:

$1 = 1(C + 0)$

and thus $C = 1$. Differentiating the solution gives:

$$\frac{dy}{dx} = e^x(2C \sin 2x - 2D \cos 2x) + e^x(C \cos 2x + D \sin 2x)$$

With $dy/dx = 2$ at $x = 0$ we have:

$2 = -2D + C$

and so $D = -\frac{1}{2}$. Thus the particular solution is:

$y = e^x(\cos 2x + \frac{1}{2} \sin 2x)$

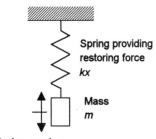

Displacement x

Figure 5.14 *Mass on a spring*

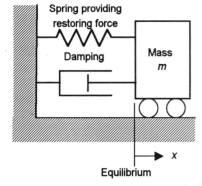

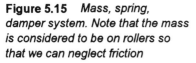

Equilibrium

Figure 5.15 *Mass, spring, damper system. Note that the mass is considered to be on rollers so that we can neglect friction*

Maths in action

System of a mass on a spring

Consider engineering systems which can be represented by a mass on a spring (Figure 5.14); we will assume there is no damping. If the mass is pulled downwards and then released, it oscillates on the spring. The force acting on the mass is just the restoring force and so:

$$F = m\frac{d^2x}{dt^2} = -kx \quad \text{or} \quad m\frac{d^2x}{dt^2} + kx = 0$$

This is a homogeneous second-order differential equation. If we try the solution $x = A\,e^t$, then we obtain:

$$mAs^2 + kA = 0$$

and so $s^2 = -k/m$ and we can write:

$$s = \pm j\sqrt{\frac{k}{m}}$$

with s an imaginary quantity. If we let $\omega = \sqrt{(k/m)}$, then the solution to the differential equation:

$$x = A\,e^{+j\omega t} + B\,e^{-j\omega t}$$

which we can write as:

$$x = C\cos \omega t + D\sin \omega t$$

If $x = 0$ when $t = 0$ then $C = 0$ and thus:

$$x = D\sin \omega t$$

The oscillations can be described by a sine function with angular frequency ω and amplitude D.

Maths in action

System of a damped mass and a spring

In a similar manner to the previous Maths in action, we may consider the oscillations of the damped system shown in Figure 5.15. The mass is constrained to move in purely a horizontal motion so we need only consider horizontal forces. There are many engineering systems which can be modelled by such a system. Later in this section we look at this system when there is an external force acting on the mass.

The force resulting from compressing the spring is proportional to the change in length x of the spring, i.e. kx with k being a constant termed the spring stiffness. The force arising from the damping is proportional to the rate at which the displacement of the piston is changing, i.e. c dx/dt with c being a constant. Thus:

net force applied to mass $= -kx - c\dfrac{dx}{dt}$

This net force will cause the mass to accelerate. Thus:

$$m\dfrac{d^2x}{dt^2} = -kx - c\dfrac{dx}{dt}$$

We can write this as:

$$m\dfrac{d^2x}{dt^2} + c\dfrac{dx}{dt} + kx = 0$$

In the absence of damping we have $m\, d^2x/dt^2 + kx = 0$ and the spring *naturally* oscillating (see Earlier Maths in action in this chapter) with an angular frequency, which we can call the natural angular frequency of ω_n given by:

$$\omega_n = \sqrt{\dfrac{k}{m}}$$

If we define a constant ζ, termed the *damping ratio*, by:

$$\zeta = \dfrac{c}{2\sqrt{mk}}$$

then we can write the second-order differential equation as:

$$\dfrac{d^2x}{dt^2} + 2\zeta\omega_n\dfrac{dx}{dt} + \omega_n^2 x = 0$$

Now, in order to solve this differential equation, we use a technique similar to that detailed in the previous Maths in action. We will try a solution of the form $x = A\,e^{st}$. This produces the auxiliary equation:

$$As^2 e^{st} + 2\zeta\omega_n sA\,e^{st} + \omega_n^2 A\,e^{st} = 0$$

and so:

$$s^2 + 2\omega_n\zeta s + \omega_n^2 = 0$$

This is a quadratic and we can use the usual equation for the roots of a quadratic to obtain:

$$s = \frac{-2\omega_n\zeta \pm \sqrt{4\omega_n^2\zeta^2 - 4\omega_n^2}}{2} = -\omega_n\zeta \pm \omega_n\sqrt{\zeta^2 - 1}$$

The general solution is thus:

$$x = A_1\,e^{s_1 t} + A_2\,e^{s_2 t}$$

The resulting oscillation of the system depends on the term inside the square root sign.

Damping ratio with a value between 0 and 1
This gives two complex roots:

$$s = -\omega_n\zeta \pm j\sqrt{1-\zeta^2}$$

If we let $\omega = \omega_n\sqrt{1-\zeta^2}$ then $s = -\omega_n\zeta \pm j\omega$ we obtain:

$$x = (A_1 e^{j\omega t} + A_2 e^{-j\omega t})\,e^{-\zeta\omega_n t}$$

By using Euler's equation (see Key point earlier in this section) we can write this as:

$$x = e^{-\zeta\omega_n t}(P\cos\omega t + Q\sin\omega t)$$

The exponential term means we have a damped oscillation. The equation can be expressed in an alternative form, since for the sine of a sum we can write $\sin(\omega t + \phi) = \sin\omega t \cos\phi + \cos\omega t \sin\phi$. If we let P and Q represent the opposite sides of a right-angled triangle of angle ϕ (Figure 5.16), then $\sin\phi = P/\sqrt{(P^2 + Q^2)}$ and $\cos\phi = Q/\sqrt{(P^2 + Q^2)}$ and so:

$$x = \sqrt{P^2 + Q^2}\;e^{-\zeta\omega_n t}\sin\left(\omega t + \phi\right)$$

$$x = C\,e^{-\zeta\omega_n t}\sin\left(\omega t + \phi\right)$$

where C is a constant and ϕ a phase difference. This describes a sinusoidal oscillation which is damped, the exponential term being the damping factor which gradually reduces the amplitude of the oscillation (Figure 5.17). Such a motion is said to be *under-damped*.

Damping ratio with the value 1
This gives two equal roots $s_1 = s_2 = -\omega_n$ and thus:

$$x = (At + B)\,e^{-\omega_n t}$$

where A and B are constants. This describes a situation where no oscillations occur but x exponentially changes with time. Such a motion is said to be *critically damped*.

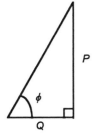

Figure 5.16   *Angle ϕ*

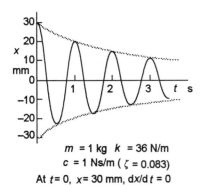

m = 1 kg k = 36 N/m
c = 1 Ns/m (ζ = 0.083)
At t = 0, x = 30 mm, dx/dt = 0

Figure 5.17   *Under-damped oscillation*

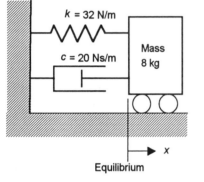

Equilibrium

Figure 5.18 *Example*

Damping ratio greater than 1
This gives two real roots $s_1 = -\omega_n\zeta + \omega_n\sqrt{(\zeta^2 - 1)}$ and $s_2 = -\omega_n\zeta - \omega_n\sqrt{(\zeta^2 - 1)}$, thus:

$$x = A_1 e^{s_1 t} + A_2 e^{s_2 t}$$

This describes a situation where no oscillations occur but x exponentially changes with time, taking longer to reach the steady-state zero displacement value than the critically damped motion. Such a motion is said to be *over-damped*.

Example

This example illustrates the discussion in the above Maths in action. For the system shown in Figure 5.18, the 8 kg mass is moved 0.2 m to the right of the equilibrium position and released from rest at time $t = 0$. Determine its displacement at time $t = 2$ s.

First we consider whether the system is underdamped, critically damped or overdamped.

$$\zeta = \frac{c}{2\sqrt{mk}} = \frac{20}{2\sqrt{8 \times 32}} = 0.625$$

Since the damping factor is less than 1 the system is underdamped. The natural frequency is:

$$\omega_n = \sqrt{\frac{k}{m}} = \sqrt{\frac{32}{8}} = 2 \text{ rad/s}$$

and so the undamped frequency is:

$$\omega = \omega_n\sqrt{1 - \zeta^2} = 2\sqrt{1 - (0.625)^2} = 1.561 \text{ rad/s}$$

The motion of the underdamped mass is described by:

$$x = C e^{-\zeta\omega_n t}\sin(\omega t + \phi) = C e^{-1.25t}\sin(1.561t + \phi)$$

At $t = 0$, we have $x = 0.2$ and so $0.2 = C \sin\phi$. Its velocity v is dx/dt:

$$\frac{dx}{dt} = -1.25C\, e^{-1.24t}\sin(1.561t + \phi)$$
$$+ 1.561C\, e^{-1.25t}\cos(1.561t + \phi)$$

At $t = 0$ the velocity is 0 and so $0 = -1.25C \sin\phi + 1.561C \cos\phi$. We can solve these two simultaneous equations to give $C = 0.256$ m and $\phi = 0.896$ rad.

> The displacement x at time t is thus given by:
>
> $x = 0.256\ e^{-1.25t} \sin(1.561t + 0.896)$
>
> Thus, at time t = 2 s we have:
>
> $x = 0.256\ e^{-2.5} \sin(3.122 + 0.896) = -0.0162$ m
>
> The minus sign indicates that the displacement is to the left of the equilibrium position.

5.3.2 Second-order non-homogeneous differential equations

Consider a non-homogeneous linear second-order differential equation with constant coefficients a_2, a_1 and a_0 with $f(x)$ being some function of x, often being referred to as the *forcing function*, applied to the system:

$$a_2 \frac{d^2 y}{dx^2} + a_1 \frac{dy}{dx} + a_0 y = f(x) \qquad [35]$$

With such a non-homogeneous differential equation there is a general solution which is equal to the sum of, what are called, the *complementary function* y_c and the *particular integral* y_p.

$y = y_c + y_p$

The complementary function is obtained by solving the equivalent homogeneous differential equation, i.e. with $f(x) = 0$, and the particular integral by considering the form of the $f(x)$ function and trying a particular solution of a similar form but which contains undetermined coefficients.

Right-hand side of non-homogeneous equation	Trial function, with A, B, C, etc. being undetermined coefficients
Constant	A
Polynomial	$A + Bx + Cx^2 + \ldots$
Exponential	$A\ e^{kx}$
Sine or cosine	$A \sin kx + B \cos kx$

Note: if the right-hand side is a sum of more than one term then the trial solution is the sum of the trial functions for these terms.

Example

Determine the general solution of the differential equation:

$$\frac{d^2y}{dx^2} - 5\frac{dy}{dx} + 6y = x^2$$

To obtain the complementary function we consider the equivalent homogeneous differential equation, i.e.

$$\frac{d^2y}{dx^2} - 5\frac{dy}{dx} + 6y = 0$$

Trying $y = A\,e^{sx}$ as a solution gives the auxiliary equation:

$$s^2 - 5s + 6 = 0$$

This can be factored as $(s - 3)(s - 2) = 0$ and so $s_1 = 3$ and $s_2 = 2$. The complementary function is thus:

$$y_c = A\,e^{3x} + B\,e^{2x}$$

To find the particular integral with x^2 we try a solution of the form $y = C + Dx + Ex^2$. This gives $dy/dx = D + 2Ex$ and $d^2y/dx^2 = 2E$. Substituting into the non-homogeneous differential equation gives:

$$2E - 5(2Ex + D) + 6(C + Dx + Ex^2) = x^2$$

Equating coefficients of x^2 gives $6E = 1$ and so $E = 1/6$. Equating coefficients of x gives $-10E + 6D = 0$ and so $D = 10/36 = 5/18$. Equating constants gives $2E - 5D + 6C = 0$ and so $C = 19/108$. Thus the particular integral is:

$$y_p = \frac{19}{108} + \frac{5}{18}x + \frac{1}{6}x^2$$

and so the general solution is:

$$y = y_c + y_p = A\,e^{3x} + B\,e^{2x} + \frac{19}{108} + \frac{5}{18}x + \frac{1}{6}x^2$$

Example

Determine the general solution of the differential equation:

$$\frac{d^2y}{dx^2} + \frac{dy}{dx} - 2x = 3\,e^{2x}$$

The corresponding homogeneous differential equation is:

$$\frac{d^2y}{dx^2} + \frac{dy}{dx} - 2y = 0$$

Trying $y = A\,e^{sx}$ as a solution gives the auxiliary equation:

$$s^2 + s - 2 = 0$$

This can be factored as $(s + 2)(s - 1) = 0$ and so the roots are $s_1 = -2$ and $s_2 = 1$ and the complementary function is:

$$y_c = A\,e^{-2x} + B\,e^x$$

For the particular integral with an exponential forcing function we try a solution of the form $y = C\,e^{kx}$. Substituting this into the non-homogeneous differential equation gives:

$$k^2C\,e^{kx} + kC\,e^{kx} - 2C\,e^{kx} = 3\,e^{2x}$$

Thus we must have $k = 2$ for equality of the exponentials and for the coefficients $(k^2 + k - 2)C = 3$ and hence $C = \frac{3}{4}$. Hence the particular integral is $y_p = \frac{3}{4}\,e^{2x}$ and the general solution is:

$$y = y_c + y_p = A\,e^{-2x} + B\,e^x + \frac{3}{4}\,e^{2x}$$

Example

Determine the general solution of the differential equation:

$$3\frac{d^2y}{dx^2} + \frac{dy}{dx} - 2y = 2\cos x$$

The corresponding homogeneous differential equation is:

$$3\frac{d^2y}{dx^2} + \frac{dy}{dx} - 2y = 0$$

Trying $y = A\,e^{sx}$ as a solution gives the auxiliary equation:

$$3s^2 + s - 2 = 0$$

This can be factored as $(3s - 2)(s + 1) = 0$ and so the roots are $s_1 = 2/3$ and $s_2 = -1$ and the complementary function is:

$$y_c = A\,e^{2x/3} + B\,e^{-x}$$

For the particular integral we try a solution of the form $y = C \cos kx + D \sin kx$. Substituting this into the non-homogeneous differential equation gives:

$$3(-C \cos kx - D \sin kx) + (-C \sin kx + D \cos kx)$$
$$- 2(C \cos kx + D \sin kx) = 2 \cos x$$

For equality of the cosines we must have $k = 1$ and $-3C + D - 2C = 2$. Equating coefficients of the sines gives $-3D - C - 2D = 0$. Thus we have $C = -5/13$ and $D = 1/13$. The particular integral is thus:

$$y_p = -\tfrac{5}{13} \cos x + \tfrac{1}{13} \sin x$$

The general solution is thus:

$$y = y_c + y_p = A\, e^{2x/3} + B\, e^{-x} - \tfrac{5}{13} \cos x + \tfrac{1}{13} \sin x$$

Exceptional cases of particular integrals

There are situations when the obvious form of function to be tried to obtain the particular integral yields no result because when it is substituted in the differential equation we obtain $0 = 0$. This occurs when the right-hand side of the non-homogeneous differential equation consists of a function that is also a term in the complementary function. To illustrate this, consider the differential equation:

$$\frac{d^2 y}{dx^2} + \frac{dy}{dx} - 2y = e^{-2x}$$

The complementary function is $y = A\, e^{-2x} + B\, e^x$. For the particular integral, if we try the solution $y = A\, e^{kx}$ we obtain:

$$4A\, e^{kx} - 2A\, e^{kx} - 2A\, e^{kx} = e^{-2x}$$

and so no solution for A. In such cases we have to try something different.

The basic rule is to multiply the trial solution by x.

Thus we try $y = Ax\, e^{kx}$. This gives, for the above differential equation:

$$(-2A\, e^{kx} + 4Ax\, e^{kx} - 2A\, e^{kx}) + (A\, e^{kx} - 2Ax\, e^{kx}) - 2Ax\, e^{kx} = e^{-2x}$$

Thus $k = -2$ and $-2A + 4Ax - 2A + A - 2Ax - 2Ax = 1$. Equating constants gives $3A = 1$, equating the x coefficients gives $0 = 0$, and so the particular integral is:

$$y_p = \tfrac{1}{3}\, e^{-2x}$$

Example

Determine the general solution of the differential equation:

$$\frac{d^2y}{dx^2} - 3\frac{dy}{dx} - 10y = 4 - e^{-2x}$$

The corresponding homogeneous differential equation is:

$$\frac{d^2y}{dx^2} - 3\frac{dy}{dx} - 10y = 0$$

Trying $y = A\,e^{sx}$ as a solution gives the auxiliary equation:

$$s^2 - 3s - 10 = 0$$

This can be factored as $(s - 5)(s + 2) = 0$ and so the complementary function is $y_c = A\,e^{5x} + B\,e^{-2x}$. The right-hand side of the non-homogeneous differential equation is the sum of two terms for which the trial functions would be C and $Dx\,e^{kx}$. We thus try the sum of these. Thus:

$$Dk^2x\,e^{kx} + Dk\,e^{kx} + Dk\,e^{kx} - 3Dkx\,e^{kx} - 3D\,e^{kx}$$
$$- 10(C + Dx\,e^{kx}) = 4 - e^{-2x}$$

Equating exponential terms gives $k = -2$, $4Dx - 2D - 2D + 6Dx - 3D - 10Dx = -1$ and so $D = 1/7$. Equating constants gives $-10C = 4$ and so $C = -4/10$. Thus the particular integral is $-(4/10) + (1/7)\,e^{2x}$. The general solution is therefore:

$$y = A\,e^{5x} + B\,e^{-2x} - \tfrac{4}{10} + \tfrac{1}{7}x\,e^{-2x}$$

Forced oscillations of elastic systems

Oscillations of elastic systems in which the system is free to adopt its own frequency of oscillation are said to be natural or free oscillations. When a system is forced to oscillate by some external force F at the frequency of this force, then the oscillations are said to be forced. In general, a simple model of such oscillations is given by a second-order differential equation of the form:

$$\frac{d^2x}{dt^2} + 2\zeta\omega_n\frac{dx}{dt} + \omega_n^2x = kF \qquad [36]$$

where F is the externally applied force, k a constant, x the system output, ω_n the natural angular frequency and ζ the damping ratio. Steady-state conditions occur when dx/dt and d^2x/dt^2 are zero and so we then have $\omega_n^2x_s = kF$.

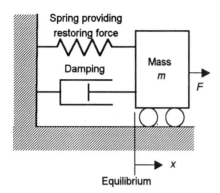

Spring providing restoring force

Damping

Mass
m

F

x

Equilibrium

Figure 5.19   *Mass, spring, damper system*

Maths in action

System of a damped mass on a spring

There are many engineering systems which can be modelled by the lumped system of damped mass on a spring and then subject to some externally applied force. An example of a measurement system which can be modelled in this way is a diaphragm pressure gauge. Figure 5.19 illustrates the basic features of such systems.

The net force applied to the mass is the applied force *F* minus the force resulting from the compressing, or stretching, of the spring and the force from the damper:

net force applied to mass = $F - kx - c\dfrac{dx}{dt}$

This net force will cause the mass to accelerate. Thus:

$$m\frac{d^2x}{dt^2} = F - kx - c\frac{dx}{dt}$$

We can write this as:

$$m\frac{d^2x}{dt^2} + c\frac{dx}{dt} + kx = F$$

In the absence of damping and a force *F*, we have $m\,d^2x/dt^2 + kx = 0$ and the spring *naturally* oscillating (see Earlier Maths in action in this chapter) with an angular frequency, which we can call the natural angular frequency of ω_n given by:

$$\omega_n = \sqrt{\frac{k}{m}}$$

If we define a constant ζ, termed the *damping ratio*, by:

$$\zeta = \frac{c}{2\sqrt{mk}}$$

then we can write the differential equation as:

$$\frac{d^2x}{dt^2} + 2\zeta\omega_n\frac{dx}{dt} + \omega_n^2x = \frac{F}{m}$$

Consider a step input such that the applied force jumps from zero to *F* at time *t* = 0. We can solve the differential equation by determining the complementary function and the particular integral. For the homogeneous form of the differential equation we try a solution of the form $x = A\,e^{st}$ (see earlier Maths in action in this chapter). We thus have the homogeneous equation solutions:

Damping ratio less than 1, i.e. underdamped

$$x = C\, e^{-\zeta \omega_n t} \sin\left(\omega t + \phi\right)$$

Damping ratio with the value 1, i.e. critically damped

$$x = (At + B)\, e^{-\omega_n t}$$

Damping ratio greater than 1, i.e. critically damped

$$x = A\, e^{s_1 t} + B\, e^{s_2 t}$$

When we have a step input then we can try for the particular integral $x = A$. Substituting this into the differential equation gives $0 + 0 + A = F/m$. Thus the particular integral is $x = F/m$ and the solutions for the different degrees of damping are:

Under-damped: $x = C\, e^{-\zeta \omega_n t} \sin\left(\omega t + \phi\right) + \dfrac{F}{m}$

Critically damped: $x = (At + B)\, e^{-\omega_n t} + \dfrac{F}{m}$

Over-damped: $x = A\, e^{s_1 t} + B\, e^{s_2 t} + \dfrac{F}{m}$

As t tends to an infinite value, in all cases the response tends to a steady-state value of F/m. Figure 5.20 shows the form the solution of the second-order differential equation takes for different values of the damping ratio.

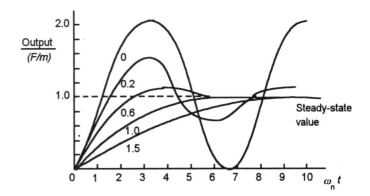

Figure 5.20 *Response of second-order system to step input for different damping factors. The output is plotted as a multiple of the steady-state value F/m. Instead of just giving the output variation with time t, the axis used is $\omega_n t$. This is because t and ω_n always appear as the product $\omega_n t$ and using this product makes the graph applicable for any value of ω_n.*

Example

The dynamic performance of a piezoelectric accelerometer is described by the following second-order differential equation:

$$\frac{d^2\theta_0}{dt^2} + 3 \times 10^3 \frac{d\theta_0}{dt} + 22.5 \times 10^9 \theta_0 = 110 \times 10^9 \theta_i$$

where θ_0 is the output charge in pC and θ_i is the input acceleration in m/s². Determine the natural angular frequency, the damping factor and the static system sensitivity.

We can compare the differential equation with the standard form of equation [36] for the oscillations of an elastic system. We thus have:

$$22.5 \times 10^9 = \omega_n{}^2$$

and so ω_n = 150 × 10³ rad/s. Since ω_n = $2\pi f_n$ then the natural frequency f_n = 150 × 10³/2π = 23.87 kHz. We also have:

$$3 \times 10^3 = 2\zeta\omega_n$$

and so ζ = 3 × 10³/(2 × 150 × 10³) = 0.01. The oscillation is thus underdamped.

Steady state occurs when $\omega_n{}^2 x_s = kF$ and so the static system sensitivity is $x_s/F = k/\omega_n{}^2$ = 110 × 10⁹/(22.5 × 10⁹) = 4.89 pC/(m/s²).

Problems 5.3

1 Determine the unique solutions for the following differential equations given the general solutions and initial conditions:

(a) $\dfrac{d^2y}{dx^2} = 3$, $y = \frac{3}{2}x^2 + Ax + B$, $y = 2$ and $\dfrac{dy}{dx} = 4$ at $x = 0$,

(b) $\dfrac{d^2y}{dx^2} + \dfrac{dy}{dx} - 6y = 0$, $y = A\,e^{2x} + B\,e^{-3x}$

$y = 1$ and $\dfrac{dy}{dx} = 0$ at $x = 0$,

(c) $\dfrac{d^2y}{dx^2} + 4\dfrac{dy}{dx} + 4y = 0$, $y = A\,e^{-2x} + Bx\,e^{-2}$

$y = 1$ and $\dfrac{dy}{dx} = 0$ at $x = 0$,

(d) $\dfrac{d^2y}{dx^2} - y = 0$, $y = A\,e^x + B\,e^{-x}$,

$y = 0$ and $\dfrac{dy}{dx} = 5$ at $x = 0$,

(e) $\dfrac{d^2y}{dx^2} - 3\dfrac{dy}{dx} + 2y = 0$, $y = A\,e^x + B\,e^{2x}$

$y = 1$ and $\dfrac{dy}{dx} = 0$ at $x = 0$,

(f) $\dfrac{d^2y}{dx^2} + 2\dfrac{dy}{dx} + y = 0$, $y = A\,e^{-x} + Bx\,e^{-x}$

$y = 2$ and $\dfrac{dy}{dx} = -1$ at $x = 0$

2 Determine the general solutions of:

(a) $2\dfrac{d^2y}{dx^2} + \dfrac{dy}{dx} - y = 0$, (b) $\dfrac{d^2y}{dx^2} - 6\dfrac{dy}{dx} + 9y = 0$,

(c) $\dfrac{d^2y}{dx^2} - 10\dfrac{dy}{dx} + 25y = 0$, (d) $\dfrac{d^2y}{dx^2} - 4\dfrac{dy}{dx} + 5y = 0$,

(e) $\dfrac{d^2y}{dx^2} - 2\dfrac{dy}{dx} + 5y = 0$, (f) $\dfrac{d^2y}{dx^2} + 3\dfrac{dy}{dx} - 10y = 0$,

(g) $\dfrac{d^2y}{dx^2} - 4\dfrac{dy}{dx} + 5y = 0$, (h) $\dfrac{d^2y}{dx^2} + 2\dfrac{dy}{dx} - 8y = 0$

3 Determine the particular solutions of:

(a) $\dfrac{d^2y}{dx^2} + 6\dfrac{dy}{dx} + 5y = 0$, $y = 0$ and $\dfrac{dy}{dx} = 3$ at $x = 0$,

(b) $\dfrac{d^2y}{dx^2} - 6\dfrac{dy}{dx} + 25y = 0$, $y = 3$ and $\dfrac{dy}{dx} = 1$ at $x = 0$,

(c) $4\dfrac{d^2y}{dx^2} + 12\dfrac{dy}{dx} + 9y = 0$, $y = 2$ and $\dfrac{dy}{dx} = 1$ at $x = 0$,

(d) $\dfrac{d^2y}{dx^2} - 7\dfrac{dy}{dx} + 12y = 0$, $y = 3$ and $\dfrac{dy}{dx} = 2$ at $x = 0$,

(e) $\dfrac{d^2y}{dx^2} + 2\dfrac{dy}{dx} + y = 0$, $y = 1$ and $\dfrac{dy}{dx} = 1$ at $x = 0$,

(f) $\dfrac{d^2y}{dx^2} - 6\dfrac{dy}{dx} + 25y = 0$, $y = 1$ and $\dfrac{dy}{dx} = 7$ at $x = 0$

4 Determine the general solutions of:

(a) $\dfrac{d^2y}{dx^2} - 4y = 2\,e^{3x}$, (b) $\dfrac{d^2y}{dx^2} + 3\dfrac{dy}{dx} + 4y = 3x + 2$,

(c) $\dfrac{d^2y}{dx^2} - 6\dfrac{dy}{dx} + 8y = 3\cos x$, (d) $\dfrac{d^2y}{dx^2} - 5\dfrac{dy}{dx} + 6y = e^x$,

(e) $\dfrac{d^2y}{dx^2} - 5\dfrac{dy}{dx} + 6y = 2e^x - 3e^{-x},$

(f) $\dfrac{d^2y}{dx^2} - \dfrac{dy}{dx} - 2y = 5\sin 2x,$

(g) $\dfrac{d^2y}{dx^2} - 3\dfrac{dy}{dx} + 2y = 3 - 2x^2$

5 Determine the particular solution of the following differential equation if $y = -2$ and $dy/dx = -3$ when $x = 0$:

$$\dfrac{d^2y}{dx^2} - 4y = 5\,e^{3x}$$

6 Determine the particular solution of the following differential equation if $y = 1$ and $dy/dx = 0$ when $x = 0$:

$$\dfrac{d^2y}{dx^2} + 9y = \sin 2x$$

7 An object of mass 1 kg is suspended from a rigid support by a vertical spring of stiffness 4 N/m. Determine how the displacement of the object varies with time when the object is pulled down from its initial position and released to freely move if the object is subject to a damping force of five times its velocity?

8 An object of mass 1 kg is suspended from a rigid support by a vertical spring of stiffness 9 N/m. The object is pulled down for an initial displacement of 0.2 m and then released with zero initial velocity. Determine how the displacement of the object varies with time when (a) there is no damping, (b) the damping is twice the velocity of the object.

9 A second-order system has a natural angular frequency of 2.0 rad/s and a damped angular frequency of 1.8 rad/s. What is the damping factor?

10 Determine the natural angular frequency and damping factor for a second-order system with input y and output x described by the following differential equation:

$$0.02\dfrac{d^2x}{dt^2} + 0.20\dfrac{dx}{dt} + 0.50x = y$$

11 A sensor can be considered to be a mass-damper-spring system with a mass of 10 g and a spring of stiffness 1.0 N/mm. Determine the natural angular frequency and the damping constant required for the damping element if the system is to be critically damped.

12 Determine whether the system described by the following differential equation is under-damped, critically damped or over-damped when subject to a step input y:

$$\dfrac{d^2x}{dt^2} + 5\dfrac{dx}{dt} + 6x = y$$

13 An object of mass 1 kg is suspended from a rigid support by a vertical spring of stiffness 9 N/m. What is the damping force per unit velocity which would be needed to give critical damping?

14 Determine the natural angular frequency and damping force per unit velocity for a system having its displacement x with time t described by the following second-order differential equation:

$$\frac{d^2x}{dt^2} + 4\frac{dx}{dt} + 7x = 0$$

15 An object of mass 1 kg is suspended from a rigid support by a vertical spring of stiffness 9 N/m. If there is a damping force of $1v$ opposing the motion of the object, where v is the velocity, determine how the displacement varies with time when the object is given an initial displacement of 0.2 m and an initial velocity of –0.3 m/s.

16 The angular displacement θ of a door controlled by a hydraulic damping mechanism is described by the differential equation:

$$\frac{d^2\theta}{dt^2} + 5\frac{d\theta}{dt} + 4\theta = 0$$

Determine how the angular displacement varies with time t when there is an initial displacement of $\pi/3$ and zero initial angular velocity.

17 An electrical circuit having resistance R, inductance L and capacitance C in series with a step voltage source V at $t = 0$ has the potential difference across the capacitor v_C described by the differential equation:

$$LC\frac{d^2v_C}{dt^2} + RC\frac{dv_C}{dt} + v_C = V$$

Show that the three possible solutions are:

$$v = A\,e^{s_1 t} + B\,e^{s_2 t} + V, \quad s = -\frac{R}{2L} \pm \sqrt{\left(\frac{R}{2L}\right)^2 - \frac{1}{LC}}$$
$$v = (A + Bt)\,e^{-Rt/2L} + V$$
$$v = e^{-Rt/2L}(A\cos\omega t + B\sin\omega t) + V, \quad \omega = \sqrt{\frac{1}{LC} - \left(\frac{R}{2L}\right)^2}$$

6 Laplace transform

Summary

In order to consider the response of engineering systems, e.g. electrical or control systems, to inputs such as step, or perhaps an impulse, we need to be able to solve the differential equation for that system with that particular form of input. As the previous chapter indicates, this can be rather laborious. A simpler method of tackling the solution is to transform a differential equation into a simple algebraic equation which we can easily solve. This is achieved by the use of the Laplace transform, the subject of this chapter.

Objectives

By the end of this chapter, the reader should be able to:

- understand what using the Laplace transform involves;
- use Laplace transform tables to convert first- and second-order differential equations into algebraic equations;
- use Laplace transform tables, and where appropriate partial fractions, to convert Laplace transform equations into real world equations;
- determine the outputs of systems to standard input signals such as step, impulse and ramp.

6.1 The Laplace transform

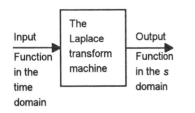

Figure 6.1 *The Laplace transform*

In this chapter a method of solving such differential equations is introduced which transforms a differential equation into an algebraic equation. This is termed the *Laplace transform*. It is widely used in engineering, in particular in control engineering and in electrical circuit analysis where it is commonplace not even to write differential equations to describe conditions but to write directly in terms of the Laplace transform.

We can think of the Laplace transform as being rather like a function machine (Figure 6.1). As input to the machine we have some function of time $f(t)$ and as output a function we represent as $F(s)$. The input is referred to as being the *time domain* while the output is said to be in the *s-domain*. Thus we take information about a system in the time domain and use our 'machine' to transform it into information in the *s*-domain. Differential equations which describe the behaviour of a system in the time domain are converted into algebraic equations in the *s*-domain, so considerably simplifying their solution. We can thus transform a

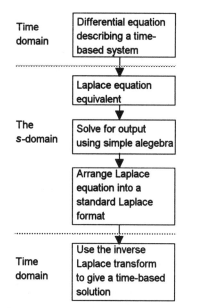

Time domain	Differential equation describing a time-based system

↓

Laplace equation equivalent

The s-domain	Solve for output using simple alegebra

↓

Arrange Laplace equation into a standard Laplace format

↓

Time domain	Use the inverse Laplace transform to give a time-based solution

Figure 6.2 *Using the Laplace transform. As an illustration, Ohm's Law gives the time-domain equation v(t) = Ri(t), both v and i being functions of time and R assumed to remain constant. In the s-domain this becomes V(s) = RI(s). After working with this equation in the s-domain we can then transform back to the time domain*

Key point

To obtain the Laplace transform of a function of time $f(t)$, multiply it by e^{-st} and integrate the product between zero and infinity.

Key point

When electrical circuits are discussed in terms of currents or voltages varying with time we use differential equations and are said to be working in the *time domain*. When we use phasors we can be said to be working in the *frequency domain*; we are no longer working with time-varying quantities. As we shall see later in this chapter, we can transform currents or voltages which vary with time into the *s-domain* by using the Laplace transform; like the transformation using phasors, we are no longer working with time-varying quantities.

differential equation into an *s*-domain equation, solve the equation and then use the 'machine' in inverse operation to transform the *s*-domain equation back into a time-domain solution (Figure 6.2).

Now this definition may look rather daunting, but do not fear, it is very likely that you will not need to use it but rather will make use of tables which other people have worked out. However, you should appreciate the basis of the transform. The *Laplace transform* of some function of time is defined by:

Multiply a given function of time f(t) by e^{-st} *and integrate the product between zero and infinity. The result, if it exists, is called the Laplace transform of f(t) and is denoted by $\mathscr{L}\{f(t)\}$ = F(s).*

$$F(s) = \mathscr{L}\{f(t)\} = \int_0^\infty e^{-st}f(t)\,dt \qquad [1]$$

Note that the integration is between 0 and $+\infty$ and so is *one-sided* and not over the full range of time from $-\infty$ to $+\infty$.

Example

Determine the Laplace transform of $f(t) = 1$.

Using equation [1]:

$$\mathscr{L}\{f(t)\} = \int_0^\infty 1\ e^{-st}\,dt = \left[\frac{e^{-st}}{-s}\right]_0^\infty = \frac{1}{s}$$

This is provided that $s > 0$ so that $e^{-st} \to 0$ as $t \to \infty$

Example

Determine the Laplace transform of $f(t) = e^{at}$.

Using equation [1]:

$$\mathscr{L}\{f(t)\} = \int_0^\infty e^{at}\ e^{-st}\,dt = \int_0^\infty e^{-(s-a)t}\,dt = \left[\frac{e^{-(s-a)t}}{-(s-a)}\right]_0^\infty = \frac{1}{s-a}$$

That is provided we have $(s - a) > 0$.

Example

Determine the Laplace transform of $f(t) = t$.

Using equation [1]:

$$\mathscr{L}\{f(t)\} = \int_0^\infty t\ e^{-st}\,dt$$

Using integration by parts:

$$\mathscr{L}\{f(t)\} = \left[-\frac{t}{s}\,e^{-st}\right]_0^\infty + \int_0^\infty \frac{1}{s}\,e^{-st}\,dt = \left[-\frac{1}{s^2}\,e^{-st}\right]_0^\infty = \frac{1}{s^2}$$

That is provided we have $s > 0$.

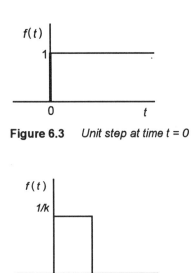

Figure 6.3 *Unit step at time t = 0*

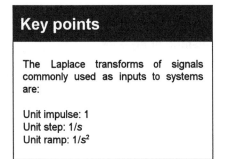

Figure 6.4 *Unit area rectangular pulse*

Key points

The Laplace transforms of signals commonly used as inputs to systems are:

Unit impulse: 1
Unit step: $1/s$
Unit ramp: $1/s^2$

Laplace transforms for step and impulse function

Consider the *unit step function* $u(t)$ shown in Figure 6.3. The Laplace transform is given by equation [1] as:

$$\mathscr{L}\{u(t)\} = \int_0^\infty 1\,e^{-st}\,dt = \left[\frac{e^{-st}}{-s}\right]_0^\infty = \frac{1}{s} \tag{2}$$

Thus a unit size step input signal to an engineering system occurring at time $t = 0$ will have a Laplace transform of $1/s$.

Now consider obtaining the *unit impulse function* (represented as $\delta(t)$). Such an impulse can be considered to be a unit area rectangular pulse which has its width k decreased to give the unit impulse in the limit when $k \to 0$. For the unit area rectangular pulse shown in Figure 6.4, the Laplace transform is:

$$\mathscr{L}\{\text{unit area pulse}\} = \int_0^\infty f(t)\,e^{-st}\,dt = \int_0^k \frac{1}{k}\,e^{-st}\,dt + \int_k^\infty 0\,e^{-st}\,dt$$

$$= \left[-\frac{1}{sk}\,e^{-st}\right]_0^k = -\frac{1}{sk}(e^{-sk} - 1)$$

We can replace the exponential by a series, thus obtaining:

$$\mathscr{L}\{\text{unit area pulse}\} = -\frac{1}{sk}\left(1 + (-sk) + \frac{(-sk)^2}{2!} + \frac{(-sk)^3}{3!} + \dots - 1\right)$$

Thus in the limit as $k \to 0$, the Laplace transform tends to the value 1 and so:

$$\mathscr{L}\{\delta(t)\} = 1 \tag{3}$$

Thus a unit size impulse input signal occurring at time $t = 0$ to an engineering system will have a Laplace transform of 1.

Standard Laplace transforms

The transforms derived above, together with others, are tabulated as a set of standard transforms so that it becomes unnecessary to derive them by the use of equation [1]. Table 6.1 gives some of the more common standard transforms. As indicated in the following section, these standard transforms can be used to derive the transforms for a wide range of functions.

Table 6.1 *Laplace transforms*

	$f(t)$	$\mathcal{L}\{f(t)\}$
1	Unit impulse $\delta(t)$	1
2	Unit step $u(t)$	$\dfrac{1}{s}$
3	Unit ramp t	$\dfrac{1}{s^2}$
4	t^n	$\dfrac{n!}{s^{n+1}}$
5	e^{-at}	$\dfrac{1}{s+a}$
6	$1 - e^{-at}$	$\dfrac{a}{s(s+a)}$
7	$t\,e^{-at}$	$\dfrac{1}{(s+a)^2}$
8	$e^{-at} - e^{-bt}$	$\dfrac{b-a}{(s+a)(s+b)}$
9	$(1 - at)\,e^{-at}$	$\dfrac{s}{(s+a)^2}$
10	$1 - \dfrac{b}{b-a}\,e^{-at} + \dfrac{a}{b-a}\,e^{-bt}$	$\dfrac{ab}{s(s+a)(s+b)}$
11	$\dfrac{e^{-at}}{(b-a)(c-a)} + \dfrac{e^{-bt}}{(c-a)(a-b)} + \dfrac{e^{-ct}}{(a-c)(b-c)}$	$\dfrac{1}{(s+a)(s+b)(s+c)}$
12	$\sin \omega t$	$\dfrac{\omega}{s^2 + \omega^2}$
13	$\cos \omega t$	$\dfrac{s}{s^2 + \omega^2}$
14	$e^{-at} \sin \omega t$	$\dfrac{\omega}{(s+a)^2 + \omega^2}$
15	$e^{-at} \cos \omega t$	$\dfrac{s+a}{(s+a)^2 + \omega^2}$
16	$\sinh \omega t$	$\dfrac{\omega}{s^2 - \omega^2}$
17	$\cosh \omega t$	$\dfrac{s}{s^2 - \omega^2}$
18	$\dfrac{\omega}{\sqrt{1 - \zeta^2}}\,e^{-\zeta \omega t} \sin \omega \sqrt{1 - \zeta^2}\ t,\ \zeta < 1$	$\dfrac{\omega^2}{s^2 + 2\zeta \omega s + \omega^2}$
19	$1 - \dfrac{1}{\sqrt{1 - \zeta^2}}\,e^{-\zeta \omega t} \sin\left(\omega \sqrt{1 - \zeta^2}\ t + \phi\right), \cos \phi = \zeta$	$\dfrac{\omega^2}{s(s^2 + 2\zeta \omega s + \omega^2)}$

Properties of Laplace transforms

The following are basic properties of Laplace transforms and can be used with the above table of standard transforms to obtain a wide range of other transforms.

Key point

If two separate time functions have Laplace transforms then the transform of the sum of the time functions is the sum of the Laplace transforms of the two functions considered separately.

- *Sum of two functions*

   If two separate time functions $f(t)$ and $g(t)$ have Laplace transforms then the transform of the sum of the time functions, i.e. $f(t) + g(t)$, is the sum of the Laplace transforms of the two functions considered separately:

   $$\mathcal{L}\{f(t) + g(t)\} = \mathcal{L}\{f(t)\} + \mathcal{L}\{g(t)\} \qquad [4]$$

This property is derived by using equation [1]:

$$\mathcal{L}\{f(t) + g(t)\} = \int_0^\infty \{f(t) + g(t)\}\ e^{-st}\ dt$$

$$= \int_0^\infty f(t)\ e^{-st}\ dt + \int_0^\infty g(t)\ e^{-st}\ dt$$

$$= \mathcal{L}\{f(t)\} + \mathcal{L}\{g(t)\}$$

Since $2f(t)$ equals $f(t) + f(t)$, then the Laplace transform of $2f(t)$ will be twice the Laplace transform of $f(t)$. Thus, in general:

$$\mathcal{L}\{af(t)\} = a\mathcal{L}\{f(t)\} \qquad\qquad [5]$$

The Laplace transform of a constant multiplying a function is the same as a the constant multiplying the Laplace transform of the function.

Key point

The Laplace transform of a constant multiplying a function is the same as a the constant multiplying the Laplace transform of the function.

Example

Determine the Laplace transform of 1 + 2*t*.

Using equations [4] and [5] and Table 6.1:

$$\mathcal{L}\{1 + 2t\} = \frac{1}{s} + \frac{2}{s^2}$$

Example

Determine the Laplace transform of 3 sin 2*t* + cos 2*t*.

Using equations [4] and [5] and Table 6.1:

$$\mathcal{L}\{3\sin 2t + \cos 2t\} = 3\frac{2}{s^2+4} + \frac{s}{s^2+4}$$

Example

Determine the Laplace transform of 3*t*² + 2 e⁻ᵗ.

Using equations [4] and [5] and Table 6.1:

$$\mathcal{L}\{3t^2 + 2\ e^{-t}\} = 3\frac{2!}{s^3} + 2\frac{1}{s+1}$$

Example

Determine the Laplace transform of $\sin(\omega t + \theta)$.

We can write $\sin(\omega t + \theta)$ as $\sin \omega t \cos \theta + \cos \omega t \sin \theta$. Thus, using equations [4] and [5] and Table 6.1:

$$\mathcal{L}\{\sin(\omega t + \theta)\} = \frac{\omega}{s^2 + \omega^2} \cos \theta + \frac{s}{s^2 + \omega^2} \sin \theta$$

Example

What is the Laplace transform of an alternating voltage which is described by 240 sin 314.16t?

We have an equation of the form constant multiplied by a sine function. Equation [12] in Table 6.1 gives for $\sin \omega t$ the transform $\omega/(s^2 + \omega^2)$. Hence:

$$\mathcal{L}\{240 \sin 314.16t\} = 240\frac{314.16}{s^2 + 314.16^2}$$

$$= \frac{75.4 \times 10^3}{s^2 + 98.7 \times 10^3}$$

- **The first shift theorem, factor e^{-at}**

  This theorem states that if $\mathcal{L}\{f(t)\} = F(s)$ then:

  $$\mathcal{L}\{e^{-at}f(t)\} = F(s + a) \qquad [6]$$

  Thus the substitution of $s + a$ for s corresponds to multiplying a time function by e^{-at}. This can be demonstrated if we consider equation [1] with such a function:

  $$\mathcal{L}\{e^{-at}f(t)\} = \int_0^\infty e^{-at}f(t)\, e^{-st}\, dt = \int_0^\infty f(t)\, e^{-(s+a)}\, dt$$

Key point

The first shift theorem: if $\mathcal{L}\{f(t)\} = F(s)$ then:

$$\mathcal{L}\{e^{-at}f(t)\} = F(s + a)$$

Example

Determine the Laplace transform of $e^{-2t} \cosh 3t$.

Using the first shift theorem and the transform for $\cosh 3t$ given by Table 39.1, the transform is that of $\cosh 3t$ with the s replaced by $s + 2$:

$$\mathcal{L}\{e^{-2t} \cosh 3t\} = \frac{s+2}{(s+2)^2 - 9}$$

> **Example**
>
> Determine the Laplace transform of $2 e^{-2t} \sin^2 t$.
>
> Since $\cos 2t = 1 - 2\sin^2 t$ we have:
>
> $$\mathcal{L}\{2\sin^2 t\} = \mathcal{L}\{1\} - \mathcal{L}\{\cos 2t\} = \frac{1}{s} - \frac{s}{s^2+4}$$
>
> Hence, using the first shift theorem and replacing the s by $s + 2$:
>
> $$\mathcal{L}\{2 e^{-2t} \sin^2 t\} = \frac{1}{s+2} - \frac{s+2}{(s+2)^2+4}$$

• *The second shift theorem, time shifting*

The second shift theorem states that if a signal is delayed by a time T then its Laplace transform is multiplied by e^{-sT}. A function $u(t)$ which is delayed is represented by $u(t - T)$, where T is the delay. Thus if $F(s)$ is the Laplace transform of $f(t)$ then:

$$\mathcal{L}\{f(t - T)u(t - T)\} = e^{-sT}F(s) \qquad [7]$$

This can be demonstrated by considering a unit step function which is delayed by a time T (Figure 6.5). Equation [1] gives for such a function:

$$\mathcal{L}\{u(t - T)\} = \int_0^\infty u(t - T)\, e^{-st}\, dt = \int_0^T 0\, dt + \int_T^\infty 1\, e^{-st}\, dt$$

$$= \left[\frac{e^{-st}}{-s}\right]_T^\infty = e^{-sT}\frac{1}{s}$$

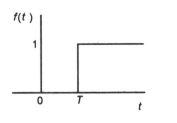

Figure 6.5 *Delayed unit step*

> **Example**
>
> Determine the Laplace transform for a unit impulse which occurs at a time of $t = 2$ s.
>
> The Laplace transform for a unit impulse at $t = 0$ is 1. Thus the transform for the delayed impulse is $1\, e^{-2s}$.

> **Example**
>
> Determine the Laplace transform of a single pulse consisting of just the first half of a sine wave (Figure 6.6).

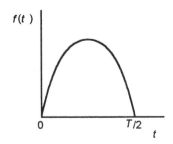

Figure 6.6 *Example*

We can think of such a function as being the sum of a sine function extending over an infinite number of cycles and a sine function that has had its start delayed by ½T. In this way all but the first half period waveform are cancelled out. Thus the Laplace transform is:

$$\frac{\omega}{s^2+\omega^2} + e^{-sT/2}\frac{\omega}{s^2+\omega^2}$$

- **_Periodic functions_**

A periodic function of period T has a Laplace transform of:

$$\frac{1}{1-e^{-sT}}F_1(s) \qquad\qquad [8]$$

where $F_1(s)$ is the Laplace transform of the function for the first period. This can be proved by considering the periodic function to be the sum of the function $f_1(t)$ describing the first period, the first period function delayed by 1 period, the first period function delayed by 2 periods, etc. The Laplace transform of the sum is thus:

$$F_1(s) + e^{-sT}F_1(s) + e^{-2sT}F_1(s) + ... = (1 + e^{-sT} + e^{-2sT} + ...)F_1(s)$$

The term in the brackets is a geometric series with the sum to infinity of $1/(1 - e^{-sT})$. Thus we obtain the equation given above.

Key point

A periodic function of period T has a Laplace transform of:

$$\frac{1}{1-e^{-sT}}F_1(s)$$

where $F_1(s)$ is the Laplace transform of the function for the first period.

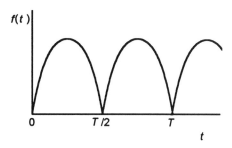

Figure 6.7 *Example*

Example

Determine the Laplace transform of a full-wave rectified sine wave (Figure 6.7).

Such a wave consists of a sequence of the pulses shown in Figure 6.6. Thus the first period function has the transform:

$$\frac{\omega}{s^2+\omega^2} + e^{-sT/2}\frac{\omega}{s^2+\omega^2}$$

Therefore the periodic wave has the Laplace transform:

$$\frac{1}{1-e^{-sT/2}}\left(\frac{\omega}{s^2+\omega^2} + e^{-sT/2}\frac{\omega}{s^2+\omega^2}\right)$$

- **_The Laplace transforms of derivatives_**

Consider the determination of the Laplace transform of the derivative of a function, i.e. $\mathcal{L}\{df(t)/dt\}$. Using equation [1]:

$$\mathcal{L}\left\{\frac{\mathrm{d}}{\mathrm{d}t}f(t)\right\} = \int_0^\infty e^{-st}\frac{\mathrm{d}}{\mathrm{d}t}f(t)\,\mathrm{d}t$$

Using integration by parts:

$$\mathcal{L}\left\{\frac{\mathrm{d}}{\mathrm{d}t}f(t)\right\} = -f(0) + s\int_0^\infty e^{-st}f(t)\,\mathrm{d}t = -f(0) + sF(s) \qquad [9]$$

where $f(0)$ is the value of $f(t)$ when $t = 0$ and $F(s)$ is the Laplace transform of $f(t)$.

For a second derivative we can similarly obtain:

$$\mathcal{L}\left\{\frac{\mathrm{d}^2}{\mathrm{d}t^2}f(t)\right\} = \int_0^\infty e^{-st}\frac{\mathrm{d}^2}{\mathrm{d}t^2}f(t)\,\mathrm{d}t$$

$$= \left[e^{-st}\frac{\mathrm{d}}{\mathrm{d}t}f(t)\right]_0^\infty + s\int_0^\infty e^{-st}\frac{\mathrm{d}}{\mathrm{d}t}f(t)\,\mathrm{d}$$

$$= -\frac{\mathrm{d}}{\mathrm{d}t}f(0) + s\{-f(0) + sF(s)\}$$

$$= s^2F(s) - sf(0) - \frac{\mathrm{d}}{\mathrm{d}t}f(0) \qquad [10]$$

where $\mathrm{d}f(0)/\mathrm{d}t$ is the value of the first derivative when $t = 0$.

Likewise for a third derivative we can obtain:

$$\mathcal{L}\left\{\frac{\mathrm{d}^3}{\mathrm{d}t^3}f(t)\right\} = s^3F(s) - s^2f(0) - s\frac{\mathrm{d}}{\mathrm{d}t}f(0) - \frac{\mathrm{d}^2}{\mathrm{d}t^2}f(0) \qquad [11]$$

where $\mathrm{d}^2f(0)/\mathrm{d}t^2$ is the value of the second derivative at $t = 0$.

Key points

$$\mathcal{L}\left\{\frac{\mathrm{d}}{\mathrm{d}t}f(t)\right\} = -f(0) + sF(s)$$

where $f(0)$ is the value of $f(t)$ when $t = 0$ and $F(s)$ is the Laplace transform of $f(t)$.

$$\mathcal{L}\left\{\frac{\mathrm{d}^2}{\mathrm{d}t^2}f(t)\right\} = s^2F(s) - sf(0) - \frac{\mathrm{d}}{\mathrm{d}t}f(0)$$

where $\mathrm{d}f(0)/\mathrm{d}t$ is the value of the first derivative when $t = 0$.

Example

Given the initial condition that $x = 2$ when $t = 0$, determine the Laplace transform of $4\dfrac{\mathrm{d}x}{\mathrm{d}t}$.

Using equation [9]:

$$\mathcal{L}\left\{4\frac{\mathrm{d}x}{\mathrm{d}t}\right\} = 4\left[sX(s) - x(0)\right] = sX(s) - 2$$

where $X(s)$ is the Laplace transform of $x(t)$.

Example

Given the initial conditions that $x = 0$ and $\mathrm{d}x/\mathrm{d}t = 0$ when $t = 0$, determine the Laplace transform of $3\dfrac{\mathrm{d}^2x}{\mathrm{d}t^2} + 2\dfrac{\mathrm{d}x}{\mathrm{d}t}$.

Using equations [9] and [10]:

$$\mathscr{L}\left\{3\frac{d^2x}{dt^2} + 2\frac{dx}{dt}\right\}$$

$$= 3\left[s^2X(s) - sx(0) - \frac{d}{dt}x(0)\right] + 2\left[sX(s) - x(0)\right]$$

$$= (3s^2 + 2s)X(s)$$

- **Laplace transform of an integral**

Consider the determination of the Laplace transform of the integral of a function, i.e.

$$\mathscr{L}\left\{\int_0^t f(t)\,dt\right\}$$

If we let $g(t) = \int_0^t f(t)\,dt$, then $\frac{d}{dt}g(t) = f(t)$. Then, using equation [7]:

$$\mathscr{L}\left\{\frac{d}{dt}g(t)\right\} = sG(s) - g(0)$$

Since $g(0) = 0$ and $G(s) = \mathscr{L}\{g(t)\}$:

$$\mathscr{L}\{f(t)\} = s\,\mathscr{L}\left\{\int_0^t f(t)\,dt\right\}$$

Thus:

$$\mathscr{L}\left\{\int_0^t f(t)\,dt\right\} = \frac{1}{s}F(s) \tag{12}$$

Key point

$$\mathscr{L}\left\{\int_0^t f(t)\,dt\right\} = \frac{1}{s}F(s)$$

Example

Determine the Laplace transform of $\int_0^t e^{-t}\,dt$.

Using equation [12]:

$$\mathscr{L}\left\{\int_0^t e^{-t}\,dt\right\} = \frac{1}{s}F(s)$$

Since:

$$F(s) = \mathscr{L}\{e^{-t}\} = \frac{1}{s+1}$$

$$\mathscr{L}\left\{\int_0^t e^{-t}\,dt\right\} = \frac{1}{s}F(s) = \frac{1}{s(s+1)}$$

6.1.1 The inverse transform

The inverse Laplace transform is the transformation of a Laplace transform into a function of time. If $\mathscr{L}\{f(t)\} = F(s)$ then $f(t)$ is the *inverse Laplace transform* of $F(s)$, the inverse being written as:

$$f(t) = \mathcal{L}^{-1}\{F(s)\} \qquad\qquad [13]$$

The inverse can generally be obtained by using standard transforms, e.g. those in Table 6.1. The basic properties of the inverse, see the following notes, can be used with the standard transforms to obtain a wider range of transforms than just those in the table. Often $F(s)$ is the ratio of two polynomials and cannot be readily identified with a standard transform. However, the use of partial fractions (see Section 4.2.3) can often convert such an expression into simple fraction terms which can then be identified with standard transforms. This is illustrated in the examples given in the next section.

Example

Determine the inverse Laplace transform of $1/s^2$.

Table 6.1 indicates that the function which has the Laplace transform of $1/s^2$ is t. Thus the inverse is t.

Basic properties of the inverse transform

The following are basic properties which aid in the obtaining of inverse transforms.

- ### Additive property
  If we have a Laplace transform as the sum of two separate terms then we can take the inverse of each separately and the sum of the two inverse transforms is the inverse of the sum:

$$\mathcal{L}^{-1}\{F(s) + G(s)\} = \mathcal{L}^{-1}\{F(s)\} + \mathcal{L}^{-1}\{G(s)\} \qquad [14]$$

  Also:

$$\mathcal{L}^{-1}\{aF(s)\} = a\mathcal{L}^{-1}\{F(s)\} \qquad [15]$$

  where a is a constant.

- ### First shift theorem
  The *first shift theorem* (see Section 6.1) can be written in inverse form as:

$$\mathcal{L}^{-1}\{F(s - a)\} = e^{at}f(t) \qquad [16]$$

  where $f(t)$ is the inverse transform of $F(s)$.

- ### Second shift theorem
  The *second shift theorem* (see Section 6.1) can be written in inverse form as:

$$\mathcal{L}^{-1}\{e^{-sT}F(s)\} = f(t - T)u(t - T) \qquad [17]$$

Key points

$f(t) = \mathcal{L}^{-1}\{F(s)\}$

A Laplace transform which is the sum of two separate terms has an inverse of the sum of the inverse transforms of each term considered separately.

A Laplace transform which is a constant multiplied by a function has an inverse of the constant multiplied by the inverse of the function.

First shift theorem:
$\mathcal{L}^{-1}\{F(s - a)\} = e^{at}f(t)$, where $f(t)$ is the inverse transform of $F(s)$.

Second shift theorem: if the inverse transform numerator contains an e^{-sT} term, we remove this term from the expression, determine the inverse transform of what remains and then substitute $(t - T)$ for t in the result.

Thus if the inverse transform numerator contains an e^{-sT} term, then we remove this term from the expression, determine the inverse transform of what remains and then substitute $(t - T)$ for t in the result.

Example

Determine the inverse Laplace transform of $\dfrac{7s}{s^2+9}$.

Table 6.1 shows the Laplace transform of $\cos \omega t$ as being $s/(s^2 + \omega^2)$. Thus:

$$\mathcal{L}^{-1}\left\{\frac{s}{s^2+\omega^2}\right\} = \cos \omega t$$

Thus, using equation [15]:

$$\mathcal{L}^{-1}\left\{\frac{7s}{s^2+9}\right\} = 7\mathcal{L}^{-1}\left\{\frac{s}{s^2+9}\right\} = 7\cos 3t$$

Example

Determine the inverse Laplace transform of $\dfrac{3s-1}{s(s-1)}$.

We can write the fraction in a simpler form by the use of partial fractions. Thus:

$$\frac{3s-1}{s(s-1)} = \frac{A}{s} + \frac{B}{s-1}$$

and so we must have $3s - 1 = A(s - 1) + Bs$. Equating coefficients of s gives $3 = A + B$ and equating numerical terms gives $-1 = -A$. Hence:

$$\frac{3s-1}{s(s-1)} = \frac{1}{s} + \frac{2}{s-1}$$

The inverse transform of $1/s$ is 1 and of $1/(s - 1)$ is e^t. Thus:

$$\mathcal{L}^{-1}\left\{\frac{3s-1}{s(s-1)}\right\} = 1 + 2\,e^t$$

Example

Determine the inverse Laplace transform of $\dfrac{6}{s^2-6s+13}$.

This fraction can be rearranged as:

$$\frac{6}{s^2 - 6s + 13} = 3\frac{2}{(s-3)^2 + 2^2}$$

The fraction term is now in the form $\omega/(s^2 + \omega^2)$, i.e. the transform of $\sin \omega t$ when s has been replaced by $s - 3$. This corresponds to a multiplication by e^{3t}. Thus, using equation [16]:

$$\mathcal{L}^{-1}\left\{\frac{5}{s^2 - 6s + 13}\right\} = 3\,e^{3t}\sin 2t$$

Example

Determine the inverse Laplace transform of $6e^{-3t}/(s + 2)$.

Using equation [17], extracting e^{-3s} from the expression gives $6/(s + 2)$. This has the inverse Laplace transform of $6\,e^{-2t}$. Thus the required inverse is $5(t - 3)\,e^{-2(t-3)}u(t - 3)$.

Initial and final values

The *initial value* of a function of time is its value at zero time, the *final value* being the value at infinite time. Often there is a need to determine the initial value and final values of systems, e.g. for an electrical circuit when there is, say, a step input. The final value in such a situation is often referred to as the *steady-state value*. The initial and final value theorems enable the initial and final values to be determined from a Laplace transform without the need to find the inverse transform.

- **The initial value theorem**

  The Laplace transform of $f(t)$ is given by equation [1] as:

  $$\mathcal{L}\{f(t)\} = \int_0^\infty e^{-st}f(t)\,dt$$

  and so:

  $$\mathcal{L}\left\{\frac{d}{dt}f(t)\right\} = \int_0^\infty e^{-st}\frac{d}{dt}f(t)\,dt \qquad [18]$$

  Integration by parts then gives:

  $$\mathcal{L}\left\{\frac{d}{dt}f(t)\right\} = \left[e^{-st}f(t)\right]_0^\infty - \int_0^\infty (-s\,e^{-st})f(t)\,dt = -f(0) + sF(s) \quad [19]$$

  As s tends to infinity then e^{-st} tends to 0. Thus we must have, as a result of equation [18], $\mathcal{L}\{df(t)/dt\}$ tending to 0 as s tends to infinity. Hence equation [19] gives:

$$\lim_{s\to\infty} [-f(0) + sF(s)] = 0$$

$$\lim_{s\to\infty} sF(s) = f(0)$$

But $f(0)$ is the initial value of the function at $t = 0$. Thus, provided a limit exists:

$$\lim_{t\to 0} f(t) = \lim_{s\to\infty} sF(s) \qquad [20]$$

This is known as the *initial value theorem*.

Key point

Initial value theorem:

$$\lim_{t\to 0} f(t) = \lim_{s\to\infty} sF(s)$$

Example

Determine the initial value of the function $f(t)$ giving the Laplace transform $4/(s + 2)$.

Applying equation [20]:

$$\lim_{t\to 0} f(t) = \lim_{s\to\infty}\left[\frac{4s}{s+2}\right] = \lim_{s\to\infty}\left[\frac{4}{1+2/s}\right] = 4$$

• **_The final value theorem_**

As with the initial value theorem, for a function $f(t)$ having a Laplace transform $F(s)$ we can write (equations [18] and [19]):

$$\mathcal{L}\left\{\frac{d}{dt}f(t)\right\} = \int_0^\infty e^{-st}\frac{d}{dt}f(t)\,dt = -f(0) + sF(s) \qquad [21]$$

As s tends to zero then e^{-st} tends to 1 and so:

$$\lim_{s\to 0}\left[\int_0^\infty e^{-st}\frac{d}{dt}f(t)\,dt\right] = \int_0^\infty \frac{d}{dt}f(t)\,dt$$

We can write this integral as:

$$\int_0^\infty \frac{d}{dt}f(t)\,dt = \lim_{t\to\infty}\int_0^t \frac{d}{dt}f(t)\,dt = \lim_{t\to\infty}\left[f(t) - f(0)\right]$$

Hence, with equation [21] we obtain:

$$\lim_{s\to 0}\left[-f(0) + sF(s)\right] = \lim_{t\to\infty}\left[f(t) - f(0)\right]$$

and so, provided a limit exists:

$$\lim_{t\to\infty} f(t) = \lim_{s\to 0} sF(s) \qquad [22]$$

This is termed the *final value theorem*.

Key point

Final value theorem:

$$\lim_{t\to\infty} f(t) = \lim_{s\to 0} sF(s)$$

Example

Determine the final value of the function which has the Laplace transform:

$$F(s) = \frac{2s+1}{(s+1)(s+3)}$$

Using equation [22]:

$$\lim_{t\to\infty} f(t) = \lim_{s\to0} sF(s) = \lim_{s\to0}\left[\frac{s(2s+1)}{(s+1)(s+3)}\right] = 0$$

Problems 6.1

1 Determine, working from first principles and the definition of the transform, the Laplace transforms of:

(a) $f(t) = t^2$, (b) $f(t) = t^3$, (c) $f(t) = \sinh at$.
(Hint: $\sinh at = \frac{1}{2}(e^{at} - e^{-at})$)

2 Determine, by the use of the transforms given in Table 6.1 and the properties of Laplace transforms, the Laplace transforms of the following functions:
(a) 4, (b) $3t - 1$, (c) e^{3t}, (d) $2t + 3\ e^t$, (e) $t^2 + 4\ e^{-2t}$,
(f) $t^2 + 2t + 1$, (g) $2\sin 3t$, (h) $5\sinh 3t$, (i) $\sin 3t\cos 3t$,
(j) $t\ e^{-3t}$, (k) $4 - 2\sin 3t + e^{2t}$, (l) $t^3\ e^{-2t}$,
(m) $(1 + e^t)(1 - e^{-t})$, (n) $e^{3t}\cos t$, (o) $(1 + t)^2\ e^{-t}$,
(p) $e^{-t}\sin^2 t$, (q) $t\cosh 3t$, (r) $t^2\cosh 3t$, (s) $t^3\ e^{-3t}$

3 Use the additive property to determine the Laplace transforms of the following functions:

(a) $t^2 + 3t + 2$, (b) $2 + 4\sin 3t$, (c) $e^{4t} + \cosh 2t$,
(d) $2 + 5\ e^{3t}$, (e) $\cos 2t + \cos 3t$, (f) $t^3 + 4\ e^{-t}$

4 Use the first shift theorem to determine the Laplace transforms of the following functions:

(a) $e^{-3t}\sin 2t$, (b) $e^{4t}t^2$, (c) $e^{2t}\cos t$

5 Use the second shift theorem to determine the Laplace transform of the following functions:

(a) a unit step function which starts at $t = 5$ s,
(b) a unit impulse which occurs at $t = 4$ s,
(c) the function described by $3(t - 10)u(t - 10)$

6 Determine the Laplace transform of the periodic function shown in Figure 6.8.

7 Determine the Laplace transform for the periodic signal shown in Figure 6.9.

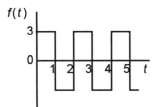

Figure 6.8 *Problem 6*

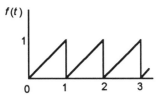

Figure 6.9 *Problem 7*

8 Determine the Laplace transform for the following periodic signals:

(a) $f(t) = 1$ for $0 \leq t < 1$ and 0 for $1 \leq t < 2$, $f(t + 2) = f(t)$,
(b) $f(t) = t$ for $0 \leq t < 1$ and 0 for $1 \leq t < 2$, $f(t + 2) = f(t)$,
(c) $f(t) = t$ for $0 \leq t < 1$ and $2 - t$ for $1 \leq t < 2$, $f(t + 2) = f(t)$

9 Determine the inverse Laplace transforms of:

(a) $\dfrac{1}{s-2}$, (b) $\dfrac{5}{s}$, (c) $\dfrac{s}{s^2+16}$, (d) $\dfrac{3}{s^2-9}$,

(e) $\dfrac{5}{(s-2)^2+25}$, (f) $\dfrac{1}{(s+3)^4}$, (g) $\dfrac{e^{-2s}}{s^2}$, (h) $\dfrac{e^{-3s}}{(s+2)^2}$

10 Determine, by the use of partial fractions, the inverse Laplace transforms of the following:

(a) $\dfrac{3s+1}{s^2-s-6}$, (b) $\dfrac{3s+3}{s^2+s-2}$, (c) $\dfrac{s-4}{(s+1)(s^2+4)}$

(d) $\dfrac{s+4}{s^2+4s+4}$

11 Determine the initial values of the functions giving the following Laplace transforms:

(a) $\dfrac{1}{s^2+1}$, (b) $\dfrac{s}{s^2+1}$

12 Determine the final values of the functions having the following Laplace transforms:

(a) $\dfrac{2}{s}$, (b) $\dfrac{1}{s+5}$

6.2 Solving differential equations

Laplace transforms offer a method of solving differential equations. The procedure adopted is:

1 Replace each term in the differential equation by its Laplace transform, inserting the given initial conditions.

2 Algebraically rearrange the equation to give the transform of the solution.

3 Invert the resulting Laplace transform to obtain the answer as a function of time.

Example
Given that $x = 0$ at $t = 0$, solve the first-order differential equation $3(dx/dt) + 2x = 4$.

Taking the Laplace transform gives:

$$3[sX(s) - x(0)] + 2X(s) = \frac{4}{s}$$

Substituting the initial condition gives:

$$3sX(s) + 2X(s) = \frac{4}{s}$$

Hence:

$$X(s) = \frac{4}{s(3s+2)}$$

Simplifying by the use of partial fractions:

$$\frac{4}{s(3s+2)} = \frac{A}{s} + \frac{B}{3s+2}$$

Hence $A(3s + 2) + Bs = 4$ and so $A = 2$ and $B = -2/3$. Thus:

$$X(s) = \frac{2}{s} - \frac{2}{3(3s+2)} = \frac{2}{s} - 2\frac{\frac{2}{3}}{\frac{2}{3}\left(s+\frac{2}{3}\right)}$$

and so $x(t) = 2 - 2\,e^{-2t/3}$.

Example

Given that $x = 0$ and $dx/dt = 1$ at $t = 0$, solve the second-order differential equation:

$$\frac{d^2x}{dt^2} - 5\frac{dx}{dt} + 6x = 2\,e^{-t}$$

Taking the Laplace transform gives:

$$s^2 X(s) - sx(0) - \frac{d}{dt}x(0) - 5[sX(s) - x(0)] + 6X(s) = \frac{2}{s+1}$$

Substituting the initial conditions:

$$s^2 X(s) - 1 - 5sX(s) + 6X(s) = \frac{2}{s+1}$$

$$X(s) = \frac{\frac{2}{s+1} + 1}{s^2 - 5s + 6} = \frac{2}{(s+1)(s-2)(s-3)} + \frac{1}{(s-2)(s-3)}$$

We can simplify the above expression by the use of partial fractions. Thus:

$$\frac{2}{(s+1)(s-2)(s-3)} = \frac{A}{s+1} + \frac{B}{s-2} + \frac{C}{s-3}$$

Hence $A(s-2)(s-3) + B(s+1)(s-3) + C(s+1)(s-2)$
$= 2$ and so $A = 1/6$, $B = -2/3$ and $C = \frac{1}{2}$.

$$\frac{1}{(s-2)(s-3)} = \frac{D}{s-2} + \frac{E}{s-3}$$

Hence $D(s-3) + E(s-2) = 1$ and so $D = -1$ and $E = 1$.
Thus:

$$X(s) = \frac{\frac{1}{6}}{s+1} + \frac{-\frac{2}{3}}{s-2} + \frac{\frac{1}{2}}{s-3} + \frac{-1}{s-2} + \frac{1}{s-3}$$

$$= \frac{\frac{1}{6}}{s+1} - \frac{\frac{5}{3}}{s-2} + \frac{\frac{3}{2}}{s-3}$$

The inverse transform is $x(t) = \frac{1}{6}e^{-t} - \frac{5}{3}e^{2t} + \frac{3}{2}e^{3t}$.

Example

Solve the following second-order differential equation:

$$\frac{d^2x}{dt^2} + 64x = 0$$

given the conditions (a) $dx/dt = 0$ and $x = 2$ when $t = 0$,
(b) $dx/dt = 2$ and $x = 0$ when $t = 0$.

Taking the Laplace transform gives:

$$s^2 X(s) - sx(0) - \frac{d}{dt}x(0) + 64X(s) = 0$$

(a) We have $dx/dt = 0$ and $x = 2$ when $t = 0$ and so:

$$s^2 X(s) - 2s - 0 + 64X(s) = 0$$

$$(s^2 + 64)X(s) = 2s$$

$$X(s) = \frac{2s}{s^2 + 64} = 2\left(\frac{s}{s^2 + 8^2}\right)$$

As Table 6.1 indicates, the bracketed term has the inverse of a cosine. Thus the solution is $x = 2\cos 8t$.

(b) We have $dx/dt = 2$ and $x = 0$ when $t = 0$, and so:

$$s^2 X(s) - 0 - 2 + 64X(s) = 0$$

$$(s^2 + 64)X(s) = 2$$

$$X(s) = \frac{2}{s^2 + 64}$$

To put this in the form $\omega/(s^2 + \omega^2)$ we multiply by 4/4:

$$X(s) = \frac{1}{4}\frac{8}{s^2 + 8^2}$$

The solution is thus $x = \frac{1}{4}\sin 8t$.

Problems 6.2

1 Solve the following differential equations:

(a) $2\frac{dx}{dt} + x = 4\,e^{2t}$,
with $x = 0$ when $t = 0$,

(b) $\frac{dx}{dt} + 5x = 2$,
with $x = 0$ when $t = 0$,

(c) $\frac{d^2x}{dt^2} + 4x = 1$,
with $x = 0$ and $dx/dt = 0$ when $t = 0$,

(d) $\frac{d^2x}{dt^2} + 2\frac{dx}{dt} + 2x = e^{-t}$,
with $x = 0$ and $dx/dt = 0$ when $t = 0$,

(e) $\frac{d^2x}{dt^2} - 2\frac{dx}{dt} + x = \sin t$, with $x = 1$
with $x = 1$ and $dx/dt = 0$ when $t = 0$,

(f) $\frac{dx}{dt} - 2x = 3$,
with $x = 0$ when $t = 0$,

(g) $\frac{dx}{dt} + 4x = \cos t$,
with $x = 0$ when $t = 0$,

(h) $\frac{d^2x}{dt^2} + x = 3$,
with $x = 0$ and $dx/dt = 1$ when $t = 0$,

(i) $\frac{d^2x}{dt^2} + \frac{dx}{dt} - 6x = 5\,e^{3t}$
with $x = 0$ and $dx/dt = 0$ when $t = 0$,

(j) $\frac{d^2x}{dt^2} - \frac{dx}{dt} - 6x = \cos 2t$
with $x = 0$ and $dx/dt = 0$ when $t = 0$,

(k) $\frac{d^2x}{dt^2} - 5\frac{dy}{dt} - 6x = t + e^{3t}$
with $x = 2$ and $dx/dt = 1$ when $t = 0$.

6.3 Transfer function

In general, when we consider inputs and outputs of systems as functions of time then the relationship between the output and input is given by a differential equation. If we have a system composed of two elements in series with each having its input–output relationships described by a differential equation, it is not easy to see how the output of the system as a whole is related to its input. We can overcome this problem by transforming the differential equations into a more convenient form by using the Laplace transform. This form is a much more convenient way of describing the relationship than a differential equation since it can be easily manipulated by the basic rules of algebra.

For a simple system we might use the term gain to relate the input and output of a system with gain G = output/input. This tells us how much bigger the output is than the input. When we are working with inputs and outputs described as functions of s we define the *transfer function* $G(s)$ as [output $Y(s)$]/[input $X(s)$] when all initial conditions before we apply the input are zero, i.e.

$$G(s) = \frac{Y(s)}{X(s)} \tag{23}$$

A transfer function can be represented as a block diagram (Figure 6.10) with $X(s)$ the input, $Y(s)$ the output and the transfer function $G(s)$ as the operator in the box that converts the input to the output. The block represents a multiplication for the input. Thus, by using the Laplace transform of inputs and outputs, we can use the transfer function as a simple multiplication factor, like the gain.

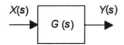

Figure 6.10 *Transfer function as the factor that multiplies the input to give the output*

Example

Determine the transfer function for an electrical system for which we have the relationship:

$$\frac{V_C(s)}{V(s)} = \frac{1}{RCs+1}$$

The transfer function $G(s)$ is thus:

$$G(s) = \frac{V_C(s)}{V(s)} = \frac{1}{RCs+1}$$

To get the output $V_C(s)$ we multiply the input $V(s)$ by $1/(RCs + 1)$.

Example

Determine the transfer function for the mechanical system having mass, stiffness and damping, and input F and output x and described by the differential equation:

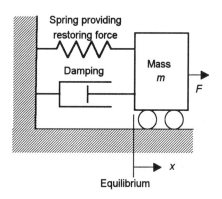

Figure 6.11 *Mass, spring, damper system*

$$F = m\frac{d^2x}{dt^2} + c\frac{dx}{dt} + kx$$

Figure 6.11 shows the type of system that would give such a differential equation.

If we now write $\mathcal{L}(x) = X(s)$ and $\mathcal{L}(F) = F(s)$, and with initial conditions zero:

$$\mathcal{L}\left\{m\frac{d^2x}{dt^2}\right\} = m\left[s^2X(s) - sx(0) - \frac{d}{dt}x(0)\right]$$

$$\mathcal{L}\left\{c\frac{dx}{dt}\right\} = c\left[sX(s) - x(0)\right]$$

$$\mathcal{L}\{k(x)\} = kX(s)$$

$$\mathcal{L}\{F\} = F(s)$$

When $t = 0$ we have $x = 0$ and $dx/dt = 0$ and so:

$$\mathcal{L}\left\{m\frac{d^2x}{dt^2}\right\} = m\left[s^2X(s)\right]$$

$$\mathcal{L}\left\{c\frac{dx}{dt}\right\} = c\left[sX(s)\right]$$

Thus, we have for the differential equation in the s-domain:

$$F(s) = ms^2X(s) + csX(s) + kX(s)$$

$$= (ms^2 + cs + k)X(s)$$

Hence the transfer function $G(s)$ of the system is:

$$G(s) = \frac{X(s)}{F(s)} = \frac{1}{ms^2 + cs + k}$$

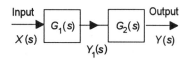

Figure 6.12 *Systems in series*

Systems in series

Consider a system of two subsystems in series (Figure 6.12). The first subsystem has an input of $X(s)$ and an output of $Y_1(s)$; thus, $G_1(s) = Y_1(s)/X(s)$. The second subsystem has an input of $Y_1(s)$ and an output of $Y(s)$; thus, $G_2(s) = Y(s)/Y_1(s)$. We thus have:

$$Y(s) = G_2(s)Y_1(s) = G_2(s)G_1(s)X(s)$$

The overall transfer function $G(s)$ of the system is $Y(s)/X(s)$ and so:

$$G_{overall}(s) = G_1(s)G_2(s) \tag{24}$$

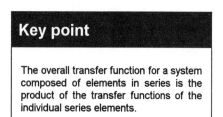

Key point

The overall transfer function for a system composed of elements in series is the product of the transfer functions of the individual series elements.

Example

Determine the overall transfer function for a system which consists of two elements in series, one having a transfer function of $1/(s + 1)$ and the other $1/(s + 2)$.

The overall transfer function is thus:

$$G_{overall}(s) = \frac{1}{s+1} \times \frac{1}{s+2} = \frac{1}{(s+1)(s+2)}$$

Key point

A simple feedback control system to, say, control the temperature of a room will have a negative feedback loop. This feeds back a measure of the output of the system which is then subtracted from the input. The input is the required temperature and the output the actual temperature. The difference between these signals, i.e. the error, is used to actuate some heating system which will continue as long as there is an error.

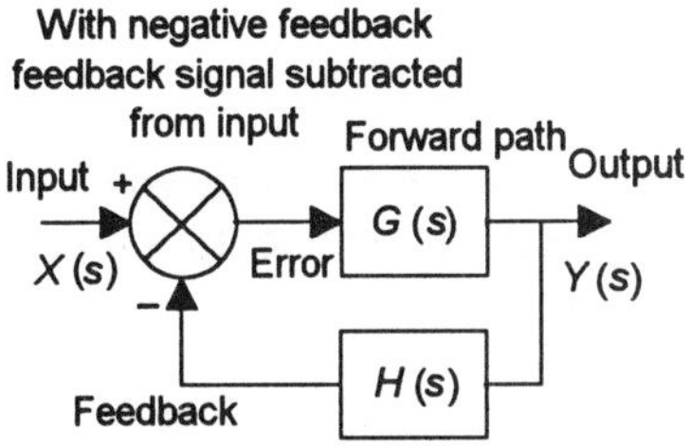

With negative feedback
feedback signal subtracted
from input

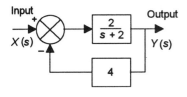

Figure 6.13 *System with negative feedback*

Key point

For a system with a negative feedback, the overall transfer function is the forward path transfer function divided by one plus the product of the forward path and feedback path transfer functions.

Systems with negative feedback

For a control system with a negative feedback loop we can have the situation shown in Figure 6.13 where the output is fed back via a system with a transfer function $H(s)$. This fed back signal subtracts from the input to the system $G(s)$ to give the error signal. The feedback system has an input of $Y(s)$ and thus an output of $H(s)Y(s)$. Thus the feedback signal is $H(s)Y(s)$. The error is the difference between the input signal $X(s)$ and the feedback signal:

$$\text{error } (s) = X(s) - H(s)Y(s)$$

This error signal is the input to the $G(s)$ system and gives an output of $Y(s)$. Thus:

$$G(s) = \frac{Y(s)}{X(s) - H(s)Y(s)}$$

and so:

$$[1 + G(s)H(s)]Y(s) = G(s)X(s)$$

which can be rearranged to give:

$$\text{overall transfer function} = \frac{Y(s)}{X(s)} = \frac{G(s)}{1 + G(s)H(s)} \qquad [25]$$

Example

Determine the overall transfer function for a control system (Figure 6.14) which has a negative feedback loop with a transfer function 4 and a forward path transfer function of $2/(s + 2)$.

The overall transfer function of the system is:

$$G_{overall}(s) = \frac{\dfrac{2}{s+2}}{1 + 4 \times \dfrac{2}{s+2}} = \frac{2}{s+10}$$

Figure 6.14 *Example*

6.3.1 Determining outputs of systems

The procedure we can use to determine how the output of a system will change with time when there is some input to the system is:

1 *Determine the output as an s function*
   In terms of the transfer function $G(s)$ we have:

   $$\text{Output } (s) = G(s) \times \text{Input } (s) \qquad [26]$$

   We can thus obtain the output of a system as an s function by multiplying its transfer function by the input s function.

2 *Determine the time function corresponding to the output s function*
   To obtain the output as a function of time we need to find the time function that will give the particular output s function that we have obtained. Tables of s functions and their corresponding time functions can be used (Table 6.1). Often, however, the s function output has to be rearranged to put it into a form given in the table.

Example

A system has a transfer function of $1/(s + 2)$. What will be its output as a function of time when it is subject to a step input of 1 V?

The step input has a Laplace transform of $(1/s)$. Thus:

$$\text{Output } (s) = G(s) \times \text{Input } (s)$$

$$= \frac{1}{s+2} \times \frac{1}{s} = \frac{1}{s(s+2)}$$

The nearest form we have in Table 6.1 to the output is item 6 as $\frac{1}{2} \times 2/[s(s + 2)]$. Thus the output, as a function of time, is $\frac{1}{2}(1 - e^{-5t})$ V.

First-order systems

A first-order system has a differential equation of the form:

$$\tau \frac{dy}{dt} + y = kx$$

As a function of s this can be written as:

$$\tau Y(s) + Y(s) = kX(s)$$

and so a transfer function of the form:

$$G(s) = \frac{Y(s)}{X(s)} = \frac{k}{\tau s + 1} \qquad [27]$$

where k is the *gain* of the system when there are steady-state conditions and τ is the *time constant* of the system.

- **Unit impulse input**

  When a first-order system is subject to a unit impulse input then $X(s) = 1$ and the output transform $Y(s)$ is:

  $$Y(s) = G(s)X(s) = \frac{k}{\tau s + 1} \times 1 = k\frac{(1/\tau)}{s + 1/\tau}$$

  Hence, since we have the transform in the form $1/(s + a)$, using item 6 in Table 6.1 gives:

  $$x = k(1/\tau)\,e^{-t/\tau} \qquad [28]$$

  Figure 6.15 shows how the output x varies with time.

- **Unit step input**

  When a first-order system is subject to a unit step input then $X(s) = 1/s$ and the output transform $Y(s)$ is:

  $$X(s) = G(s)Y(s) = \frac{k}{s(\tau s + 1)} = k\frac{(1/\tau)}{s(s + 1/\tau)}$$

  Hence, since we have the transform in the form $a/s(s + a)$, using item 6 in Table 6.1 gives:

  $$x = k(1 - e^{-t/\tau}) \qquad [29]$$

  Figure 6.16 shows how the output x varies with time.

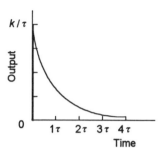

Figure 6.15 *Output with a unit impulse input to a first-order system*

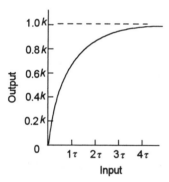

Figure 6.16 *Behaviour of a first-order system when subject to a unit step input*

Example

A circuit has a resistance R in series with a capacitance C. The differential equation relating the input v and output v_C, i.e. the voltage across the capacitor, is:

$$v = RC\frac{dv_C}{dt} + v_C$$

Determine the output of the system when there is a 2 V impulse input.

As a function of s the differential equation becomes:

$$V(s) = RCsV_C(s) + V_C(s)$$

Hence the transfer function is

$$G(s) = \frac{V_C(s)}{V(s)} = \frac{1}{RCs+1}$$

The output when there is 2 V impulse input is:

$$V_C(s) = G(s)V(s) = \frac{1}{RCs+1} \times 2 = \frac{2/RC}{s+1/RC}$$

Hence, since we have the transform in the form $1/(s + a)$, using item 6 in Table 6.1 gives:

$v = (2/RC)\, e^{-t/RC}$

Example

A thermocouple which has a transfer function linking its voltage output V and temperature input of:

$$G(s) = \frac{30 \times 10^{-6}}{10s+1} \text{ V/°C}$$

Determine the response of the system when it is suddenly immersed in a water bath at 100°C.

The output as an s function is:

$V(s) = G(s) \times \text{input } (s)$

The sudden immersion of the thermometer gives a step input of size 100°C and so the input as an s function is $100/s$. Thus:

$$V(s) = \frac{30 \times 10^{-6}}{10s+1} \times \frac{100}{s} = \frac{30 \times 10^{-4}}{10s(s+0.1)}$$

$$= 30 \times 10^{-4} \frac{0.1}{s(s+0.1)}$$

The fraction element is of the form $a/s(s + a)$, item 6 in Table 6.1, and so the output as a function of time is:

$V = 30 \times 10^{-4}\, (1 - e^{-0.1t})$ V

Second-order systems

The differential equation for a second-order system can be written as:

$$\frac{d^2y}{dt^2} + 2\zeta\omega_n\frac{dy}{dt} + \omega_n^2 y = kx$$

where x is the input and y the output (see Chapter 5, equation [36]). Since the steady-state output occurs when $\omega_n^2 y = kx$, a more usual way of writing the standard form of the equation, so that the steady-state value occurs when $y = kx$, is as:

$$\frac{d^2y}{dt^2} + 2\zeta\omega_n\frac{dy}{dt} + \omega_n^2 y = k\omega_n^2 x \qquad [30]$$

where ω_n is the natural angular frequency with which the system oscillates and ζ is the damping ratio. Hence we have:

$$s^2 Y(s) + 2\zeta\omega_n s Y(s) + \omega_n^2 Y(s) = k\omega_n^2 X(s)$$

and so a transfer function of:

$$G(s) = \frac{Y(s)}{X(s)} = \frac{k\omega_n^2}{s^2 + 2\zeta\omega_n s + \omega_n^2} \qquad [31]$$

When a second-order system is subject to a unit step input, i.e. $X(s) = 1/s$, then the output transform is:

$$Y(s) = G(s)X(s) = \frac{k\omega_n^2}{s(s^2 + 2\zeta\omega_n s + \omega_n^2)}$$

There are three different forms of answer to this equation for the way the output varies with time; these depending on the value of the damping constant and whether it gives an overdamped, critically damped or underdamped system. We can determine the condition for these three forms of output by putting the equation in the form:

$$Y(s) = \frac{k\omega_n^2}{s(s + p_1)(s + p_2)} \qquad [32]$$

where p_1 and p_2 are the roots of the quadratic term:

$$s^2 + 2\zeta\omega_n s + \omega_n^2 = 0 \qquad [33]$$

Hence, if we use the equation to determine the roots of a quadratic equation, we obtain:

$$p = \frac{-2\zeta\omega_n \pm \sqrt{4\zeta^2\omega_n^2 - 4\omega_n^2}}{2}$$

and so the two roots are given by:

$$p_1 = -\zeta\omega_n + \omega_n\sqrt{\zeta^2 - 1} \quad \text{and} \quad p_2 = -\zeta\omega_n - \omega_n\sqrt{\zeta^2 - 1} \qquad [34]$$

Key point

In general, we can write a transfer function as:

$$G(s) = \frac{K(s - z_1)(s - z_2)...(s - z_m)}{(s - p_1)(s - p_2)...(s - p_n)}$$

with the values of s that make $G(s)$ zero being termed zeros and so correspond to $s = z_1, z_2, ... z_m$. The values of s that make $G(s)$ infinite are known as poles and so correspond to $s = p1, p2, ... p_n$. As will be apparent from the discussion on this page and the next, poles and zeros can be real or complex. We can thus plot them on an Argand diagram with the vertical axis being the imaginary element and the horizontal axis the real part. Such a plot is said to have the poles and zeros plotted on an s-plane diagram.

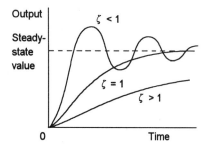

Figure 6.17 *Behaviour of a second-order system when subject to a unit step input signal*

The important issue in determining the form of the roots is the value of the square root term and this is determined by the value of the damping factor (Figure 6.17).

- ### Damping factor $\zeta > 1$

  With the damping factor ζ greater than 1 the square root term is real and will factorise. To find the inverse transform we can either use partial fractions to break the expression down into a number of simple fractions or use item 10 in Table 6.1. The output is thus:

$$y = \frac{k\omega_n^2}{p_1 p_2}\left[1 - \frac{p_2}{p_2 - p_1}\,\mathrm{e}^{-p_1 t} + \frac{p_1}{p_2 - p_1}\,\mathrm{e}^{-p_2 t}\right] \qquad [35]$$

This describes an output which does not oscillate but dies away with time and thus the system is *overdamped*. As the time t tends to infinity then the exponential terms tend to zero and the output becomes the steady value of $k\omega_n^2/(p_1 p_2)$. Since $p_1 p_2 = \omega_n^2$, the steady value is k.

- ### Damping factor $\zeta = 1$

  With $\zeta = 1$ the square root term is zero and so $p_1 = p_2 = -\omega_n$; both roots are real and both the same. The output equation then becomes:

$$Y(s) = \frac{k\omega_n^2}{s(s + \omega_n)^2}$$

This equation can be expanded by means of partial fractions to give:

$$Y(s) = k\left[\frac{1}{s} - \frac{1}{s + \omega_n} - \frac{\omega_n}{(s + \omega_n)^2}\right]$$

Hence:

$$y = k[1 - \mathrm{e}^{-\omega_n t} - \omega_n t\,\mathrm{e}^{-\omega_n t}] \qquad [36]$$

This is the critically damped condition and describes an output which does not oscillate but dies away with time. As the time t tends to infinity then the exponential terms tend to zero and the output tends to the steady state value of k.

- ### Damping factor $\zeta < 1$

  With $\zeta < 1$ the square root term does not have a real value. Using item 19 in Table 6.1 then gives:

$$y = k\left[1 - \frac{\mathrm{e}^{-\zeta\omega_n t}}{\sqrt{1 - \zeta^2}}\,\sin\left(\omega_n\sqrt{(1 - \zeta^2)}\,t + \phi\right)\right] \qquad [37]$$

where $\cos\phi = \zeta$. This is an under-damped oscillation. The angular frequency ω of the damped oscillation is:

$$\omega = \omega_n \sqrt{1 - \zeta^2} \qquad\qquad [38]$$

Only when the damping is very small does the angular frequency of the oscillation become nearly the natural angular frequency ω_n. As the time t tends to infinity then the exponential term tends to zero and so the output tends to the value k.

Example

What will be the state of damping of a system having the following transfer function and subject to a unit step input?

$$G(s) = \frac{1}{s^2 + 8s + 16}$$

The output $Y(s)$ from such a system is given by:

$$Y(s) = G(s)X(s)$$

For a unit step input $X(s) = 1/s$ and so the output is given by:

$$Y(s) = \frac{1}{s(s^2 + 8s + 16)} = \frac{1}{s(s+4)(s+4)}$$

The roots of $s^2 + 8s + 16$ are $p_1 = p_2 = -4$. Both the roots are real and the same, hence we have critical damping.

Example

A system has an output y related to the input x by the differential equation:

$$\frac{d^2y}{dt^2} + 5\frac{dy}{dt} + 6y = x$$

What will be the output from the system when it is subject to a unit step input? Initially both the output and input are zero.

We can write the Laplace transform of the equation as:

$$s^2Y(s) + 5sY(s) + 6Y(s) = X(s)$$

The transfer function is thus:

$$G(s) = \frac{Y(s)}{X(s)} = \frac{1}{s^2 + 5s + 6}$$

For a unit step input the output is given by:

$$Y(s) = \frac{1}{s(s^2 + 5s + 6)} = \frac{1}{s(s+3)(s+2)}$$

Because the quadratic term has two real roots, the system is overdamped. We can directly use one of the standard forms given in Table 6.1 or partial fractions to first simplify the expression before using Table 6.1. Using partial fractions:

$$\frac{1}{s(s+3)(s+2)} = \frac{A}{s} + \frac{B}{s+3} + \frac{C}{s+2}$$

Thus, we have $1 = A(s + 3)(s + 2) + Bs(s + 2) + Cs(s + 3)$. When $s = 0$ then $1 = 6A$ and so $A = 1/6$. When $s = -3$ then $1 = 3B$ and so $B = 1/3$. When $s = -2$ then $1 = -2C$ and so $C = -1/2$. Hence we can write the output in the form:

$$Y(s) = \frac{1}{6s} + \frac{1}{3(s+3)} - \frac{1}{2(s+2)}$$

Hence, using Table 6.1 gives:

$$y = 0.17 + 0.33\,e^{-3t} - 0.5\,e^{-2t}$$

Example

A system has the transfer function:

$$G(s) = \frac{9}{s^2 + 3.6s + 9}$$

Determine its natural frequency, the damping ratio and the frequency of the damped oscillation.

If we compare the transfer function with that given in equation [31], i.e.

$$G(s) = \frac{k\omega_n^2}{s^2 + 2\zeta\omega_n s + \omega_n^2}$$

we are led to conclude that $\omega_n^2 = 9$ and so $\omega_n = 3$ rad/s and $f_n = \omega_n/2\pi = 3/2\pi = 0.48$ Hz. The damping ratio is given by $2\zeta\omega_n = 3.6$ and so $\zeta = 3.6/(2 \times 3) = 0.6$; the system is underdamped. Using equation [38], the angular frequency of the undamped oscillation is given by:

$$\omega = \omega_n\sqrt{1 - \zeta^2} = 3\sqrt{1 - 0.6^2} = 2.4 \text{ rad/s}$$

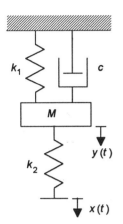

Figure 6.18 *Example*

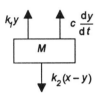

Figure 6.19 *Example*

Example

For the system shown in Figure 6.18, determine its transfer function if M = 50 kg, $k_1 = k_2$ = 400 N/m and c = 180 Ns/m. What will be the damped frequency of its oscillation when subject to a unit step input?

Considering the free-body diagram of the mass (Figure 6.19), and applying Newton's second law, we have:

$$k_2(x - y) - k_1 y - c\frac{dy}{dt} = ma = m\frac{d^2y}{dt^2}$$

The Laplace transform of this equation, with zero initial conditions, is:

$$k_2\left[X(s) - Y(s)\right] - k_1 Y(s) - csY(s) = ms^2 Y(s)$$

$$k_2 X(s) = ms^2 Y(s) + k_2 Y(s) + k_1 Y(s) + csY(s)$$

$$G(s) = \frac{Y(s)}{X(s)} = \frac{k_2}{ms^2 + cs + k_1 + k_2}$$

For comparison with the standard form of the transfer function equation, we write the above equation as:

$$G(s) = \frac{(k_2/m)}{s^2 + (c/m)s + (k_1 + k_2)/m}$$

Hence, with the given data:

$$G(s) = \frac{400/50}{s^2 + (180/50)s + (400 + 400)/50} = \frac{8}{s^2 + 3.6s + 16}$$

Comparing this with the standard form of transfer function for a second-order system [31]:

$$G(s) = \frac{k\omega_n^2}{s^2 + 2\zeta\omega_n s + \omega_n^2}$$

then the natural angular frequency ω_n^2 = 16 and ω_n = 4 rad/s. The damping ratio ζ is given by $2\zeta\omega_n$ = 3.6 and so ζ = 3.6/(2 × 4) = 0.45. The oscillation is underdamped.

Using equation [38], the angular frequency of the undamped oscillation is given by:

$$\omega = \omega_n\sqrt{1 - \zeta^2} = 4\sqrt{1 - 0.45^2} = 3.57 \text{ rad/s}$$

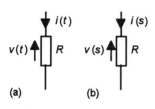

Figure 6.20   *Resistance:*
(a) time,   (b) s-domain

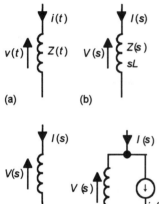

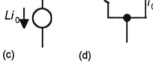

Figure 6.21   *Inductance:*
(a) time, (b), (c), (d) s-domain

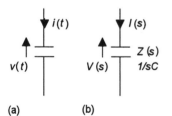

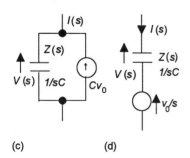

Figure 6.22   *Capacitance:*
(a) time domain, (b), (c), (d) s-domain

6.3.2 Electrical circuit analysis

While we could write differential equations to represent electrical circuits and then solve them by the use of the Laplace transform, a simpler method is to replace time-domain components by their equivalents in the *s*-domain.

Resistance R in the time domain is defined as $v(t)/i(t)$. Taking the Laplace transform of this equation gives a definition of resistance in the *s*-domain (Figure 6.20) as:

$$R = \frac{V(s)}{I(s)} \qquad [39]$$

Inductance L in the time domain (Figure 6.21(a)) is defined by:

$$v(t) = L\frac{di(t)}{dt}$$

The Laplace transform of this equation is $V(s) = L[sI(s) - i(0)]$. With zero initial current then $V(s) = sLI(s)$. Impedance in the *s*-domain $Z(s)$ is defined as $V(s)/I(s)$, thus for inductance (Figure 6.21(b)):

$$Z(s) = \frac{V(s)}{I(s)} = sL \qquad [40]$$

If the current was not initially zero but $i(0) = i_0$, then $V(s) = sLI(s) - Li_0$. This equation can be considered to describe two series elements (Figure 6.21(c)). The first term then represents the potential difference across the inductance L, being $Z(s)I(s)$, and the second term a voltage generator of $(-Li_0)$. Alternatively we can rearrange equation $V(s) = sLI(s) - Li_0$ in a form to represent two parallel elements (Figure 6.21(d)):

$$I(s) = \frac{V(s) + Li_0}{sL} = \frac{V(s)}{sL} + \frac{i_0}{s} \qquad [41]$$

$I(s)$ is the current into the system, $V(s)/sL = V(s)/Z(s)$ can be considered to be the current through the inductance and i_0/s a parallel current source.

Capacitance C in the time domain (Figure 6.22(a)) is defined by:

$$i(t) = C\frac{dv(t)}{dt}$$

The Laplace transform of this equation is $I(s) = C[sV(s) - v(0)]$. If we have $v(0) = 0$ then (Figure 6.22(b)):

$$Z(s) = \frac{V(s)}{I(s)} = \frac{1}{sC} \qquad [42]$$

If $v(0) = v_0$ then $I(s) = CsV(s) - Cv(0) = CsV(s) - Cv_0$. We can think of this representing $I(s)$ entering a parallel arrangement

(Figure 6.22(c)) of a capacitor, and giving a current through it is $V(s)/Z(s) = CsV(s)$, and a current source $(-Cv_0)$. Alternatively we can rearrange the equation as:

$$V(s) = \frac{1}{sC}I(s) + \frac{v_0}{s}$$ [43]

This equation now represents a capacitor in series with a voltage source of v_0/s (Figure 6.22(d)).

Example

Determine the impedance and equivalent series circuit in the *s*-domain of an inductance of 50 mH if there is a current of 0.1 A at time *t* = 0.

The impedance in the *s*-domain is given by equation [40] as $0.050s$ Ω. Its equivalent series circuit with the initial condition $i(0) = 0.1$ A is of a voltage source of $-0.050 \times 0.1 = 0.005$ V in series with the impedance of $0.050s$ Ω.

Example

Determine the impedance in the *s*-domain of a capacitance of 0.1 μF and its equivalent series circuit when the capacitor has been charged to 5 V at time *t* = 0.

The impedance in the *s*-domain is given by equation [42] as $1/sC = 1/(0.1 \times 10^{-6}s)$ Ω, and its equivalent series circuit with the initial condition $v(0) = 5$ V is of a voltage source of $-5/s$ in series with the impedance of $10^7/s$ Ω.

Using Kirchhoff's laws

Because of the additive property of the Laplace transform, the transform of a number of time-domain functions is the sum of the transforms of each separate function. Thus with *Kirchhoff's current law*, the algebraic sum of the time-domain currents at a junction is zero and so the sum of the transformed currents is also zero. With *Kirchhoff's voltage law*, the sum of the time-domain voltages around a closed loop is zero and thus the sum of the transformed voltages is also zero. A consequence of this is that:

All the techniques developed for use in the analysis of circuits in the time domain can be used in the s-domain.

The following examples illustrate this.

Example

Determine the impedance in the s-domain of a 10 Ω resistor in (a) series and (b) parallel with a 1 mH inductor.

(a) For impedances in series $Z(s) = Z_1(s) + Z_2(s) = 10 + 0.001s\ \Omega$.
(b) For impedances in parallel we have:

$$\frac{1}{Z(s)} = \frac{1}{Z_1(s)} + \frac{1}{Z_2(s)} = \frac{1}{10} + \frac{1}{0.001s} = \frac{0.001s+10}{0.01s}$$

Hence $Z(s) = 0.01s/(0.001s + 10)\ \Omega$.

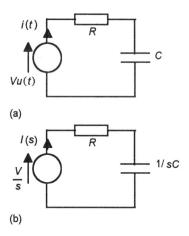

(a)

(b)

Figure 6.23 *Example*

Example

Determine how the circuit current varies with time for a circuit having a resistance R in series with an initially uncharged capacitance C when the input to the circuit is a step voltage V at time $t = 0$.

Figure 6.23(a) shows the circuit in the time domain and Figure 6.23(b) the equivalent circuit in the s-domain. A unit step at $t = 0$ has the Laplace transform $1/s$ and thus a voltage step of V has a transform of V/s. The impedance of the capacitance is $1/sC$. Thus, applying Kirchhoff's voltage law to the circuit:

$$\frac{V}{s} = RI(s) + \frac{1}{sC}I(s)$$

$$I(s) = \frac{V}{Rs + 1/C} = \frac{V(1/R)}{s + (1/RC)}$$

This is a constant multiplied by $1/(s + a)$, thus:

$$i(t) = \frac{V}{R}\,e^{-t/RC}$$

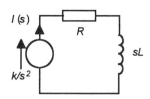

Figure 6.24 *Example*

Example

A ramp voltage of $v = kt$ is applied at time $t = 0$ to a circuit consisting of an inductance L in series with a resistance R. If initially at $t = 0$ there is no current in the circuit, determine how the circuit current varies with time.

The Laplace transform of kt is k/s^2. The inductance has an impedance in the s-domain of sL. Thus the circuit in the s-domain is as shown in Figure 6.24. Applying Kirchhoff's voltage law to the circuit gives:

$$\frac{k}{s^2} = sLI(s) + RI(s)$$

and so:

$$I(s) = \frac{k}{s^2(sL+R)} = \frac{(k/R)(R/L)}{s^2(s+R/L)}$$

This can be simplified by partial fractions, writing a for R/L:

$$\frac{a}{s^2(s+a)} = \frac{A}{s^2} + \frac{B}{s} + \frac{C}{s+a} = \frac{A(s+a) + Bs(s+a) + Cs^2}{s^2(s+a)}$$

Hence $A = 1$, $B = -1/a$ and $C = 1/a$. Thus:

$$I(s) = \frac{k}{R}\left(\frac{1}{s^2} - \frac{1}{(R/L)s} + \frac{1}{(R/L)(s+R/L)}\right)$$

Hence:

$$i(t) = \frac{k}{R}\left(t - \frac{1}{R/L} + \frac{e^{-Rt/L}}{R/L}\right)$$

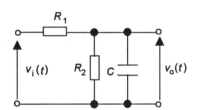

Figure 6.25 *Example*

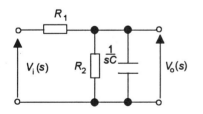

Figure 6.26 *Example*

Example

Determine the transfer function of the circuit shown in Figure 6.25 and the output $v_o(t)$ resulting from a unit step input, given that $R_1 = 10 \text{ k}\Omega$, $R_2 = 22 \text{ k}\Omega$ and $C = 1 \text{ }\mu\text{F}$.

The Laplace equivalent circuit is shown in Figure 6.26. The impedance $Z_p(s)$ for the parallel arrangement of R_2 and the capacitor is given by:

$$\frac{1}{Z_p(s)} = \frac{1}{R_2} + \frac{1}{1/sC}$$

$$Z_p(s) = \frac{R_2}{1+sCR_2}$$

We have a potential divided circuit and so:

$$V_0(s) = V_i(s)\frac{\left(\dfrac{R_2}{1+scR_2}\right)}{\left(R_1 + \dfrac{R_2}{1+sCR_2}\right)} = V_i(s)\frac{R_2}{R_1(1+sCR_2)+R_2}$$

The transfer function $G(s)$ of the system is thus:

$$G(s) = \frac{V_0(s)}{V_i(s)} = \frac{R_2}{R_1(1+sCR_2)+R_2}$$

Using the given values:

$$G(s) = \frac{22 \times 10^3}{220s + 32 \times 10^3}$$

For a unit step input we have $V_i(s) = 1/s$ and so:

$$V_0(s) = \frac{22 \times 10^3}{220s + 32 \times 10^3} \times \frac{1}{s} = \frac{100}{s(s + 145.45)}$$

We can use equation [6] in Table 6.1 or partial fractions to obtain the inverse. Partial fractions give:

$$V_0(s) = \frac{0.687}{s} - \frac{0.687}{s + 145.45}$$

and so:

$$v_0(t) = 0.687(1 - e^{-145.45t})$$

Problems 6.3

1 A system has an input of a voltage of 3 V which is suddenly applied by a switch being closed. What is the input as an s function?

2 A system has an input of a voltage impulse of 2 V. What is the input as an s function?

3 A system has an input of a voltage of a ramp voltage which increases at 5 V per second. What is the input as an s function?

4 A system gives an output of $1/(s + 5)$ $V(s)$. What is the output as a function of time?

5 A system has a transfer function of $5/(s + 3)$. What will be its output as a function of time when subject to (a) a unit step input of 1 V, (b) a unit impulse input of 1 V?

6 A system has a transfer function of $2/(s + 1)$. What will be its output as a function of time when subject to (a) a step input of 3 V, (b) an impulse input of 3 V?

7 A system has a transfer function of $1/(s + 2)$. What will be its output as a function of time when subject to (a) a step input of 4 V, (b) a ramp input unit impulse of 1 V/s?

8 Use partial fractions to simplify the following expressions:

(a) $\dfrac{s-6}{(s-1)(s-2)}$, (b) $\dfrac{s+5}{s^2+3s+2}$, (c) $\dfrac{2s-1}{(s+1)^2}$

9 A system has a transfer function of:

$$\frac{8(s+3)(s+8)}{(s+2)(s+4)}$$

What will be the output as a time function when it is subject to a unit step input? Hint: use partial fractions.

10 A system has a transfer function of:

$$G(s) = \frac{8(s+1)}{(s+2)^2}$$

What will be the output from the system when it is subject to a unit impulse input? Hint: use partial fractions.

11 What will be the state of damping of systems having the following transfer functions and subject to a unit step input?

(a) $\dfrac{1}{s^2+2s+1}$, (b) $\dfrac{1}{s^2+7s+12}$, (c) $\dfrac{1}{s^2+s+1}$

12 The input x and output y of a system are described by the differential equation:

$$\frac{dy}{dt}+2y=x$$

Determine how the output will vary with time when there is an input which starts at zero time and then increases at the constant rate of 6 units/s. The initial output is zero.

13 The input x and output y of a system are described by the differential equation:

$$\frac{d^2y}{dt^2}+3\frac{dy}{dt}+2y=x$$

If initially the input and output are zero, what will be the output when there is a unit step input?

14 The input x and output y of a system are described by the differential equation:

$$\frac{d^2y}{dt^2}+4\frac{dy}{dt}+3y=x$$

If initially the input and output are zero, what will be the output when there is a unit impulse input?

15 A control system has a forward path transfer function of $2/(s + 2)$ and a negative feedback loop with transfer function 4. What will be the response of the system to a unit step input?

16 A system has a transfer function of $100/(s^2 + s + 100)$. What will be its natural frequency ω_n and its damping ratio ζ?

17 A system has a transfer function of $10/(s^2 + 4s + 9)$. Is the system under-damped, critically damped or over-damped?

18 A system has a transfer function of $3/(s^2 + 6s + 9)$. Is the system under-damped, critically damped or over-damped?

19 A system has a forward path transfer function of $10/(s + 3)$ and a negative feedback loop with transfer function 5. What is the time constant of the resulting first-order system?

20 Determine the series and parallel models in the s-domain for (a) an inductance of 10 mH when $i(0) = 0.2$ A, (b) a capacitance of 2 μF when $v(0) = 5$ V.

21 Determine the impedance in the s-domain of a resistance of 10 Ω in (a) series, (b) parallel with a 2 mH inductance.

22 Determine how the current varies with time when a charged capacitor, with a potential difference of v_0, is allowed to discharge through a resistance R.

23 Determine how the current varies with time when a step voltage $Vu(t)$ is applied to a circuit consisting of a resistance R in series with an inductance L, there being no initial current in the circuit.

24 Determine how the current varies with time when a 1 V impulse is applied at time $t = 0$ to a circuit consisting of a resistance R in series with a capacitance C, there being no initial potential difference across the capacitor.

7 Sequences and series

Summary

This chapter introduces the idea of sequences, such concepts proving particularly relevant in considerations of digital signals which can be thought of as sequences of pulses. The main aspect of the chapter is, however, series and the use of the Fourier series to represent non-sinusoidal signals.

Objectives

By the end of this chapter, the reader should be able to:

- understand what is meant by a sequence and uses the idea to describe digital signals;
- recognise arithmetic and geometric series;
- recognise that some series can converge to a limit, determining the sums of such series;
- recognise the binomial series and uses it in engineering problems;
- represents waveforms by Fourier series and applies the series in the analysis of electrical circuit problems involving non-sinusoidal signals.

7.1 Sequences and series

This section considers what is meant by sets and sequences, considering some commonly encountered forms and their relevance to engineering.

7.1.1 Sequences

Consider the numbers, 1, 3, 5, 7, 9. Such a set of numbers is termed a *sequence* because the numbers are stated in a definite order, 1 followed by 3 followed by 5, etc. Another sequence might be $1, \frac{1}{2}, \frac{1}{4}, \frac{1}{8}, \frac{1}{16}$. These sequences have a finite number of terms but often we can meet ones involving an infinite number of terms, e.g. 2, 4, 6, 8, 10, 12, ..., etc.

The term sequence is used for a set of quantities stated in a definite order.

In general we can write a sequence as:

Key point

A sequence is a set of quantities stated in a definite order.

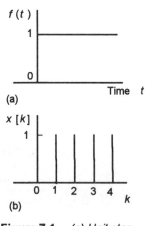

Figure 7.1 *(a) Unit step,*

(b) unit step sequence

Key points

An arithmetic sequence has each term formed from the previous term by simply adding on a constant value.

$x[k] = a + (k - 1)d$

A geometric sequence has each term formed from the previous term by multiplying it by a constant factor, e.g. 3, 6, 12, 24, ...

$x[k] = ar^{k-1}$

---

first value of variable, second value of variable, third value of variable, ..., etc.

or, if x is the variable:

$x[1]$, $x[2]$, $x[3]$, ..., etc.

This is usually more compactly written as $x[k]$, where $k = 1, 2, 3$, ..., etc. Such a form of notation is commonly encountered in signal processing when perhaps an analogue signal is sampled at a number of sequential points and the resulting sequence of digital signal values processed. For example, if an analogue unit step signal is sampled the sampled data output might be expressed as $x[k] = 0$ for $k < 0$, $x[k] = 1$ for $k \geq 0$ with $k = 0, 1, 2, 3, 4$, etc. Figure 7.1 shows graphs of the unit step input and the sampled output.

Sometimes it is possible to describe a sequence by giving a rule for the kth term, common forms being the arithmetic and geometric sequences.

- **Arithmetic sequence**

  An arithmetic sequence has each term formed from the previous term by simply adding on a constant value. If a is the first term and d the common difference between successive terms, the terms are:

  a, $(a + d)$, $(a + 2d)$, $(a + 3d)$, ..., etc.　　　　[1]

  The kth term is $a + (k - 1)d$, with $k = 1, 2, 3, 4$, ..., etc. (note that if k has the values 0, 1, 2, etc. the kth term is $a + kd$). Thus for such a sequence we can write:

  $x[k] = a + (k - 1)d$　　　　[2]

- **Geometric sequence**

  A geometric sequence has each term formed from the previous term by multiplying it by a constant factor, e.g. 3, 6, 12, 24, ... If a is the first term and r the common ratio between successive terms, the terms are:

  a, ar, ar^2, $ar^3 + ...$, etc.　　　　[3]

  The kth term is ar^{k-1}, with $k = 1, 2, 3, 4$, ..., etc. Thus for such a sequence we can write:

  $x[k] = ar^{k-1}$　　　　[4]

- **Harmonic sequence**

  The sequence $1, \frac{1}{2}, \frac{1}{3}, \frac{1}{4}, ...$ is termed the *harmonic sequence* and defined for $k = 1, 2, 3$, etc. by:

$$x[k] = \frac{1}{k} \qquad\qquad [5]$$

Sequences can be generated by other rules. For example, the sequence 1, 2, 5, 10, 17, ... is generated by $x[k] = 1 + (k - 1)^2$, where $k = 1, 2, 3, ...$. This sequence is neither an arithmetic nor a geometric sequence.

Example

Write down the first five terms of the sequence $x[k]$ defined by $x[k] = \frac{1}{2}k^2 + k$ when $k \geq 0$.

When $k = 0$ we have $0 + 0$, when $k = 1$ we have $0.5 + 1$, when $k = 2$ we have $2 + 2$, and so on. The sequence is thus 0, 1.5, 4, 7.5, 12.

7.1.2 Series

A *series* is formed by adding the terms of a sequence. Thus $1 + 3 + 5 + 7 + 9 + ...$, etc. is a series.

A series is the sum of the terms of a sequence.

The sum of n terms of a series is written using *sigma notation* as:

$$S_n = \sum_{k=1}^{n} x[k] \qquad\qquad [6]$$

The first and the last values of k are shown below and above the sigma. For example, the series $1 + 3 + 5 + 7 + 9$ would have the sum, over the five terms, written as:

$$S_5 = \sum_{k=1}^{5} (2k - 1)$$

Common series are:

- *Arithmetic series*

  An arithmetic series has each term formed from the previous term by simply adding on a constant value. Such a series can be written in the general form as:

  $$a + \{a + d\} + \{a + 2d\} + \{a + 3d\} + ... + \{a + (n - 1)d\} \qquad [7]$$

  The sum to k terms is:

  $$S_k = \{a\} + \{(a + d)\} + \{(a + 2d)\} + \{(a + 3d)\} + ... \\ + \{a + (n - 1)d\}$$

  If we write this back to front then:

$$S_k = \{a + (n-1)d\} + \{a + (n-2)d\} + \{a + (n-3)d\} + \ldots \{a\}$$

Adding these two equations gives first term plus first term, second term plus second term, etc. and we obtain:

$$2S_k = \{2a + (n-1)d\} + \{2a + (n-1)d\} + \{2a + (n-1)d\} + \ldots$$

for k terms. Thus $2S_k = n\{2a + (n-1)d\}$ and so:

$$S_k = \tfrac{1}{2}n\{2a + (n-1)d\} \tag{8}$$

- *Geometric series*

  A geometric series has each term formed from the previous term by multiplying it by a constant factor. Such a series can be written in the general form as:

  $$a + ar + ar^2 + ar^3 + \ldots + ar^{n-1} \tag{9}$$

  The sum to the kth terms is:

  $$S_k = a + ar + ar^2 + ar^3 + \ldots + ar^{n-1}$$

  Multiplying by r gives:

  $$rS_k = ar + ar^2 + ar^3 + ar^4 + \ldots + ar^n$$

  Hence $S_k - rS_k = a - ar^n$, and so, provided $k \neq 1$:

  $$S_k = \frac{a(1 - r^n)}{1 - r} \tag{10}$$

Example

Determine the sum of the arithmetic series $1 + 5 + 9 + \ldots$ if it contains 10 terms.

Such a series has a first term a of 1 and a common difference d of 4. Thus, using equation [8]:

$$S_k = \tfrac{1}{2}\{2a + (k-1)d\} = \tfrac{1}{2} \times 10\{2 + 9 \times 4\} = 190$$

Example

Determine the sum of the geometric series $4 + 6 + 9 + \ldots$ if it contains 10 terms.

Such a series has a first term of 4 and a common ratio of 3/2. Thus, using equation [10]:

$$S_k = \frac{a(1 - r^k)}{1 - r} = \frac{4(1 - 1.5^{10})}{1 - 1.5} = 453.3$$

Convergent and divergent series

So far we have considered the sums of series with a finite number of terms. What about the sum when we have a series with an infinite number of terms?

> *A series in which the sum of the series tends to a definite value as the number of terms tends to infinity is called a convergent series.*

Consider an *arithmetic series* $a + (a + d) + (a + 2d) + ...$ for an infinite number of terms. For k terms we have the sum (equation [8]) of:

$$S_k = \tfrac{1}{2}n\{2a + (n - 1)d\}$$

As k tends to infinity then n tends to infinity and so the sum tends to infinity. The sum of an infinite arithmetic series is infinity. The series is said to be *divergent*.

Consider a *geometric series* $a + ar + ar^2 + ...$ for an infinite number of terms. For k terms we have the sum (equation [10]) of:

$$S_k = \frac{a(1 - r^n)}{1 - r} = \frac{a}{1 - r} - \frac{ar^n}{1 - r} \qquad [11]$$

Suppose we have $-1 < r < 1$, as n tends to infinity then r^n tends to 0. Thus the second term converges to zero and we are left with just the first term. Thus such a series converges to the sum:

$$S_\infty = \frac{a}{1 - r} \text{ for } -1 < r < 1 \qquad [12]$$

Thus the geometric series $x[k] = 3^{1/2}$ converges to the sum 6. However, if we had the geometric series $x[k] = 3^2$ then the sum is given by equation [15] as $-3 + 3 \times 2^n$ and thus as n tends to infinity the sum tends to infinity. For $|r| \geq 1$ the geometric series does not converge.

There are a number of ways that are used to determine whether a series will converge:

- *Comparison test*

  A series of positive terms is convergent if its terms are less than the corresponding terms of a positive series which is known to converge. Similarly, the series is divergent if its terms are greater than the corresponding terms of a series which is known to be divergent. As an example, consider:

  $$1 + \frac{1}{2^2} + \frac{1}{3^3} + \frac{1}{4^4} + ...$$

Suppose we know that the series:

$$1 + \frac{1}{2^2} + \frac{1}{2^3} + \frac{1}{2^4} + \ldots$$

converges (it is a geometric series with $r = \frac{1}{2}$), then if, after the first two terms, we compare terms we find that every term in our convergent series is greater than the one we are considering. Thus the series must also converge.

Key point

The size of a real number x is called its modulus and denoted by $|x|$.

- ## D'Alembert's ratio test

An infinite series is convergent if, as k tends to infinity, the ratio of each term u_{n+1} to the preceding term u_n is numerically less than 1 and divergent if greater than 1, i.e.

$$\lim_{n \to \infty} \left| \frac{u_{n+1}}{u_n} \right| < 1, \text{ the series converges,}$$

$$\lim_{n \to \infty} \left| \frac{u_{n+1}}{u_n} \right| > 1, \text{ the series diverges,}$$

$$\lim_{n \to \infty} \left| \frac{u_{n+1}}{u_n} \right| = 1, \text{ the series may converge or diverge.}$$

Consider the series:

$$1 - \frac{1}{2!} + \frac{1}{3!} - \frac{1}{4!} + \ldots$$

Key point

The symbol ! appearing after a number means that it is multiplied by all the integers between it and 1, e.g. $5! = 5 \times 4 \times 3 \times 2 \times 1$.

The nth term u_n is $|1/n!|$ and the $(n + 1)$th term u_{n+1} is $|1/(n + 1)!|$. Therefore:

$$\frac{u_{n+1}}{u_n} = \left| \frac{\frac{1}{(n+1)!}}{\frac{1}{n!}} \right| = \left| \frac{1}{n+1} \right|$$

As n tends to infinity then:

$$\lim_{n \to \infty} \left| \frac{u_{n+1}}{u_n} \right| < 1$$

and so the series converges.

Example

Find the sum to infinity of the series $4 + 2 + 1 + \frac{1}{2} + \ldots$.

This is a geometric series with $a = 4$ and $r = \frac{1}{2}$. Using equation [12]:

$$S_\infty = \frac{a}{1-r} = \frac{4}{1-\frac{1}{2}} = 8$$

Example

Determine, using the comparison test, whether the series $x[k] = 1/n^n$, i.e. $1 + 1/2^2 + 1/3^3 + 1/4^4 + ...$, is convergent.

If we exclude the first two terms we can compare it with the geometric series $1/2^3 + 1/2^4 + 1/2^5 + ...$ which is known to be convergent. Each term in this series being tested is smaller than the comparable term in the comparison series. Thus it must be convergent.

Example

Determine, using d'Alembert's ratio test, whether the series $1 + x + x^2/2! + x^3/3! + ...$ is convergent.

Using d'Alembert's ratio test, since $u_n = x^{n-1}/(n-1)!$ and $u_{n+1} = x^n/n!$:

$$\frac{u_{n+1}}{u_n} = \frac{\dfrac{x^{n-1}}{(n-1)!}}{\dfrac{x^n}{n!}} = \frac{x}{n}$$

In the limit as n tends to infinity then the ratio tends to 0. Thus the series is convergent.

Power series

A series of the type:

$$a_0 + a_1x + a_2x^2 + a_3x^3 + ... + a_nx^n + ...$$

is known as a *power series*. If we apply d'Alembert's ratio test then the series will be convergent when:

$$\lim_{n\to\infty} \left| \frac{a_{n+1}x^{n+1}}{a_nx^n} \right| < 1$$

This can be written as:

$$|x| \lim_{n\to\infty} \left| \frac{a_{n+1}}{a_n} \right| < 1$$

or:

$$|x| < \lim_{n\to\infty} \left| \frac{a_{n+1}}{a_n} \right| \qquad [13]$$

Thus there are conditions attached to the value of x if the series is to converge. Examples are given later in this chapter.

Example

For what values of x is the series $x[k] = x^n/n$ convergent?

Here $a_n = 1/n$ and $a_{n+1} = 1/(n + 1)$. Thus $|a_{n+1}/a_n| = (n + 1)/n = 1 + 1/n$ and so in the limit we have the value of 1 for the limit. Thus the condition for convergence is that $|x| < 1$ or $-1 < x < +1$.

Binomial series

For $(1 + x)^2$ we can readily show that it can be written as $1 + 2x + x^2$. If we multiply this by $(1 + x)$ we obtain $1 + 3x + 3x^2 + x^3$. Multiplying by repeated factors of $(1 + x)$ enables expansions of higher powers of $(1 + x)$ to be generated. This is, however, rather cumbersome if, say, we wanted the expansion of $(1 + x)^{10}$. There is, however, a pattern in the results:

$$(1 + x)^1 = \qquad\quad 1 + 1x$$
$$(1 + x)^2 = \quad\ 1 + 2x + 1x^2$$
$$(1 + x)^3 = \ 1 + 3x + 3x^2 + 1x^3$$
$$(1 + x)^4 = 1 + 4x + 6x^2 + 4x^3 + 1x^4$$

If we just write the coefficients the pattern is more readily discerned:

$$1 \quad 1$$
$$1 \quad 2 \quad 1$$
$$1 \quad 3 \quad 3 \quad 1$$
$$1 \quad 4 \quad 6 \quad 4 \quad 1$$

Every coefficient is obtained by adding the two either side of it in the row above. Thus, for example, we have:

The above pattern is known as *Pascal's triangle*.

However, we can show that the above pattern can be given by:

$$(1 + x)^n = 1 + nx + \frac{n(n-1)}{2!}x^2 + \frac{n(n-1)(n-2)}{3!}x^3 + \qquad [14]$$

This is known as the *Binomial theorem*. The theorem can be used for both positive and negative values of n and fractional values. With n a positive number the series will eventually terminate. With n a negative number, the series does not terminate. The series converges if we have $-1 < x < 1$.

Example

Expand by the binomial theorem $(1 + x)^6$.

$$(1+x)^6 = 1 + 6x + \frac{6 \times 5}{2!}x^2 + \frac{6 \times 5 \times 4}{3!}x^3 + \frac{6 \times 5 \times 4 \times 3}{4!}x$$

$$+ \frac{6 \times 5 \times 4 \times 3 \times 2}{5!}x^5 + \frac{6 \times 5 \times 4 \times 3 \times 2 \times 1}{6!}x^6$$

$$= 1 + 6x + 15x^2 + 20x^3 + 15x^4 + 6x^5 + x^6$$

Example

Write the first four terms in the expansion of $(1 + x)^{1/2}$.

$$(1+x)^{1/2} = 1 + \tfrac{1}{2}x + \frac{\tfrac{1}{2}\left(-\tfrac{1}{2}\right)}{2!}x^2 + \frac{\tfrac{1}{2}\left(-\tfrac{1}{2}\right)\left(-\tfrac{3}{2}\right)}{3!}x^3 + \dots$$

$$= 1 + \tfrac{1}{2}x - \tfrac{1}{8}x^2 + \tfrac{1}{16}x^3 + \dots$$

Maths in action

Making approximations

A common use of the Binomial theorem in engineering is for making approximations. For example, we might want to determine the change in the second moment of area of a rectangle which was given by $bL^3/12$ if b is increased by 3% and L reduced by 2%. The new second moment of area is:

$$I = \tfrac{1}{12}(1 + 0.03)b\left[(1 - 0.02)L\right]^3$$

$$= \frac{bL^3}{12}(1 + 0.03)(1 - 0.02)^3$$

Using the Binomial theorem for the cubed term and neglecting, since they will be very small, all terms involving powers of 0.02:

$$I = \frac{bL^3}{12}(1 + 0.03)(1 - 0.06 + ..) = \frac{bL^3}{12}(1 + 0.03 - 0.06 + ..)$$

Hence, $I = 0.97bL^3/12$ and so the percentage change is a reduction by approximately 3%.

Useful power series

Table 7.1 gives some commonly met functions and their series expansions.

Table 7.1 *Power series*

Function	Series	Validity
$\sin x$	$x - \dfrac{x^3}{3!} + \dfrac{x^5}{5!} - \dfrac{x^7}{7!} + \ldots$	For all x
$\cos x$	$x - \dfrac{x^2}{2!} + \dfrac{x^4}{4!} - \dfrac{x^6}{6!} + \ldots$	For all x
$\tan x$	$x + \dfrac{x^3}{3!} + \dfrac{2x^5}{15} + \dfrac{17x^7}{315} + \ldots$	$-\pi/2 < x < \pi/2$
e^x	$1 + x + \dfrac{x^2}{2!} + \dfrac{x^3}{3!} + \dfrac{x^4}{4!} + \ldots$	For all x
$\sinh x$	$x + \dfrac{x^3}{3!} + \dfrac{x^5}{5!} + \dfrac{x^7}{7!} + \ldots$	For all x
$\cosh x$	$1 + \dfrac{x^2}{2!} + \dfrac{x^4}{4!} + \dfrac{x^6}{6!} + \ldots$	For all x
$\ln(1 + x)$	$x - \dfrac{x^2}{2!} + \dfrac{x^3}{3!} - \dfrac{x^4}{4!} + \ldots$	$-1 < x < 1$

Example

Using series given in Table 7.1, determine the series expansion of the function $e^x \sin x$.

Table 7.1 gives:

$$\sin x = x - \frac{x^3}{3!} + \frac{x^5}{5!} + \ldots, \text{valid for all values of } x$$

$$e^x = 1 + \frac{x}{1!} + \frac{x^2}{2!} + \frac{x^3}{3!} + \ldots, \text{valid for all values of } x$$

We can multiply these two series to give

$$e^x \sin x = \left(1 + \frac{x}{1!} + \frac{x^2}{2!} + \frac{x^3}{3!} + \ldots\right)\left(x - \frac{x^3}{3!} + \frac{x^5}{5!} + \ldots\right)$$

$$= x + x^2 + \left(\frac{1}{2} - \frac{1}{6}\right)x^3 + \left(\frac{1}{6} - \frac{1}{6}\right)x^4$$
$$+ \left(\frac{1}{120} + \frac{1}{24} - \frac{1}{12}\right)x^5 + \ldots$$

$$= x + x^2 + \frac{1}{3}x^3 - \frac{1}{30}x^5 + \ldots$$

Example

Using Table 6.1, determine the series, as far as the x^3 term, for the function $y = e^{4x}$.

Table 6.1 gives:

$$e^x = 1 + \frac{x}{1!} + \frac{x^2}{2!} + \frac{x^3}{3!} + \dots$$

If we substitute $4x$ for x then we obtain:

$$e^{4x} = 1 + \frac{4x}{1!} + \frac{16x^2}{2!} + \frac{64x^3}{3!} + \dots = 1 + 4x + 8x^2 + \frac{32}{3}x^3 + \dots$$

Problems 7.1

1 A sinusoidal signal $f(t) = \sin t$ is sampled every quarter period starting when $t = 0$. State the sequence of sampled values.

2 Write down the first five terms of the sequence $x[k]$ defined, for $k \geq 0$, by (a) $x[k] = k$, (b) $x[k] = e^{-k}$.

3 State the fifth term of (a) the arithmetic sequence given by 4, 7, 10, ..., (b) the geometric sequence given by 12, 6, 3, ...

4 Write an equation for the kth term, where $k = 1, 2, 3, ...,$ for the following sequences (a) 1, −1, 1, −1, ..., (b) 5, 10, 15, 20, ..., (c) 2, 1, 5, 1, 0.5, ...

5 Write down the first five terms of the sequence $x[k]$ defined, for $k \geq 0$, by (a) $x[k] = k^2$, (b) $x[k] = e^k$, (c) $x[k] = \frac{1}{2}k^2 + 2k$.

6 State the fifth term of the arithmetic progression given by 5, 7, 9, ...

7 State the fifth term of the geometric progression given by 8, 4, 2, ...

8 Write an equation for the kth term for the following sequences: (a) $\frac{1}{4}$, $\frac{1}{16}$, $\frac{1}{64}$, ..., (b) −2, +2, −2, ..., (c) 3.1, 3.01, 3.001, ...

9 State the first three terms of the sequences given by:

(a) $(0.1)^k$, (b) $5 + (0.1)^k$, (c) $(-1)^k$

10 Determine the sums of the following series if each contains 12 terms:

(a) $2 + 5 + 8 + ...$, (b) $5 + \frac{5}{2} + \frac{5}{4} + ..$,
(c) $4 + 3.6 + 3.24 + ...$

11 Determine the sums of the following arithmetic or geometric series if each contains 10 terms:

(a) $3 + 2.5 + 2.0 + ...$, (b) $12 + 6 + 3 + ...$,
(c) $1 + 2 + 4 + 8 + ...$

12 Find the sum to infinity of the series:

(a) $6 + 3 + 1.5 + ...$, (b) $4 + 3 + 2.25 + ...$,
(c) $12 + 3 + 0.75 + ...$

13 Using the comparison test, determine whether the following series are convergent or divergent:

(a) $x[k] = 1/3^n$ (compare with $1/2^n$), (b) $x[k] = 1.5^n$ (compare with 1^n)

14 Using d'Alembert's ratio test, determine whether the following series are convergent or divergent:

(a) $x - \dfrac{x^2}{2} + \dfrac{x^3}{3} + ... + \dfrac{(-1)^{n-1}x^n}{n} + ...$

(b) $3 + \dfrac{3^2}{2} + \dfrac{3^3}{3} + ... + \dfrac{3^n}{n} + ...$

15 Determine which of the following series is convergent and which divergent:

(a) $-1 + 1 - 1 + ... (-1)^n + ...$,

(b) $1\,e^{-1} + 2\,e^{-2} + 3\,e^{-3} + ... n\,e^{-n} + ...$,

(c) $\dfrac{2 \times 1 + 1}{2} + \dfrac{2 \times 2 + 1}{2^2} + \dfrac{2 \times 3 + 1}{2^3} + ... + \dfrac{2n + 1}{2^n} + ...$,

(d) $\displaystyle\sum_{k=0}^{\infty} \dfrac{10^n}{n!}$, (e) $\displaystyle\sum_{k=1}^{\infty} \dfrac{1}{n^2 + 2^2}$, (f) $\displaystyle\sum_{k=1}^{\infty} \dfrac{2^{n-1}}{10 + (n-1)}$

16 Expand by the binomial theorem:

(a) $(1 + x)^4$, (b) $(1 + x)^{3/2}$, (c) $(1 - x)^{-5/2}$,

(d) $(1 + 0.25)^{-1}$ for four terms, (e) $(4 + x)^{1/2}$ for four terms.

17 Use the binomial theorem to write the first four terms of:

(a) $(1 + x)^{12}$, (b) $(1 - 2x)^{-2}$, (c) $(3 - 2x)^{2/5}$, (d) $\dfrac{1}{1-x}$,

(e) $(1 + 3x)^{-1/2}$, (f) $\dfrac{1}{(1 - x^3)^2}$

18 By using the binomial theorem, determine the cube root of 1.04 to four decimal places. Hint: write 1.04 as 1 + 0.04.

19 The transverse deflection δ of a column of length L when subject to a vertical load F and a horizontal load H at the top is given by:

$$\delta = \frac{HL}{F}\left(\frac{\tan aL}{aL} - 1\right)$$

where $a^2 = F/EI$. Show that as F tends to a zero value that δ tends to $HL^3/3EI$.

20 Determine the series expansion for cosh x using the relationship cosh $x = \frac{1}{2}(e^x + e^{-x})$.

21 Determine the series expansion for tan x using tan $x =$ sin x/cos x.

22 Using Table 7.1, determine the series for the following functions:

(a) $y = e^{2x}$, (b) $y = e^x \cos x$, (c) $y = (1 + x)^{-1/2}$,

(d) $y = e^x \ln(1 + x)$, (e) $y = \sec x$, (f) $y = \cos^2 x$

23 Show that, if x is small:

$$\frac{1}{1+x} - (1-2x)^{1/2} \approx \tfrac{3}{2}x^2$$

24 For a continuous belt passing round two wheels, diameters d and D, with centres a distance x apart, the length L of belt required, if there is no sag, is:

$$L = 2x \cos a + \tfrac{1}{2}\pi(D+d)+(D-d)a$$

where sin $a = (D - d)/2x$. Show that:

$$L \approx 2x + \tfrac{1}{2}\pi(D+d) + \frac{(D-d)^2}{4x}$$

25 The displacement x of the slider of a reciprocating mechanism depends on the crankshaft angle θ, being related by

$$x = r\cos\theta + L\sqrt{1 - \frac{r^2}{L^2}\sin^2\theta}$$

where r is the radius of the crankshaft and L the length of the connecting rod. Show, when r/L is considerably smaller than 1, that:

$$x \approx r\cos\theta + L - \frac{r^2}{2L}\sin^2\theta$$

26 Determine the approximate percentage change in the volume of a cylinder if its radius is reduced by 4% and its height increased by 2%.

27 The resonant frequency of an electrical circuit containing capacitance C and inductance L is given by $1/[2\pi\sqrt{(LC)}]$. Determine the approximate percentage change in the frequency if the capacitance is increased by 2% and the inductance decreased by 1%.

7.2 Fourier series

Alternating waveforms in, say, electrical circuits are not always sinusoidal. For example, many voltages which might initially have been sinusoidal have their waveforms 'distorted' by being applied to some non-linear device and thus we need to be able to considered the behaviour of such a waveform with an electrical circuit. In other cases we might have a rectangular waveform rather than a sinusoidal one. This section is a consideration of how we can use a series to describe such waveforms.

7.2.1 Fourier series

In 1822 Jean Baptiste Fourier proposed that any periodic waveform could be made up of a combination of sinusoidal waveforms, i.e.

$$y = A_0 + A_1 \sin(1\omega t + \phi_1) + A_2 \sin(2\omega t + \phi_2) + A_3 \sin(3\omega t + \phi_3) + \ldots$$
[15]

This is termed the *Fourier* series. where A_0 is a non-alternating component, e.g. a d.c. component. The waveform element with the $1\omega t$ frequency is called the *fundamental frequency* or the *first harmonic*, the element with the $2\omega t$ frequency the *second harmonic*, the element with the $3\omega t$ the third harmonic, and so on. A_1, A_2, A_3 are the amplitudes of the components and ϕ_1, ϕ_2, ϕ_3, etc. their phases.

As an illustration, consider the waveform produced by having just sine terms with the fundamental and the third harmonic and $A_3 = A_1/3$, i.e.

$$y = A_1 \sin 1\omega t + \tfrac{1}{3}A_1 \sin 3\omega t$$

Figure 7.2 shows graphs of the two terms and the waveform obtained by adding the two, ordinate by ordinate.

> **Key point**
>
> *Fourier series*:
> Any periodic waveform can be represented by a constant d.c. signal term plus terms involving sinusoidal waveforms of multiples of a basic frequency.
>
> The Fourier series is concisely expressed as:
>
> $$y = A_0 + \sum_{n=1}^{\infty} A_n \sin\left(n\omega t + \phi_n\right)$$

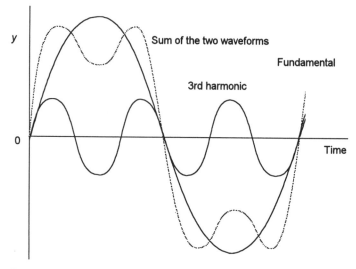

Figure 7.2 *Adding two waveforms*

The result of adding the two waveforms is something that begins to look a bit like a rectangular waveform. The addition of a d.c. term shifts the waveform up or down. If we add a d.c. term of $0.79\,A_1$ then Figure 7.2 becomes transformed to Figure 7.3:

$$y = 0.79A_1 + A_1 \sin 1\omega t + \tfrac{1}{3}A_1 \sin 3\omega t$$

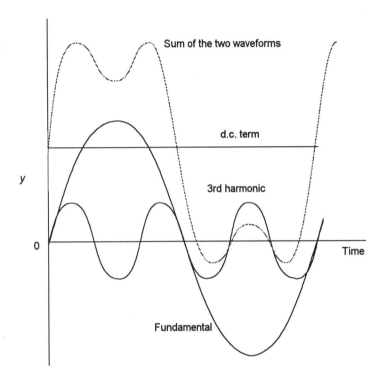

Figure 7.3 *Adding a d.c. term*

A better approximation to a rectangular waveform is given by adding more terms:

$$y = 0.79A_1 + A_1 \sin 1\omega t + \tfrac{1}{3}A_1 \sin 3\omega t + \tfrac{1}{5}A_1 \sin 5\omega t \\ + \tfrac{1}{7}A_1 \sin 7\omega t + \dots$$

We then obtain a rectangular waveform which approximates to a periodic sequence of pulses (Figure 7.4).

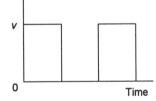

Figure 7.4 *Rectangular waveform*

Alternative way of writing the Fourier series

There is an alternative, simpler, way of writing equation [15]. Since $\sin(A + B) = \sin A \cos B + \cos A \sin B$ we can write:

$$A_1 \sin(1\omega t + \phi_1) = A_1 \sin \phi_1 \cos 1\omega t + A_1 \cos \phi_1 \sin 1\omega t$$

If we represent the non-time varying terms $A_1 \sin \phi_1$ by a constant a_1 and $A_1 \cos \phi_1$ by b_1, then:

$A_1 \sin(1\omega t + \phi_1) = a_1 \cos 1\omega t + b_1 \sin 1\omega t$

Likewise we can write:

$A_2 \sin(2\omega t + \phi_2) = a_2 \cos 2\omega t + b_2 \sin 2\omega t$

$A_3 \sin(3\omega t + \phi_3) = a_3 \cos 3\omega t + b_3 \sin 3\omega t$

and so on. If, for convenience we choose to write $\frac{1}{2}a_0$ for A_0, equation [10] can be written as:

$$y = \tfrac{1}{2}a_0 + a_1 \cos 1\omega t + a_2 \cos 2\omega t + a_3 \cos 3\omega t + \ldots$$
$$+ b_1 \sin 1\omega t + b_2 \sin 2\omega t + b_3 \sin 3\omega t + \ldots \qquad [16]$$

Hence we can write the Fourier series equation as:

$$y = \tfrac{1}{2}a_0 + \sum_{n=1}^{\infty} a_n \cos n\omega t + \sum_{n=1}^{\infty} b_n \sin n\omega t \qquad [17]$$

The a and b terms are called the *Fourier coefficients*.

Since we have $a_n = A_n \sin \phi_n$ and $b_n = A_n \cos \phi_n$ then:

$$\phi_n = \tan^{-1}\left(\frac{a_n}{b_n}\right) \qquad [18]$$

and, since:

$$a_n^2 + b_n^2 = A_n^2 \sin^2\phi_n + A_n^2 \cos^2\phi_n = A_n^2$$

we have:

$$A_n = \sqrt{a_n^2 + b_n^2} \qquad [19]$$

7.2.2 Fourier coefficients

Now consider how we can establish the Fourier coefficients for a waveform. Suppose we have the Fourier series in the form of equation [16]:

$$y = \tfrac{1}{2}a_0 + a_1 \cos \omega t + a_2 \cos 2\omega t + \ldots + a_n \cos n\omega t$$
$$+ b_1 \sin \omega t + b_2 \sin 2\omega t + \ldots + b_n \sin n\omega t$$

If we integrate both sides of the equation over one period T of the fundamental, the integral for each cosine and sine term will be the area under the graph of that expression for one cycle and thus zero. A consequence of this is that the only term which is not zero when we integrate the equation is the integral of the a_0 term. Thus, integrating over one period T gives:

$$\int_0^T y \, dt = \int_0^T \tfrac{1}{2}a_0 \, dt = \tfrac{1}{2}a_0 T$$

> **Key point**
>
> Note: in Figure 7.3 the addition of a d.c. term of 0.79 to the waveform results in an average value of this waveform over one cycle of 0.79. The term $\frac{1}{2}a_0$ in the Fourier series thus represents the average value of the waveform over a cycle.

and so:

$$a_0 = \frac{2}{T} \int_0^T y \, \mathrm{d}t \qquad\qquad [20]$$

We can obtain the a_1 term by multiplying the equation by $\cos \omega t$ and then integrating over one period. Thus the equation becomes:

$$y \cos \omega t = \tfrac{1}{2} a_0 \cos \omega t + a_1 \cos \omega t \cos \omega t + a_2 \cos \omega t \cos 2\omega t$$
$$+ \dots + b_1 \cos \omega t \sin \omega t + b_2 \cos \omega t \sin 2\omega t + \dots$$

$$= \tfrac{1}{2} a_0 \cos \omega t + a_1 \cos^2 \omega t + a_2 \cos \omega t \cos 2\omega t$$
$$+ \dots + b_1 \cos \omega t \sin \omega t + b_2 \cos \omega t \sin 2\omega t + \dots$$

The integration over a period T of all the terms involving $\sin \omega t$ and $\cos \omega t$ will be zero. Thus we are only left with the $\cos^2 \omega t$ term and so, using equation [13]:

$$\int_0^T y \cos \omega t \, \mathrm{d}t = \int_0^T a_1 \cos^2 \omega t \, \mathrm{d}t$$

$$= \tfrac{1}{2} a_1 \int_0^T (1 + \cos 2\omega t) \, \mathrm{d}t$$

$$= \tfrac{1}{2} a_1 \left[t + \frac{\sin 2\omega t}{2\omega} \right]_0^T = \tfrac{1}{2} a_1 T$$

and so we have:

$$a_1 = \frac{2}{T} \int_0^T y \cos \omega t \, \mathrm{d}t \qquad\qquad [21]$$

In general, multiplying the equation by $\cos n\omega t$ gives:

$$a_n = \frac{2}{T} \int_0^T y \cos n\omega t \, \mathrm{d}t \qquad\qquad [22]$$

This equation gives for $n = 0$ the equation given earlier for a_0. This would not have been the case if the first term in the Fourier series had been written as a_0 instead of $a_0/2$.

In a similar way, multiplying the equation by $\sin \omega t$ and integrating over a period enables us to obtain the b coefficients. Thus:

$$y \sin \omega t = \tfrac{1}{2} a_0 \sin \omega t + a_1 \sin \omega t \cos \omega t + a_2 \sin \omega t \cos 2\omega t$$
$$+ \dots + b_1 \sin \omega t \sin \omega t + b_2 \sin \omega t \sin 2\omega t + \dots$$

$$= \tfrac{1}{2} a_0 \sin \omega t + a_1 \sin \omega t \cos \omega t + a_2 \sin \omega t \cos 2\omega t$$
$$+ \dots + b_1 \sin^2 \omega t + b_2 \sin \omega t \sin 2\omega t + \dots$$

The integration over a period T of all the terms involving $\sin \omega t$ and $\cos \omega t$ will be zero and so:

$$\int_0^T y \sin \omega t \, \mathrm{d}t = \int_0^T b_1 \sin^2 \omega t \, \mathrm{d}t$$

$$= \tfrac{1}{2} \int_0^T b_1 (1 - \cos 2\omega t)\, dt$$

$$= \tfrac{1}{2} b_1 \left[t - \frac{\sin 2\omega t}{2\omega} \right]_0^T = \tfrac{1}{2} b_1 T$$

Hence:

$$b_1 = \frac{2}{T} \int_0^T y \sin \omega t\, dt \qquad\qquad [23]$$

In general, multiplying the equation by $\sin n\omega t$ and integrating gives:

$$b_n = \frac{2}{T} \int_0^T y \sin n\omega t\, dt \qquad\qquad [24]$$

The following illustrates how the Fourier series can be established for a number of common waveforms.

Rectangular waveform

Consider the *rectangular waveform* shown in Figure 7.4. It can be described as:

$y = A$ for $0 \leq t < T/2$, and $y = 0$ for $T/2 \leq t < T$, period T

Now consider the determination of the coefficients. Equation [20] for a_0:

$$a_0 = \frac{2}{T} \int_0^T y\, dt$$

has an integral which is the area under the graph of y against t for the period T. Since this area is $AT/2$, we have $a_0 = A$. To obtain a_n we use equation [22]:

$$a_n = \frac{2}{T} \int_0^T y \cos n\omega t\, dt$$

Since y has the value A up to $T/2$ and is zero thereafter, we can write the above equation in two parts as:

$$a_n = \frac{2}{T} \int_0^{T/2} A \cos n\omega t\, dt + \frac{2}{T} \int_{T/2}^T 0 \cos n\omega t\, dt$$

The value of the second integral is 0 and so:

$$a_n = \frac{2}{T} \left[\frac{A}{n\omega} \sin n\omega t \right]_0^{T/2}$$

Since $\omega = 2\pi/T$ then the sine term is $\sin 2n\pi t/T$. Thus with $t = T/2$ we have $\sin n\pi$ which is zero and since $\sin 0 = 0$, we have $a_n = 0$.

For the b_n terms we use equation [24]:

$$b_n = \frac{2}{T} \int_0^T y \sin n\omega t\, dt$$

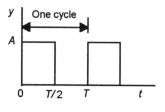

Figure 7.4 *Rectangular waveform*

Since we have $y = A$ from 0 to $T/2$ and then $y = 0$ for the remainder of the period, this equation can be written in two parts as:

$$b_n = \frac{2}{T} \int_0^{T/2} A \sin n\omega t\, dt + \frac{2}{T} \int_{T/2}^T 0 \sin n\omega t\, dt$$

The value of the second integral is 0 and so:

$$b_n = \frac{2}{T}\left[-\frac{A}{n\omega} \cos n\omega t \right]_0^{T/2} = \frac{A}{\pi n}(1 - \cos n\pi)$$

Hence:

$$b_1 = \frac{A}{\pi}(1 - \cos \pi) = \frac{2A}{\pi}, \quad b_2 = \frac{A}{2\pi}(1 - \cos 2\pi) = 0$$

$$b_3 = \frac{A}{3\pi}(1 - \cos 3\pi) = \frac{2A}{3\pi}, \quad \text{etc.}$$

Thus the Fourier series for the rectangular waveform can be written as:

$$y = A\left(\frac{1}{2} + \frac{2}{\pi} \sin \omega t + \frac{2}{3\pi} \sin 3\omega t + \dots \right) \qquad [25]$$

Note that only odd harmonics are present.

Sawtooth waveform

Consider the *sawtooth waveform* shown in Figure 7.5. It can be described by:

$$y = At/T \text{ for } 0 \le t < T, \text{ period } T$$

To determine a_0 we use equation [20]:

$$a_0 = \frac{2}{T} \int_0^T y\, dt$$

The integral is the area under the graph of y against t between 0 and time T. This is $AT/2$ and so $a_0 = A$. To obtain a_n we use equation [22]:

$$a_n = \frac{2}{T} \int_0^T y \cos n\omega t\, dt$$

Since $\omega = 2\pi/T$ and $y = At/T$ then:

$$a_n = \frac{2}{T} \int_0^T \frac{At}{T} \cos \frac{2\pi n t}{T}\, dt$$

Using integration by parts gives:

$$a_n = \frac{2}{T}\left[\frac{At}{2\pi n} \sin \frac{2\pi n t}{T} + \frac{At}{4\pi^2 n^2} \cos \frac{2\pi n t}{T} \right]_0^T$$

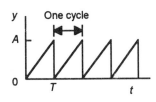

Figure 7.5   *Sawtooth waveform*

$$= \frac{2A}{T}\left[\frac{T}{4\pi^2 n^2} - \frac{T}{4\pi^2 n^2}\right] = 0$$

The values of a_n are zero for all values other than a_0. The values of b_n can be found by using equation [24]:

$$b_n = \frac{2}{T}\int_0^T y \sin n\omega t \, dt = \frac{2}{T}\int_0^T \frac{At}{T} \sin \frac{2\pi n t}{T} \, dt$$

Integration by parts gives:

$$b_n = \frac{2}{T}\left[-\frac{At}{2\pi n}\cos\frac{2\pi n t}{T} + \frac{At}{4\pi^2 n^2}\sin\frac{2\pi n t}{T}\right]_0^T$$

$$= \frac{2A}{T}\left[-\frac{T}{2\pi n}\right]_0^T = -\frac{A}{\pi n}$$

The Fourier series for the sawtooth waveform is thus:

$$y = \frac{A}{2} - \frac{A}{\pi}\sin\omega t - \frac{A}{2\pi}\sin 2\omega t - \frac{A}{3\pi}\sin 3\omega t - \dots \qquad [26]$$

We can write this as:

$$y = \frac{A}{2} + \frac{A}{\pi}\cos(\omega t + \frac{\pi}{2}) + \frac{A}{2\pi}\cos(2\omega t + \frac{\pi}{2})$$
$$+ \frac{A}{3\pi}\cos(3\omega t + \frac{\pi}{2}) + \dots$$

Half-wave rectified sinusoid

Consider a half-rectified sinusoidal waveform of period T (Figure 7.6). This can be described by:

$$y = A \sin \omega t = A \sin 2\pi t/T \text{ for } 0 \le t \le T/2, \ y = 0 \text{ for } T/2 \le t < T$$

We can determine a_0 by using equation [20]:

$$a_0 = \frac{2}{T}\int_0^T y \, dt = \frac{2}{T}\left(\int_0^{T/2} A \sin \omega t \, dt + \int_{T/2}^T 0 \, dt\right)$$

$$= -\frac{2A}{T\omega}[\cos \omega t]_0^{T/2} = \frac{2A}{\pi}$$

Equation [22] can be used to determine a_n:

$$a_n = \frac{2}{T}\int_0^T y \cos n\omega t \, dt = \frac{2}{T}\left(\int_0^{T/2} A \sin \omega t \cos n\omega t \, dt + \int_{T/2}^T 0 \, dt\right)$$

Since $2 \sin A \cos B = \sin(A + B) + \sin(A - B)$:

$$a_n = \frac{A}{T}\int_0^{T/2}\left[\sin(1+n)\omega t + \sin(1-n)\omega t\right] dt$$

For $n = 1$ we have:

$$a_1 = \frac{A}{T}\int_0^{T/2} \sin(1+1)\omega t \, dt = -\frac{A}{2T\omega}[\cos \omega t]_0^{T/2} = 0$$

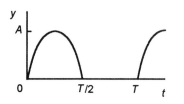

Figure 7.6 *Half-wave rectified sinusoid*

For $n > 1$ we have:

$$a_n = \frac{A}{T}\left[-\frac{1}{(1+n)\omega}\cos(1+n)\omega t - \frac{1}{(1-n)\omega}\cos(1-n)\omega t\right]_0^{T/2}$$

For even values of n we have $\cos(1+n)\pi = -1$ and $\cos(1-n)\pi = -1$ and so:

$$a_{n\,even} = \frac{A}{T}\left(\frac{1}{(1+n)\omega} + \frac{1}{(1+n)\omega} + \frac{1}{(1-n)\omega} + \frac{1}{(1-n)\omega}\right.$$

$$= \frac{A}{\pi}\left(\frac{2}{1+n} + \frac{2}{1-n}\right)$$

$$= \frac{2A}{\pi(1-n^2)}$$

For odd values, other than 1, of n we have $\cos(1+n)\pi = 1$ and $\cos(1-n)\pi = 1$. This gives:

$$a_{n\,odd} = \frac{A}{T}\left(-\frac{1}{(1+n)\omega} + \frac{1}{(1+n)\omega} - \frac{1}{(1-n)\omega} + \frac{1}{(1-n)\omega}\right) = 0$$

The values of b_n can be found using equation [24]:

$$b_n = \frac{2}{T}\int_0^T y\sin n\omega t\, dt = \frac{2}{T}\int_0^{T/2} A\sin\omega t\sin n\omega t\, dt$$

Since $2\sin A\sin B = \cos(A-B) - \cos(A+B)$:

$$b_n = \frac{A}{T}\int_0^{T/2}\left[\cos(1-n)\omega t - \cos(1+n)\omega t\right]dt$$

For $n = 1$ we have:

$$b_n = \frac{A}{T}\int_0^{T/2}\left[1 - \cos(1+n)\omega t\right]dt$$

$$= \frac{A}{t}\left[t - \frac{1}{2\omega}\sin 2\omega t\right]_0^{T/2} = \frac{A}{2}$$

For $n > 1$ we have:

$$b_n = \frac{A}{T}\left[\frac{1}{(1-n)\omega}\sin(1-n)\omega t - \frac{1}{(1+n)\omega}\sin(1+n)\omega t\right]_0^{T/2}$$

Since $\sin(1-n)\pi = 0$ and $\sin(1+n)\pi = 0$, we have $b_n = 0$ for all values of n other than 1.

The Fourier series for the half-wave rectified sinusoid is thus:

$$y = \frac{A}{\pi} - \frac{2A}{3\pi}\cos 2\omega t - \frac{2A}{15\pi}\cos 4\omega t + \ldots + \frac{A}{2}\sin\omega t \qquad [27]$$

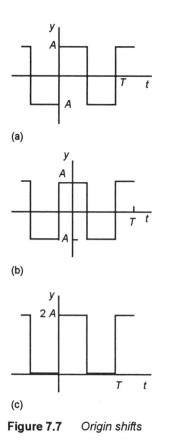

(a)

(b)

(c)

Figure 7.7   *Origin shifts*

Shift of origin

The Fourier series for the rectangular waveform shown in Figure 7.7(a) is:

$$y = \frac{4A}{\pi}\left[\sin \omega t + \tfrac{1}{3} \sin 3\omega t + \tfrac{1}{5} \sin 5\omega t + ...\right]$$ [28]

Now consider the waveform in Figure 33.7(b). This is the waveform in (a) with the time origin shifted to the right by $\pi/2$. If we work out the Fourier series for this waveform we find that it is equation [28] with t replaced by $(t + \pi/2)$.

$$y = \frac{4A}{\pi}\left[\sin\left(\omega t + \frac{\pi}{2}\right) + \tfrac{1}{3} \sin 3\omega\left(t + \frac{\pi}{2}\right) + \tfrac{1}{5} \sin 5\omega\left(t + \frac{\pi}{2}\right) + ...\right]$$

and so:

$$y = \frac{4A}{\pi}\left[\cos \omega t - \tfrac{1}{3} \cos 3\omega t + \tfrac{1}{5} \cos 5\omega t - ...\right]$$ [29]

Thus we have the rule:

Shifting the time origin of a waveform to the right by θ means replacing t by (t + θ) in the Fourier series. Shifting the time origin to the left by h means replacing t by (t – θ).

Now consider the waveform in Figure 33.7(c). This is that in (a) shifted vertically by A, i.e. the waveform in (a) plus A. The Fourier series is then that of (a) plus A:

$$y = A + \frac{4A}{\pi}\left[\sin \omega t + \tfrac{1}{3} \sin 3\omega t + \tfrac{1}{5} \sin 5\omega t + ...\right]$$ [30]

This gives the rule:

Shifting the time axis vertically downwards adds to the Fourier series the amount of the shift, shifting upwards subtracts.

7.2.3 Odd and even symmetry

As will be apparent from the above examples, certain terms are not always present in a Fourier series. Consideration of whether functions have odd or even symmetry about the origin enables us to determine the presence or otherwise of terms.

- *Odd symmetry*

  A function with odd symmetry is defined as having $f(-t) = -f(t)$. This means that the function value for a particular positive value of time is equal in magnitude but of opposite sign to that for the corresponding negative value of that time. Thus $y = f(x) = x^3$ is an odd function since $f(-2) = -8 = -f(2)$. For every point on the waveform for positive times there is a corresponding point on the waveform on a straight line drawn through the origin and equidistant from it (Figure 7.8).

- **Even symmetry**

  A function with even symmetry is defined as having $f(-t) = f(t)$. This means that the function value for a particular positive value of time is identical to that for the corresponding negative value of that time. Thus $y = f(x) = x^2$ is an even function since $f(-2) = 4 = f(2)$. If the y-axis was a plane mirror then the reflection of the positive time values for the waveform would give the negative time values (Figure 7.9).

In determining Fourier coefficients it is necessary to consider the odd or even nature of products of two odd or even functions.

- **Product of two even functions**

  Consider $f(x)$ and $g(x)$ and the product $F(x) = f(x)g(x)$. We can write $F(-x) = f(-x)g(-x)$. Thus if $f(x)$ and $g(x)$ are both even we must have $F(-x) = f(x)g(x)$ and so $F(-x) = F(x)$. The product of two even functions is an even function.

- **Product of two odd functions**

  Consider $f(x)$ and $g(x)$ and the product $F(x) = f(x)g(x)$. We can write $F(-x) = f(-x)g(-x)$. Thus if $f(x)$ and $g(x)$ are both odd we must have $F(-x) = \{-f(x)\}\{-g(x)\}$ and so $F(-x) = F(x)$. The product of two odd functions is an even function.

- **Product of an odd and an even function**

  Consider $f(x)$ and $g(x)$ and the product $F(x) = f(x)g(x)$. We can write $F(-x) = f(-x)g(-x)$. Thus if $f(x)$ is even and $g(x)$ is odd we must have $F(-x) = f(x)\{-g(x)\} = -f(x)g(x)$ and so $F(-x) = -F(x)$. The product of an even and an odd function is an odd function.

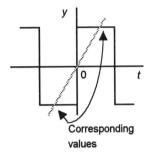

Figure 7.8   *Odd symmetry*

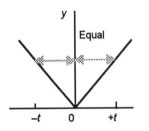

Figure 7.9   *Even symmetry*

Example

Determine whether (a) x^2, (b) cos $2x$ and (c) x^2 cos $2x$ are even or odd functions.

(a) $y = f(x) = x^2$ is an even function since if we consider some particular value of x, say -2, we have $f(-2) = 4 = f(2)$.

(b) $y = f(x) = $ cos $2x$ is an even function since if we consider some particular value of x, say $-\pi/2$ we have $f(-\pi/2) = 0 = f(\pi/2)$.

(c) Since the product of two even functions is even, x^2 cos $2x$ is an even function.

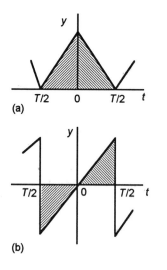

(a)

(b)

Figure 7.10 *(a) Even, (b) odd*

Fourier coefficients for odd/even symmetry

Consider the coefficients for a Fourier series for functions showing odd or even symmetry.

- **a_0 coefficients**

  a_0 is given by equation [20] as:

  $$a_0 = \frac{2}{T} \int_0^T f(t)\, dt = \frac{2}{T} \int_{-T/2}^{T/2} f(t)\, dt$$

  $$= \frac{2}{T} \int_0^{T/2} f(t)\, dt + \frac{2}{T} \int_{-T/2}^0 f(t)\, dt$$

  For a function with even symmetry we have the areas under the waveform on each side of the y-axis equal in both size and sign. Figure 7.10(a) illustrates this. A consequence of this is:

  $$a_0 = 2 \times \frac{2}{T} \int_0^{T/2} f(t)\, dt \qquad [31]$$

  But for an odd function (Figure 7.10(b)) the areas under the waveform on each side of the y-axis are equal in size but opposite in sign. A consequence of this is that there can be no a_0 term:

  $$a_0 = \int_0^{T/2} f(t)\, dt + \int_{-T/2}^0 f(t)\, dt = 0 \qquad [32]$$

  We can look at this issue in another way. The mean value over one cycle of a waveform is $a_0/2$. Thus for an odd function the mean value is 0 because the mean value is 0.

- **a_n coefficients**

  For the a_n coefficients equation [22] gives:

  $$a_n = \frac{2}{T} \int_0^T f(t) \cos n\omega t\, dt = \frac{2}{T} \int_{-T/2}^{T/2} f(t) \cos n\omega t\, dt$$

  Since $\cos n\omega t$ is an even function, if $f(t)$ is even then the product is even. Hence we have, on the basis of the discussion used for a_0:

  $$a_n = 2 \times \frac{2}{T} \int_0^{T/2} f(t) \cos n\omega t\, dt \qquad [33]$$

  If $f(t)$ is odd then the product is odd. Thus on the basis of the discussion used for a_0:

  $$a_n = \frac{2}{T} \int_{-T/2}^{T/2} f(t) \cos n\omega t\, dt = 0 \qquad [34]$$

- **b_n coefficients**

  For the b_n coefficients equation [24] gives:

  $$b_n = \frac{2}{T} \int_0^T f(t) \sin n\omega t\, dt = \frac{2}{T} \int_{-T/2}^{T/2} f(t) \sin n\omega t\, dt$$

Since $\sin n\omega t$ is an odd function, if $f(t)$ is even then the product is odd. Thus, on the basis of the discussion used for a_0:

$$b_n = \frac{2}{T} \int_{-T/2}^{T/2} f(t) \sin n\omega t \, dt = 0 \qquad [35]$$

If $f(t)$ is odd then the product is even. Thus, on the basis of the discussion used for a_0:

$$b_n = \frac{2}{T} \int_{-T/2}^{T/2} f(t) \sin n\omega t \, dt = 2 \times \frac{2}{T} \int_{0}^{T/2} f(t) \sin n\omega t \, dt \qquad [36]$$

To summarise:

If f(t) is an even function then the Fourier series contains an a_0 term and only cosine terms. If f(t) is an odd function then the Fourier series contains no a_0 term and only sine terms.

Key points

If f(t) is an even function then the Fourier series contains an a_0 term and only cosine terms. If f(t) is an odd function then the Fourier series contains no a_0 term and only sine terms.

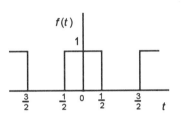

Figure 7.11   *Example*

Example

Determine the Fourier series for the function shown in Figure 7.11.

The function is an even function and so the b coefficients are all zero. The period is 2 and so $\omega = \pi$. Thus, using equation [31]:

$$a_0 = 2 \times \frac{2}{T} \int_{0}^{T/2} f(t) \, dt = 2\left(\int_{0}^{1/2} 1 \, dt + \int_{1/2}^{1} 0 \, dt\right) = 2[t]_{0}^{1/2} = 1$$

Using equation [33]:

$$a_n = 2 \times \frac{2}{T} \int_{0}^{T/2} f(t) \cos n\omega t \, dt$$

$$= 2\left(\int_{0}^{1/2} 1 \cos n\pi t \, dt + \int_{1/2}^{1} 0 \, dt\right)$$

$$= 2\left[\frac{\sin n\pi t}{n\pi}\right]_{0}^{1/2} = \frac{2}{n\pi} \sin \tfrac{1}{2} n\pi$$

Thus the Fourier series is:

$$y = \tfrac{1}{2} + \frac{2}{\pi}\left(\cos \pi t - \tfrac{1}{3} \cos 3\pi t + \tfrac{1}{5} \cos 5\pi t + \ldots\right)$$

Half-wave symmetry

It is often possible by considering the symmetry of successive half-cycle waves within a waveform to recognise whether it will contain odd or even harmonics.

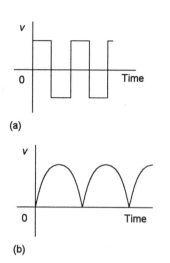

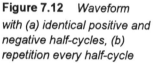

(b)

Figure 7.12 *Waveform with (a) identical positive and negative half-cycles, (b) repetition every half-cycle*

Key point

Any complex waveform which has a negative half-cycle which is just the positive cycle inverted will contain only odd harmonics, such a form of symmetry being termed *half-wave inversion*.

Waveforms which repeat themselves after each half-cycle of the fundamental frequency will have just even harmonics, such a form of symmetry being termed *half-wave repetition*.

- **Half-wave inversion**

  Any complex waveform which has a negative half-cycle which is just the positive cycle inverted will contain only odd harmonics, such a form of symmetry being termed *half-wave inversion*. Thus Figure 7.12(a) shows a waveform which has negative half-cycles which are just the positive half-cycles inverted and so does not contain any even harmonics.

- **Half-wave repetition**

  Waveforms which repeat themselves after each half-cycle of the fundamental frequency will have just even harmonics, such a form of symmetry being termed *half-wave repetition*. Figure 7.12(b) show a waveform which repeats itself after each half-cycle and so has just even harmonics.

We can see why the above statements occur by considering the conditions that are necessary for a Fourier series to give the required symmetry. Thus if we have the series describing the waveform at time t:

$$v \text{ at } t = \tfrac{1}{2}a_0 + a_1 \cos 1\omega t + a_2 \cos 2\omega t + a_3 \cos 3\omega t + ...$$
$$+ b_1 \sin 1\omega t + b_2 \sin 2\omega t + b_3 \sin 3\omega t + ... \qquad [37]$$

To obtain the value of the waveform after half a cycle, i.e. at time $t + \pi$, we put this value of time into equation [37]:

$$v \text{ at } (t + \pi) = \tfrac{1}{2}a_0 + a_1 \cos 1\omega(t + \pi) + a_2 \cos 2\omega(t + \pi)$$
$$+ a_3 \cos 3\omega(t + \pi) + ... + b_1 \sin 1\omega(t + \pi)$$
$$+ b_2 \sin 2\omega(t + \pi) + b_3 \sin 3\omega(t + \pi) + ...$$

$$= \tfrac{1}{2}a_0 - a_1 \cos 1\omega t + a_2 \cos 2\omega t - a_3 \cos 3\omega t + ...$$
$$- b_1 \sin 1\omega t + b_2 \sin 2\omega t - b_3 \sin 3\omega t + ... \qquad [38]$$

If the waveform is to have negative half-cycles which are just the positive half-cycles inverted we must have the waveform after half a cycle, i.e. at time $t + \pi$, which is $-v$ at t. Thus we must have:

$$\tfrac{1}{2}a_0 + a_1 \cos 1\omega t + a_2 \cos 2\omega t + a_3 \cos 3\omega t + ...$$
$$+ b_1 \sin 1\omega t + b_2 \sin 2\omega t + b_3 \sin 3\omega t + ...$$
$$= -(\tfrac{1}{2}a_0 - a_1 \cos 1\omega t + a_2 \cos 2\omega t - a_3 \cos 3\omega t + ...$$
$$- b_1 \sin 1\omega t + b_2 \sin 2\omega t - b_3 \sin 3\omega t + ...) \qquad [39]$$

This can only occur if $a_0 = 0$, $a_2 = 0$, and all even harmonics are 0.

If the waveform is to have waveforms which repeat themselves after half a cycle then we must have the waveform at time $t + \pi$ equal to v at time t. Thus we must have:

$$\tfrac{1}{2}a_0 + a_1 \cos 1\omega t + a_2 \cos 2\omega t + a_3 \cos 3\omega t + ...$$
$$+ b_1 \sin 1\omega t + b_2 \sin 2\omega t + b_3 \sin 3\omega t + ...$$
$$= \tfrac{1}{2}a_0 - a_1 \cos 1\omega t + a_2 \cos 2\omega t - a_3 \cos 3\omega t + ...$$
$$- b_1 \sin 1\omega t + b_2 \sin 2\omega t - b_3 \sin 3\omega t + ... \qquad [40]$$

This can only occur if $a_1 = 0$, $a_3 = 0$ and all odd harmonics are 0.

Key point

Equations [41] and [42] are for the Fourier series in the form:

$y = A_0 + A_1 \sin (\omega t + \phi_1)$
$\qquad + A_2 \sin (2\omega t + \phi_2) + ...$

Note that if we want to refer to a series in the form:

$y = A_0 + A_1 \cos (\omega t + \phi_1)$
$\qquad + A_2 \cos (2\omega t + \phi_2) + ...$

then, in order to take account of $\sin \omega t = \cos (\omega t - \pi/2)$, i.e. the phase difference of $-90°$ between the cosine and sine, equation [42] becomes:

$\phi = \tan^{-1}\left(\dfrac{-b_n}{a_n}\right)$

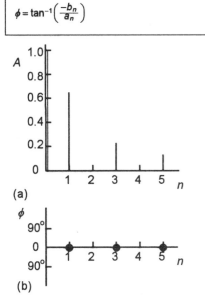

(a)

(b)

Figure 7.13   *Frequency spectrum*

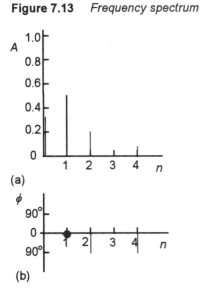

(a)

(b)

Figure 7.14   *Frequency spectrum*

7.2.4 Frequency spectrum

The *frequency spectrum* comprises an *amplitude spectrum*, which is a graph of the amplitudes of each of the constituent sinusoidal components in the Fourier series plotted against frequency, and a *phase spectrum* which is their phases. The amplitudes are given from the Fourier coefficients by equation [19]:

$$A_n = \sqrt{a_n^2 + b_n^2} \qquad [41]$$

and the phases of sinusoidal components by equation [18] as:

$$\phi_n = \tan^{-1}\left(\frac{a_n}{b_n}\right) \qquad [42]$$

Example

Determine the frequency spectrum of the rectangular waveform with $a_0 = 1$, $a_n = 0$ and $b_n = (1 - \cos n\pi)/n\pi$.

We have $b_1 = 2/\pi = 0.64$, $b_2 = 0$, $b_3 = 2/3\pi = 0.21$, $b_4 = 0$, $b_5 = 2/5\pi = 0.13$, etc. The A_0 term is 1. Using equation [41], the A_1 term is 0.64, the A_2 term 0, the A_3 term 0.21, the A_4 term 0, the A_5 term 0.13, etc. The phases, when referred to a sine wave, are $0°$ for all components. When referred to a cosine wave they are $-90°$. Figure 7.13(a) shows the resulting amplitude spectrum and Figure 7.13(b) the phase spectrum when referring to sinusoidal components.

Example

Determine the frequency spectrum for a half-wave rectified sinusoidal waveform if it has the Fourier series:

$$y = \frac{1}{\pi} - \frac{2}{3\pi}\cos 2\omega t - \frac{2}{15\pi}\cos 4\omega t + ... + \frac{1}{2}\sin \omega t$$

The A_0 term is $1/\pi = 0.32$. Using equation [41], the A_1 term is 0.5, the A_2 term $2/3\pi = 0.21$, the A_3 term 0 and the A_4 term $2/15\pi = 0.04$. The phases, when referred to a sine wave are $\phi_1 = 0$ and since $-\cos \omega t = \sin(\omega t - 90°)$, $\phi_2 = -90°$ and $\phi_4 = -90°$. Figure 7.14(a) shows the amplitude spectrum and Figure 7.14(b) the phase spectrum when referring to sinusoidal components.

Maths in action

Electric circuit analysis

Often in considering electrical systems the input is not a simple d.c. or sinusoidal a.c. signal but perhaps a square wave periodic signal or a distorted sinusoidal signal or a half-wave rectified sinusoid. Such problems can be tackled by representing the waveform as a Fourier series and using the *principle of superposition*; we find the overall effect of the waveform by summing the effects due to each term in the Fourier series considered alone. Thus if we have a voltage waveform:

$$v = V_0 + V_1 \sin \omega t + V_2 \sin 2\omega t + V_3 \sin 3\omega t + \ldots + V_n \sin n\omega t$$

then we can consider the effects of each element taken alone. Thus we can calculate the current due to the voltage V_0, that due to $V_1 \sin \omega t$, that due to $V_2 \sin 2\omega t$, that due to $V_3 \sin 3\omega t$, and so on for all the terms in the series. We then add these currents to obtain the overall current due to the waveform.

Consider the application to a pure resistance R. Since $i = v/R$ and resistance R is independent of frequency, then the current due to the V_0 term is V_0/R, that due to the first harmonic term is $(V_1 \sin \omega t)/R$, that due to the second harmonic term is $(V_2 \sin 2\omega t)/R$ and so on. Thus the resulting current waveform is:

$$i = \frac{V_0}{R} + \frac{V_1}{R} \sin \omega t + \frac{V_2}{R} \sin 2\omega t + \ldots + \frac{V_n}{R} \sin n\omega t$$

Because the resistance is the same for each harmonic, the amplitude of each voltage harmonic is reduced by the same factor, i.e. the resistance. The phases of each harmonic are not changed. The current waveform is thus the same shape as the voltage waveform.

Consider the application to a pure inductance L. The impedance of a pure inductance depends on the frequency, i.e. its reactance $X_L = \omega L$. Also the current lags the voltage by 90°. The impedance is 0 when the frequency is 0 and thus the current due to the V_0 term will be 0. The current due to the first harmonic will be the voltage of that harmonic divided by the impedance at that frequency and so $V_1 \sin (\omega t - 90°)/\omega L$. The current due to the second harmonic will be the voltage of that harmonic divided by the impedance at that frequency and so $V_1 \sin (2\omega t - 90°)/2\omega L$. Thus the current waveform will be:

$$i = \frac{V_1}{\omega L} \sin(\omega L - 90°) + \frac{V_2}{2\omega L} \sin(2\omega L - 90°) + \dots$$

$$+ \frac{V_n}{n\omega L} \sin(n\omega L - 90°)$$

Each of the voltage terms has its amplitude altered by a different amount; the phase, however, is changed by the same amount. The result is that the shape of the current waveform is different to that of the voltage waveform.

Consider a pure capacitor capacitance C. The impedance of a pure capacitor depends on the frequency, i.e. its reactance $X_C = 1/\omega C$, and the current leads the voltage by 90°. The impedance is 0 when the frequency is 0 and thus the current due to the V_0 term will be 0. The current due to the first harmonic will be the voltage of that harmonic divided by the impedance at that frequency and so $V_1 \sin (\omega t + 90°)/(1/\omega C)$. For the second harmonic the current will be the voltage of that harmonic divided by the impedance at that frequency and so $V_1 \sin (2\omega t + 90°)/(1/2\omega C)$. Thus the current waveform will be:

$$i = \omega C V_1 \sin(\omega t + 90°) + 2\omega C V_2 \sin(2\omega t + 90°) + \dots$$

$$+ n\omega C V_n \sin(n\omega t + 90°)$$

Each of the voltage terms has had their amplitude altered by a different amount but the phase changed by the same amount. The result of this is that the shape of the current waveform is different to that of the voltage waveform.

Example

A voltage of 2.5 + 3.2 sin 100t + 1.6 sin 200t V is applied across a resistor having a resistance of 100 Ω. Determine the current through the resistor.

The complex current will be the sum of the currents due to each of the voltage terms in the complex voltage. Since the resistance is the same at all frequencies, the complex current will be:

i = 0.025 + 0.032 sin 100t + 0.016 sin 200t A

Thus, each of the elements has the same phase as the corresponding voltage element.

Example

A complex voltage of 2.5 + 3.2 sin 100t + 1.6 sin 200t V is applied across a pure inductor having an inductance of 100 mH. Determine the current through the inductor.

The impedance is 0 when the frequency is 0 and thus the current due to the 2.5 V term will be 0. For the second term, the reactance is 100 × 0.100 = 10 Ω and the current lags the voltage by 90° and so the current due to this harmonic is 0.32 sin (100t − 90°) A. For the third term, the reactance is 200 × 0.100 = 20 Ω and the current lags the voltage by 90° and so the current due to this harmonic is 0.08 sin (100t − 90°) A. Thus the current waveform is:

i = 0.32 sin (100t − 90°) + 0.08 sin (100t − 90°) A

Example

Determine the waveform of the current occurring when a 2μF capacitor has connected across it the half-wave rectified sinusoidal voltage v = 0.32 + 0.5 cos 100t + 0.21 cos 200t V.

There will be no current arising from the d.c. term. For the first harmonic the reactance is 1/(2 × 10⁻⁶ × 100) Ω and so we have a current of 0.5 × 2 × 10⁻⁶ × 100 cos (100t + 90°) A. For the second harmonic the reactance is 1/(2 × 10⁻⁶ × 200) Ω and so the current is 0.21 × 2 × 10⁻⁶ × 200 cos (200t + 90°). Thus the resulting current is:

i = 2 × 10⁻⁶ × 0.5 × 100 cos (100t + 90°)
      + 2 × 10⁻⁶ × 0.21 × 200 cos (200t + 90°) A

Example

A voltage of v = 100 cos 314t + 50 sin(5 × 314t − π/6) V is applied to a series circuit consisting of a 10 Ω resistor, a 0.02 H inductor and a 50 μF capacitor. Determine the circuit current.

For the first harmonic, the resistance is 10 Ω, the inductive reactance is ωL = 314 × 0.02 = 6.28 Ω and the capacitive reactance is 1/ωC = 1/(314 × 50 × 10⁻⁶) = 63.8 Ω. Thus the total impedance is:

Z_1 = 10 + j6.28 − j63.8 = 10 − j57.52

$$= \sqrt{10^2 + 57.52^2} \angle \tan^{-1} \frac{-57.52}{10} = 58.4 \angle (-80.1°) \, \Omega$$

Thus the current due to the first harmonic is:

$$i_1 = \frac{100 \angle 0°}{58.4 \angle (-80.1°)} = 1.71 \angle 80.1° \text{ A}$$

For the fifth harmonic, the resistance is 10 Ω, the inductive reactance is $5\omega L = 5 \times 314 \times 0.02 = 31.4$ Ω and the capacitive reactance is $1/5\omega C = 1/(5 \times 314 \times 50 \times 10^{-6}) = 12.76$ Ω. Thus the total impedance is:

$$Z_5 = 10 + j31.4 - j12.76 = 10 + j18.64$$

$$= \sqrt{10^2 + 18.64^2} \angle \tan^{-1} \frac{18.64}{10} = 21.2 \angle 61.8° \, \Omega$$

Thus the current due to the fifth harmonic is:

$$i_5 = \frac{50 \angle (-30°)}{21.2 \angle 61.8°} = 2.36 \angle (-91.8°) \text{ A}$$

Thus the current waveform is:

$$i = 1.71 \cos(314t + 80.1°) + 2.36 \cos(3 \times 314t - 91.8°) \text{ A}$$

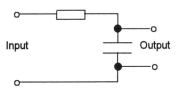

Figure 7.15 *Example*

Input
Output

Example

A half-wave rectified sinusoidal voltage:

$$v = 0.32 + 0.5 \sin \pi t - 0.21 \cos 2\pi t - 0.04 \cos 4\pi t \text{ V}$$

is applied to a circuit consisting of a 1 Ω resistor in series with a 1 F capacitor. Determine the waveform of the voltage output across the capacitor.

Figure 7.15 shows the circuit. The output is the fraction of the input voltage that is across the capacitor. Thus, using phasors and the component values given:

$$V_{out} = \frac{1/jn\omega C}{(1/jn\omega C) + R} V_{in} = \frac{1}{1 + jn\omega CR} V_{in} = \frac{1}{1 + jn\omega} V_{in}$$

where $\omega = \pi$ is the fundamental frequency. For the d.c. component, with $\omega = 0$, we have $V_{out\,0} = V_{in\,0} = 0.32$ V. For the first harmonic we have $V_{in\,1} = -j0.5$ V and thus the output due to this term is:

$$V_{out\,1} = \frac{1}{1 + j\pi}(-j0.5) = \frac{1 - j\pi}{(1 + j\pi)(1 - j\pi)}(-j0.5) = \frac{-0.5\pi - j0.}{1 + \pi^2}$$

$$= \sqrt{\frac{0.5^2\pi^2 + 0.5^2}{(1+\pi^2)^2}} \angle \tan^{-1}\frac{-0.5}{-0.5\pi} = 0.15\angle(-162.3°) \text{ V}$$

For the second harmonic we have $\mathbf{V}_{\text{in 2}} = -j0.21$ V and thus the output due to this term is:

$$\mathbf{V}_{\text{out 2}} = \frac{1}{1+j2\pi}(-j0.21) = \frac{-0.42\pi - j0.21}{1+4\pi^2} = 0.033\angle(-189°)$$

For the fourth harmonic we have $\mathbf{V}_{\text{in 4}} = -j0.04$ V and thus the output due to this term is:

$$\mathbf{V}_{\text{out 4}} = \frac{1}{1+j4\pi}(-j0.04) = \frac{-0.16\pi - j0.04}{1+16\pi^2} = 0.003\,2\angle(-184.5°)$$

Thus the output is:

$$\mathbf{V}_{\text{out}} = 0.32 + 0.15 \sin(\pi t - 162.3°) + 0.033 \cos(2\pi t - 189°)$$
$$+ 0.003\,2 \cos(4\pi t - 184.5°) \text{ V}$$

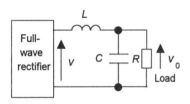

Figure 7.16 *Full-wave recifier with filter*

Maths in action

Rectifier filter circuit

A full-wave rectifier produces a far-from-smooth output and relies on the use of a *LC* filter in order to give an output which reasonably approximates to a smooth d.c. voltage. Figure 7.16 shows the circuit. The output from the full-wave rectifier can be described by the Fourier series:

$$v = \frac{2V_m}{\pi}\left(1 - \tfrac{2}{3}\cos 2\omega t - \tfrac{2}{15}\cos 4\omega t - \tfrac{2}{35}\cos 6\omega t - ...\right)$$

The first term is a constant and so represents a d.c. component. The second, and succeeding terms, represent alternating voltages which can be considered to be superimposed on the d.c. voltage.

The output voltage from filter circuit is across the resistive load. Assuming ideal components, we have a d.c. voltage of $2V_m/\pi$ across the load resistor. For the capacitor to provide effective smoothing of the output, its reactance must be low compared with the load resistance so as to divert most of the a.c. element away from the load resistor. For the a.c. elements, the circuit is effectively just a pure inductance in series with a pure capacitor. It is a voltage-divider circuit, thus for the *n*th harmonic:

$$\frac{\mathbf{V_0}}{\mathbf{V}} = \frac{-jX_C}{jX_L - jX_C}$$

and so:

$$\frac{v_0}{v} = \frac{X_C}{X_C - X_L}$$

For the 2nd harmonic we have $X_C = 1/2\omega C$ and $X_L = 2\omega L$. Thus:

$$\frac{v_0}{v} = \frac{1/2\omega C}{(1/2\omega C) - 2\omega L} = \frac{1}{1 - 4\omega^2 LC}$$

Since $4\omega^2 LC$ will be much greater than 1, the equation approximates to:

$$\frac{v_0}{v} = -\frac{1}{4\omega^2 LC}$$

For the 2nd harmonic $v = -(2V_m/\pi)(2/3) \cos 2\omega t$ and so:

$$v_0 = \frac{V_m}{3\pi\omega^2 LC} \cos 2\omega t$$

This will give a ripple on the output d.c. voltage. The size of the ripple is the peak-to-peak value of the alternating component and so is 2 times the maximum amplitude of the ripple:

$$\text{size of ripple disturbance} = \frac{2V_m}{3\pi\omega^2 LC}$$

A measure of the smoothness of the d.c. output is provided by the *ripple factor r*. This can be defined as:

$$r = \frac{\text{ripple voltage at load}}{\text{average load voltage}}$$

Thus, since we have a d.c. voltage component of $2V_m/\pi$:

$$r = \frac{1}{3\omega^2 LC}$$

As an illustration, consider the inductance needed with such a filter circuit to give a 1% ripple factor for a frequency of 50 Hz and a smoothing capacitor of 10 μF. Using these values in the above equation:

$$L = \frac{1}{3r\omega^2 C} = \frac{1}{3 \times 0.01 \times 4\pi^2 50^2 \times 10 \times 10^{-6}} = 3.38 \text{ H}$$

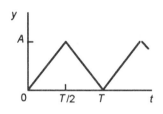

Figure 7.17 *Problem 2*

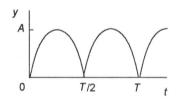

Figure 7.18 *Problem 3*

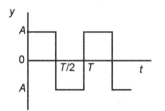

Figure 7.19 *Problem 4*

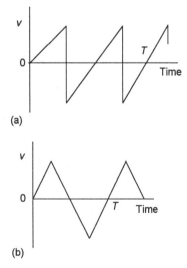

Figure 7.20 *Problem 5*

Problems 7.2

1 What harmonics are present in the waveform given by v = $1.0 - 0.67 \cos 2\omega t - 0.13 \cos 4\omega t$?

2 Determine the Fourier series for the waveform shown in Figure 7.17.

3 Determine the Fourier series for the waveform in Figure 7.18.

4 Determine the Fourier series for the full-wave rectified sinusoid (Figure 7.19).

5 Determine the nature of the terms within the Fourier series for the waveforms shown in Figure 7.20. T is the periodic time for a cycle.

6 Given that the Fourier series for the waveform in Figure 7.21(a) is:

$$y = \frac{4A}{\pi}\left[\sin \omega t + \tfrac{1}{3}\sin 3\omega t + \tfrac{1}{5}\sin 5\omega t + \ldots\right]$$

Determine, by considering the shift of origin, the Fourier series for the waveform shown in Figure 7.21(b).

7 What terms will be present in the Fourier series for the waveforms shown in Figure 7.22?

8 From considerations of the mean values of the waveforms in Figure 7.23, what will be the values of a_0?

9 Determine whether the following are even or odd functions:

(a) $\sin x$, (b) x, (c) $x \sin x$, (d) $x \cos 2x$, (e) $x^3 \cos 2x$

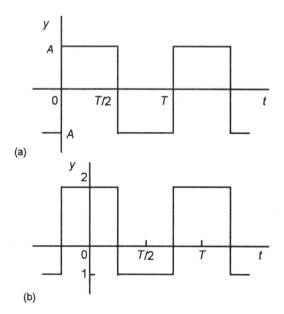

Figure 7.21 *Problem 6*

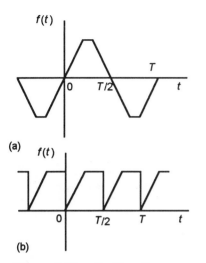

(a)

(b)

Figure 7.22 *Problem 7*

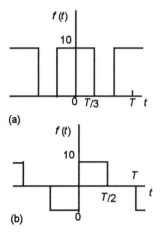

Figure 7.23 *Problem 8*

10 Determine what terms the following waveforms will contain in their Fourier series:

(a) $f(t) = 3t$ for $-\pi \le t < \pi$, period 2π,

(b) $f(t) = \cos t$ for $-\pi \le t < \pi$, period 2π,

(c) $f(t) = t^2 \cos t$ for $-\pi \le t < \pi$, period 2π

11 Determine the Fourier series for the waveform described by $f(t) = t$ for $-\pi \le t < 0$ with a period of 2π.

12 Determine the amplitude and phase (referred to a sine) elements for the frequency spectrum of the waveforms giving the following Fourier series:

(a) $y = \dfrac{1}{2} - \dfrac{1}{\pi} \sin \omega t - \dfrac{1}{2\pi} \sin 2\omega t - \dfrac{1}{3\pi} \sin 3\omega t - \dots$,

(b) $a_0 = \pi/2$, $a_n = 0$ for n even and $-2/n^2\pi$ for n odd, $b_n = -(-1)^n/n$

13 Determine the waveform of the current occurring when a resistor of resistance 1 kΩ has connected across it the half-wave rectified sinusoidal voltage $v = 0.32 + 0.5 \cos 100t + 0.21 \cos 200t$ V.

14 Determine the waveform of the current when a pure inductor of inductance 10 mH has connected across it the half-wave rectified sinusoidal voltage $v = 0.32 + 0.5 \cos 100t + 0.21 \cos 200t$ V.

15 A voltage of $2.5 + 3.2 \sin 100t + 1.6 \sin 200t$ V is applied across a 10 μF capacitor. Determine the current.

8 Logic gates

Summary

In digital circuits extensive use is made of switching circuits. A switch is either on or off with these states being denoted by the digits 1 or 0. A logic circuit can be considered as a collection of switching circuits. In this chapter the basic mathematics necessary to analyse and synthesise such circuits is introduced. The mathematics involved is named after George Boole (1815–64) who first developed the modern ideas of the mathematics concerned with the manipulation of logic statements. In this chapter, Boolean algebra is approached by means of the analysis of switching circuits.

Objectives

By the end of this chapter, the reader should be able to:

- represent switching systems by logic gates;
- represent the action of such gates by truth tables;
- describe switching logic by Boolean statements;
- manipulate Boolean statements by the use of the rules of Boolean algebra.

8.1 Logic gates

Digital electronic logic gates are relatively cheap and readily available as integrated circuits. Such gates find a wide range of applications. For example, they might be used to determine when an input signal control system is to be allowed to give an output, as in an alarm system. Such logic gates are essentially just switching devices and this section considers the basics of such devices.

8.1.1 Switching circuits

Consider a simple on-off switch (Figure 8.1). If we denote a closed contact by a 1 and an open contact by a 0 then the switch has just two possible states: 1 or 0.

Suppose we have two switches a and b in series. Each switch has two possible states, 0 and 1. Figure 8.2 shows the various possibilities for switches. In (a) both switches are open, in (b) a is open and b is closed, in (c) a is closed and b is open and in (d) a and b are both closed. With (a) the effect of both switches being

Switch open: 0

Switch closed: 1

Figure 8.1   *The two states*
of a switch

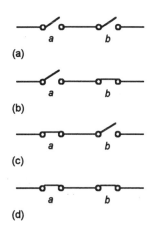

(a)

(b)

(c)

(d)

Figure 8.2 *Switches in series*

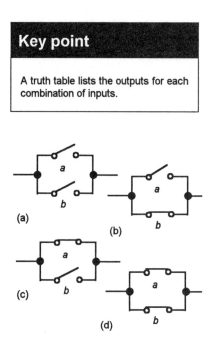

(a)

(b)

(c)

(d)

Figure 8.3 *Parallel switches*

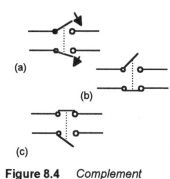

(a)

(b)

(c)

Figure 8.4 *Complement*

open is the same as would be obtained by a single open switch; (b) and (c) likewise are equivalent to a single open switch but (d) is equivalent to a single closed switch. Thus we can say that the two elements are equivalent to 0 for (a), (b) and (c) but 1 for (d). In tabular form we can represent the state of the circuit by Table 8.1:

Table 8.1 *Truth table for a AND b*

a	b	Output
0	0	0
0	1	0
1	0	0
1	1	1

Such a table is known as a *truth table*. If a AND b are 1 then the result is 1. Such an arrangement is known as an AND gate since both a and b have to be 1 for the output to be 1.

Consider two switches in parallel. Figure 8.3 shows the various possibilities for switches. In (a) both switches are open, in (b) a is open and b is closed, in (c) a is closed and b is open and in (d) a and b are both closed. With (a) the effect of both switches being open is the same as would be obtained by a single open switch; (b), (c) and (d) are equivalent to a single closed switch. Thus we can say that the two elements are equivalent to 0 for (a), and 1 for (b), (c) and (d). In tabular form we can represent the state of the circuit by the truth table (Table 8.2):

Table 8.2 *Truth table for a OR b*

a	b	Output
0	0	0
0	1	1
1	0	1
1	1	1

Such an arrangement is known as an OR gate since if a or b is 1 then the result is 1.

Another possible form of switch circuit is where two switches are connected together so that the closing of one switch results in the opening of the other. Figure 8.4(a) illustrates the switch action with (b) showing the upper switch open when the lower switch is closed and (c) the upper switch closed when the lower switch is open. The lower switch is said to give the *complement* of the upper switch. Table 8.3 is the truth table:

Table 8.3 *Truth table for NOT*

Upper switch	Lower switch
0	1
1	0

Such an arrangement constitutes a NOT switching circuit, since if one switch is 1 then the other switch is not 1.

Logic gates

With a mechanical switch we can represent the two logical states of 0 or 1 as the switch being open and closed. With electronic switches, 0 is taken to be a low voltage level and 1 a high voltage level for what is called *positive logic*, although the opposite convention (*negative logic*) can be used with 0 being represented by a high voltage level and 1 by a low voltage level. The 0 and the 1 do <u>not</u> represent actual numbers but the state of the voltage or current. The term *logic level* is often used with the voltage being said to be at logic level 0 or logic level 1.

The basic building blocks of digital electronic circuits are called *logic gates*. A logic gate is an electronic block which has one or more inputs and an output. The output can be either high or low depending on the digital levels at the input terminals. The following sections take a look at the logic gates: AND, OR, INVERT/NOT, NAND, NOR and XOR. Different sets of standard circuit symbols have been developed in Britain, Europe and the United States; an international standard (IEEE/ANSI) has, however, been developed based on squares. In this text, both the IEEE/ANSI form and the older United States form are shown.

- **AND** *gate*

    The AND gate gives an output 1 when both input A <u>and</u> input B are 1. Figure 8.5 shows the symbol, the associated truth table being given in Table 8.4.

Table 8.4 AND *gate*

A	B	Output
0	0	0
0	1	0
1	0	0
1	1	1

- **OR** *gate*

    The OR gate gives an output 1 when either input A <u>or</u> input B is 1. Figure 8.6 shows the symbol and Table 8.5 the truth table.

Table 8.5 OR *gate*

A	B	Output
0	0	0
0	1	1
1	0	1
1	1	1

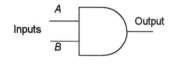

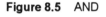

Figure 8.5   AND

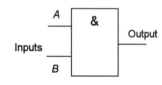

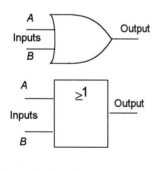

Figure 8.6   OR

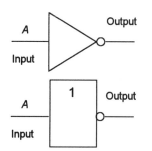

Figure 8.7   NOT

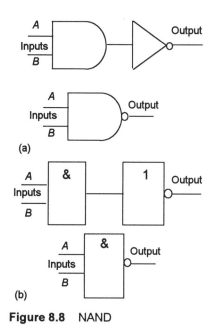

(a)

(b)

Figure 8.8   NAND

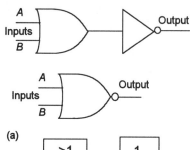

(a)

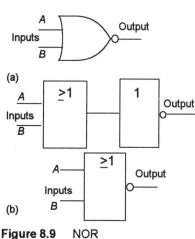

(b)

Figure 8.9   NOR

• **INVERT/NOT** *gate*

The INVERT or NOT gate has a single input and gives a 1 output when the input is 0. The gate inverts the input, giving a 1 when the input is 0 and a 0 when the input is 1. Figure 8.7 shows the gate symbol and Table 8.6 gives the truth table.

Table 8.6 NOT *gate*

A	Output
0	1
1	0

• **NAND** *gate*

This gate (Figure 8.8) is logically equivalent to a NOT gate in series with an AND gate, NAND standing for NotAND. The symbol for the gate is the AND symbol followed by a small circle, the small circle being used to indicate negation. The gate has the truth table shown in Table 8.7. There is a 1 output when A and B are both not 1, i.e. are both 0.

Table 8.7 NAND *gate*

A	B	Output
0	0	1
0	1	1
1	0	1
1	1	0

• **NOR** *gate*

This gate (Figure 8.9) is logically equivalent to a NOT gate in series with an OR gate. It is represented by the OR gate symbol followed by a small circle to indicate negation. Table 8.8 gives the truth table, there being a 1 output when neither A nor B is 1.

Table 8.8 NOR *gate*

A	B	Output
0	0	1
0	1	0
1	0	0
1	1	0

• **EXCLUSIVE OR (XOR)** *gate*

The OR gate gives an output 1 when either input A or input B is 1 or both A and B are 1. The EXCLUSIVE OR gate gives an output 1 when either input A or input B is 1 but not when both are. Figure 8.10 shows the gate symbol and Table 8.9 the truth table.

Table 8.9 XOR *gate*

A	B	Output
0	0	0
0	1	1
1	0	1
1	1	0

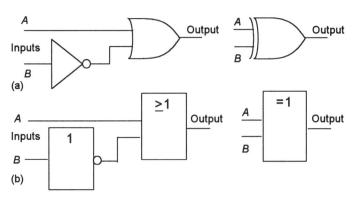

Figure 8.10 XOR

By combining gates it is possible to generate other switching operations. This is illustrated in the following example and discussed later in this chapter.

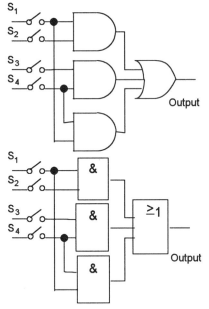

Figure 8.11   *Example*

Example

Suppose we wanted to design a switching circuit in order to operate a relay from a combination of four switches so that the relay is energised when switch 1 and switch 2 are both closed, or when switch 3 and switch 4 are both closed, or when switch 1 and switch 3 are both closed. Design a system of logic gates which would give this.

The output required is when we have (S1 and S2) or (S3 and S4) or (S1 and S3). Figure 8.11 shows how this may be realised with gates.

Maths in action

Ladder programming with PLCs

Programmable logic controllers (PLCs) use a simple form of programming in order to exercise control functions. This program involves drawing each step in a program as the rung on a ladder, each rung then being taken in turn from top to bottom. Each rung can execute logic switching functions such as AND and OR. Figure 8.12(a) shows the symbols used to represent normally open switches, normally closed switches and output devices. Figure 8.12(b) shows three rungs in a ladder program. With rung 1 we have an AND gate situation in that both A and B have to be on for there to be an output. With rung 2 either A or B have to be on for there to be an output and so we have an OR gate. Rung 3 shows a NOT gate in that when A has an input it switches the output off.

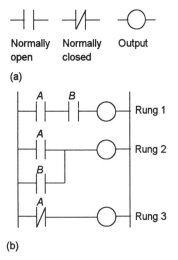

Figure 8.12   *Ladder programming*

8.2 Boolean algebra

In this section we look at how we can develop algebraic notation and rules to describe and manipulate logic gate arrangements.

Notation

For the AND operation, i.e. the series connections of switches a and b, a is considered to be *multiplied* by b. Generally $\cdot$ is used for the multiplication symbol. From truth table 8.4 we thus have the rules:

$$0 \cdot 0 = 0, \qquad 0 \cdot 1 = 0, \qquad 1 \cdot 0 = 0, \qquad 1 \cdot 1 = 1 \qquad [1]$$

For the OR operation, i.e. the parallel connection of switches a and b, a is considered to be *added* to b. From truth table 8.5 we have the rules:

$$0 + 0 = 0, \quad 0 + 1 = 1, \quad 1 + 0 = 1, \quad 1 + 1 = 1 \qquad [2]$$

For the NOT operation, i.e. the complement with the switches, we use a bar over a symbol to indicate NOT. Thus truth table 8.6 gives the rules:

$$\overline{0} = 1, \qquad \overline{1} = 0 \qquad [3]$$

Boolean algebra

The binary digits 1 and 0 are the *Boolean variables* and, together with the operations $\cdot$, $+$ and the complement, form what is known as *Boolean algebra*. By constructing the appropriate truth tables the following laws can be derived:

- **Anything ORed with itself is equal to itself**
  See Table 8.10.

  $$a + a = a \qquad [4]$$

- **Anything ANDed with itself is equal to itself**
  See Table 8.11.

  $$a \cdot a = a \qquad [5]$$

- **Input sequence for OR and AND**
  It does not matter in which order we take the inputs for OR and AND gates, the output is the same. This is illustrated by Table 8.12 for OR.

  $$a + b = b + a \qquad [6]$$

  $$a \cdot b = b \cdot a \qquad [7]$$

- **Handling bracketed terms**

Table 8.10 OR

a	a	$a + a$
0	0	0
1	1	1

Table 8.11 AND

a	a	$a \cdot a$
0	0	0
1	1	1

Table 8.12 OR

a	b	$a + b$
0	0	0
0	1	1
1	0	1
1	1	1

As Table 8.13(a) indicates:

$$a + (b \cdot c) = (a + b) \cdot (a + c) \qquad [8]$$

As Table 8.13(b) indicates.

$$a \cdot (b + c) = a \cdot b + a \cdot c \qquad [9]$$

Table 8.13(a)

a	b	c	$b \cdot c$	$a + b \cdot c$	$a + b$	$a + c$	$(a + b) \cdot (a + c)$
0	0	0	0	0	0	0	0
0	0	1	0	0	0	1	0
0	1	0	0	0	1	0	0
0	1	1	1	1	1	1	1
1	0	0	0	1	1	1	1
1	0	1	0	1	1	1	1
1	1	0	0	1	1	1	1
1	1	1	1	1	1	1	1

Table 8.13(b)

a	b	c	$b + c$	$a \cdot (b + c)$	$a \cdot b$	$a \cdot c$	$a \cdot b + a \cdot c$
0	0	0	0	0	0	0	0
0	0	1	1	0	0	0	0
0	1	0	1	0	0	0	0
0	1	1	1	0	0	0	0
1	0	0	0	0	0	0	0
1	0	1	1	1	0	1	1
1	1	0	1	1	1	0	1
1	1	1	1	1	1	1	1

Table 8.14

a	$\overline{a}$	$a + \overline{a}$
0	1	1
1	0	1

Table 8.15

a	$\overline{a}$	$a \cdot \overline{a}$
0	1	0
1	0	0

Table 8.16

a	$a + 0$	$a + 1$
0	0	1
1	1	1

- *Complementary law*
  Anything ORed with its own negative is 1. See Table 8.14.

$$a + \bar{a} = 1 \qquad [10]$$

Anything ANDed with its own negative is 0. See Table 8.15.

$$a \cdot \bar{a} = 0 \qquad [11]$$

- **OR*ing with 0 or 1***
  Anything ORed with a 0 is equal to itself, anything ORed with a 1 is equal to 1. See Table 8.16.

$$a + 0 = a, \quad a + 1 = 1 \qquad [12]$$

- **AND*ing with 0 or 1***

Table 8.17

a	$a \cdot 0$	$a \cdot 1$
0	0	0
1	0	1

Anything ANDed with a 0 is equal to 0, any thing ANDed with a 1 is equal to itself. See Table 8.17.

$$a \cdot 1 = a, \quad a \cdot 0 = 0 \qquad [13]$$

De Morgan laws:

- The complement of the outcome of switches a and b in parallel, i.e. an OR situation, is the same as when the complements of a and b are separately combined in series, i.e. the AND situation. Table 8.18 shows the validity of this.

$$\overline{a+b} = \overline{a} \cdot \overline{b} \qquad [14]$$

- The complement of the outcome of switches a and b in series, i.e. the AND situation, is the same as when the complements of a and b are separately considered in parallel, i.e. the OR situation. Table 8.19 shows the validity of this.

$$\overline{a \cdot b} = \overline{a} + \overline{b} \qquad [15]$$

Key points

De Morgan laws:

$\overline{a+b} = \overline{a} \cdot \overline{b}$

$\overline{a \cdot b} = \overline{a} + \overline{b}$

Table 8.18

a	b	$a+b$	$\overline{a+b}$	$\overline{a}$	$\overline{b}$	$\overline{a} \cdot \overline{b}$
0	0	0	1	1	1	1
0	1	1	0	1	0	0
1	0	1	0	0	1	0
1	1	1	0	0	0	0

Table 8.19

a	b	$a \cdot b$	$\overline{a \cdot b}$	$\overline{a}$	$\overline{b}$	$\overline{a} + \overline{b}$
0	0	0	1	1	1	1
0	1	0	1	1	0	1
1	0	0	1	0	1	1
1	1	1	0	0	0	0

Using the rules given above, complicated switching circuits can be reduced to simpler equivalent circuits.

Example

Simplify the following Boolean function:

$f = a \cdot c + (a+b) \cdot \overline{c}$.

Using equation [9]: gives $a \cdot (b + c) = a \cdot b + a \cdot c$. Since:

$$(a+b) \cdot \overline{c} = a \cdot \overline{c} + b \cdot \overline{c}$$

we can write:

$$f = a \cdot c + a \cdot \overline{c} + b \cdot \overline{c}$$

Using equation [9] for the first two terms gives:

$$f = a \cdot (c + \overline{c}) + b \cdot \overline{c}$$

Then using equations [7] and [10] gives:

$$f = a \cdot 1 + b \cdot c = a + b \cdot \overline{c}$$

Example

Simplify the function: $f = a + a \cdot b \cdot c + \overline{a} \cdot \overline{c}$.

Using equation [13] we can replace a by $a \cdot 1$. The function can then be written as:

$$f = a \cdot 1 + a \cdot (b \cdot c) + \overline{a} \cdot \overline{c}$$

Then using equation [9]:

$$f = a \cdot (1 + (b \cdot c)) + a \cdot \overline{c}$$

Using the second of the equations in [12] gives $1 + (b \cdot c) = 1$ and so:

$$f = a \cdot 1 + \overline{a} \cdot \overline{c}$$

Since $a \cdot 1 = a$ (equation [10]), and applying equation [8]:

$$f = a + \overline{a} \cdot \overline{c} = (a + \overline{a}) \cdot (a + \overline{c})$$

But: $a + \overline{a} = 1$

and so, using equation [13]:

$$f = a + \overline{c}$$

Example

The operation of an output relay controlled by a PLC program is given by the Boolean expression:

Y007 = (X001 · X002 · M000 · M002)

    + (M002 · X001 · $\overline{X004}$ · M000)
      + (X003 · M003 · M002 · X002)
        + (M003 · X003 · M002 · X006)

(a) Represent this expression as rungs in a PLC ladder program, with a rung for each part of the expression.
(b) Simplify the ladder program and hence write another Boolean expression which describes the simplified program.

(a) Each bracketed term can be represented by a rung in a ladder program and so give the program shown in Figure 8.13(a).
(b) Figure 8.13(b) shows how we can simplify the ladder program and still give the same outcome. Such a program can then be described by the Boolean expression:

Y007 = ((X002 + X004) · X001 · M000)
    + ((X002 + X006) · X003 · M003)) · M002

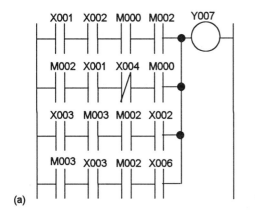

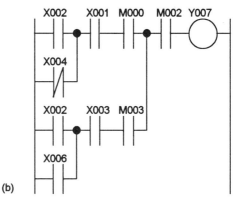

Note: X is an input device, Y is an output device and M is effectively an internal relay (really just a piece of software)

Figure 8.13 *Example*

Problems 8.2

1 Complete the following:

    (a) 1 + 0 = ?, (b) 1 · 1 = ?, (c) $\overline{1}$ = ?

2 Simplify the following Boolean functions:

    (a) $a(\overline{a} + a \cdot b)$, (b) $a + b + c + \overline{a} \cdot b$, (c) $(a + b) \cdot (a + b)$,

    (d) $a \cdot \overline{b} \cdot c + a \cdot \overline{b} \cdot \overline{c}$, (e) $a + \overline{a} \cdot \overline{b}$

8.3 Logic gate systems

The operations ·, + and the complement can be used to write the Boolean functions for complex switching circuits, the states of such circuits being determined by developing the truth table to indicate all the various switching possibilities. Boolean algebra might then be used to simplify the switching circuits.

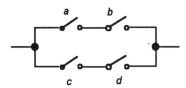

Figure 8.14   *Example*

Example

Write, for the circuit shown in Figure 8.14, (a) the truth table and (b) the Boolean function to describe that truth table.

(a) *a* and *b* are in series, and in parallel with the series arrangement of *c* and *d*. The result of using the switches is that only when either *a* and *b* are closed or *c* and *d* are closed will there be an output. Table 8.20 shows the truth table.

(b) The Boolean function for the two switches *a* and *b* in series is $a \cdot b$, the AND function, and thus, since the function for two items in parallel is OR, the function for the circuit as a whole is:

$$a \cdot b + c \cdot d$$

Table 8.20 *Example*

a	*b*	*c*	*d*	Result
0	0	0	0	0
0	0	0	1	0
0	0	1	0	0
0	0	1	1	1
0	1	0	0	0
0	1	0	1	0
0	1	1	0	0
0	1	1	1	0
1	0	0	0	0
1	0	0	1	0
1	0	1	0	0
1	0	1	1	1
1	1	0	0	1
1	1	0	1	1
1	1	1	0	1
1	1	1	1	1

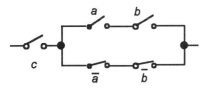

Figure 8.15 *Example*

Example

Derive the Boolean function for the switching circuit shown in Figure 8.15.

In the upper parallel arm of the circuit, the switches *a* and *b* are in series and so have a Boolean expression of $a \cdot b$. In the lower arm the complements of *a* and *b* are in series. Thus the Boolean expression for that part of the circuit is $\overline{a} \cdot \overline{b}$.

Because the two arms are in parallel the expression for the parallel part of the circuit is $a \cdot b + \overline{a} \cdot \overline{b}$.

In series with this is switch *c*. Thus the Boolean function for the circuit is: $c \cdot (a \cdot b + \overline{a} \cdot \overline{b})$.

Combining gates

By combining logic gates it is possible to represent other Boolean functions and use of Boolean algebra can often be used to simplify the arrangement.

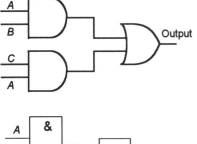

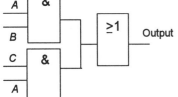

Figure 8.16 *Example*

Example

Determine the Boolean function describing the relation between the output from the logic circuit shown in Figure 8.16. Hence, consider how the circuit could be simplified.

This might be a circuit used with a car warning buzzer so that it sounds when the key is in the ignition (*A*) and a car door is opened (*B*) or the headlights are on (*C*) and a car door is opened (*A*). We have two AND gates and an OR gate. The output from the top AND gate is $A \cdot B$, and from the lower AND gate $C \cdot A$. These outputs are the inputs to the OR gate and thus the output is

output $= A \cdot B + C \cdot A$

The circuit can be simplified by considering the Boolean algebra. Using equation [9] the Boolean function can be written as:

$A \cdot B + C \cdot A = A \cdot (B + C)$

We now have *A* and *B* or *C*. This function now describes a logic circuit with just two gates, an OR gate and an AND gate. Figure 8.17 shows the circuit.

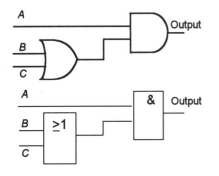

Figure 8.17 *Example*

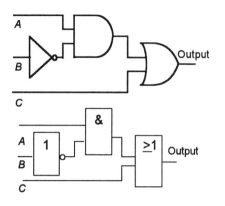

Figure 8.18 *Example*

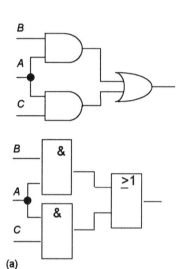

(a)

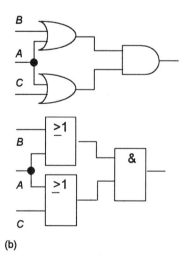

(b)

Figure 8.19 *(a) Sum of products, (b) product of sums*

Example

Devise a logic gate system to generate the Boolean function $A \cdot \overline{B} + C$.

$A \cdot B$ requires an AND gate, but as the B input has to be inverted we precede the input from B to the AND gate by a NOT gate. We then require an OR gate for the output from the AND gate and C. Figure 8.18 shows the gate system.

Boolean function generation from truth tables

Often the requirements for a system are specified in terms of a truth table and the problem then becomes one of determining how a logic gate system can be devised, using the minimum number of gates, to give that truth table. The forms to which most are minimised are an AND gate driving a single OR gate or vice versa.

- **Sum of products**

  Two AND gates driving a single OR gate (Figure 8.19(a)) give, what is termed, the sum of products form:

  $$A \cdot B + A \cdot C$$

- **Product of sums**

  Two OR gates driving a single AND gate (Figure 8.19(b)) give, what is termed the product of sums form:

  $$(A + B) \cdot (A + C)$$

The usual procedure to find the minimum logic gate system is thus to find the sum of products or the product of sums form that fits the data. Generally the sum of products form is used. The procedure used is:

1 Consider each row of the truth table in turn that generates a 1 output and find the product that would fit a row. Only a row of a truth table that has an output of 1 need be considered, since the rows with 0 output do not contribute to the final expression. For example, suppose we have a row in a truth table of: $A = 1$, $B = 0$ and output = 1. When A is 1 and B is not 1 then the output is 1, thus the product which fits this is:

$$Q = A \cdot \overline{B}$$

2 The overall result is the sum of all the products for the rows giving 1 output.

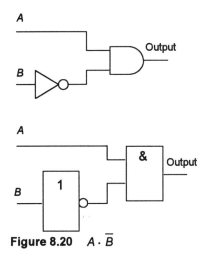

Figure 8.20 $A \cdot \overline{B}$

Example

Determine a logic gate system to give the following truth table.

A	B	Output	Products
0	0	0	
0	1	0	
1	0	1	$A \cdot \overline{B}$
1	1	0	

We only need consider the third row, thus the result is:

$$Q = A \cdot \overline{B}$$

The logic gate system that will give this truth table is thus that shown in Figure 8.20.

Example

Determine a logic gate system which will give the following truth table.

A	B	C	Output	Products
0	0	0	1	$\overline{A} \cdot \overline{B} \cdot \overline{C}$
0	0	1	0	
0	1	0	1	$\overline{A} \cdot B \cdot \overline{C}$
0	1	1	0	
1	0	0	0	
1	0	1	0	
1	1	0	0	
1	1	1	0	

There are two rows for which we need to find a product. Thus the sum of products which fits this table is:

$$Q = \overline{A} \cdot \overline{B} \cdot \overline{C} + \overline{A} \cdot B \cdot \overline{C}$$

This can be simplified to give:

$$Q = \overline{A} \cdot \overline{C} \cdot (\overline{B} + B) = \overline{A} \cdot \overline{C}$$

The truth table can thus be generated by just a NAND gate.

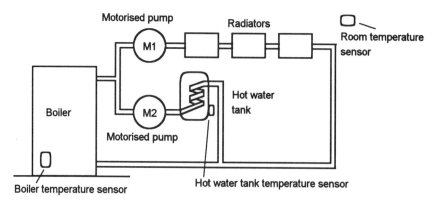

Figure 8.21 *Central heating system*

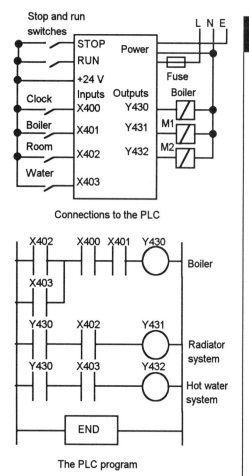

Connections to the PLC

The PLC program

Figure 8.22 *Central heating system*

Maths in action

A PLC and a central heating system

Consider a domestic central heating system (Figure 8.21) and its control by a PLC (see earlier Maths in Action in this chapter). The central heating boiler is to be thermostatically controlled and supply hot water to the radiator system in the house and also to the hot water tank to provide hot water from the taps in the house. Pump motors have to be switched on to direct the hot water from the boiler to either, or both, of the radiator and hot water systems according to whether the temperature sensors for the room temperature and the hot water tank indicate that the radiators or tank need heating. The entire system is to be controlled by a clock so that it only operates for certain hours of the day. Figure 8.22 shows how a PLC might be used and its ladder program.

The boiler, output Y430, is switched on if X400 and X401 and either X402 or X403 are switched on. This means if the clock switched is on, the boiler temperature sensor gives an on input, and either the room temperature sensor or the water temperature sensors give on inputs. The motorised valve M1, output Y431, is switched on if the boiler, Y430, is on and if the room temperature sensor X402 gives an on input. The motorised valve M2, output Y432, is switched on if the boiler, Y430, is on and if the water temperature sensor gives an on input.

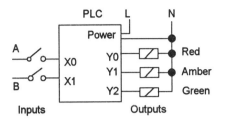

Figure 8.23   *Example*

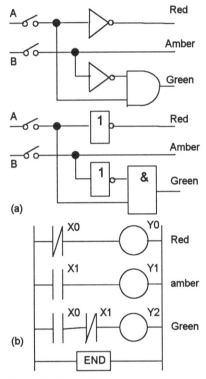

(a)

(b)

Figure 8.24   *Example*

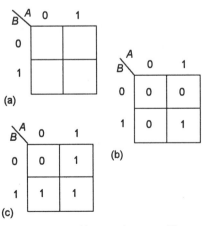

Figure 8.25   *Karnaugh map with a four-cell box*

Example

Design a PLC ladder program that will control a simple red–amber–green traffic light sequence for two inputs X0 and X1 to give the outputs Y0, Y1 and Y2 (Figure 8.23) shown in the following table:

Inputs		Outputs		
X0	X1	Y0	Y1	Y2
0	0	1	0	0
0	1	1	1	0
1	0	0	0	1
1	1	0	1	0

Note: logic 0 defines an open switch or a light turned OFF, logic 1 defines a closed switch and a light turned ON.

Figure 8.24(a) shows how we can represent the above truth table by a logic gate system and Figure 8.24(b) by rungs in a ladder program.

When there is no input to X0 then the red light is ON. Thus, when the input to X0 is 0 then, as the switch is normally closed, the output Y1 is ON; when the input is 1, to open the switch, then the output is OFF.

Karnaugh maps

The *Karnaugh map* is a graphical way of representing a truth table and a method by which simplified Boolean expressions can be obtained from sums of products. The Karnaugh map is drawn as a rectangular array of cells with each cell corresponding to the output for a particular combination of inputs, i.e. a particular product value. Thus, Figure 8.25(a) shows the four-cell box corresponding to two input variables A and B, this giving four product terms. We then insert the function for each input combination, Figure 8.25(b) showing this for an AND gate and Figure 8.25(c) for an OR gate. Figure 8.26 shows how we can represent such maps with input labels A and B for 1 entries and not A and B for 0 entries.

Karnaugh maps not only pictorially represent truth tables but also can be used for minimisation. Suppose we have the following truth table:

A	B	Output	Products
0	0	0	$\overline{A} \cdot \overline{B}$
0	1	0	$\overline{A} \cdot B$
1	0	1	$A \cdot \overline{B}$
1	1	0	$A \cdot B$

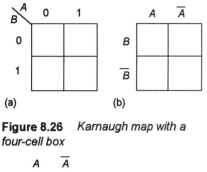

Figure 8.26 *Karnaugh map with a four-cell box*

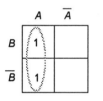

Figure 8.27 *Karnaugh map*

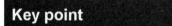

Figure 8.28 *Karnaugh map*

Key point

Simplification using Karnaugh maps

1 Construct the Karnaugh map and place 1s in those squares which correspond to the 1s in the truth table.
2 Examine the map for adjacent 1s and loop them.
3 Form the OR sum of all those terms generated by each loop.

Figure 8.27 shows the Karnaugh map for this truth table with just the 1 output shown. On the map this entry is just the cell with the coordinates $A = 1$, $B = 0$ and so gives the indicated product. The Karnaugh map enables the minimisation to be spotted visually.

As a further example, consider the following truth table:

A	B	Output	Products
0	0	0	$\overline{A} \cdot \overline{B}$
0	1	0	$\overline{A} \cdot B$
1	0	1	$A \cdot \overline{B}$
1	1	1	$A \cdot B$

Figure 8.28 shows the Karnaugh map for this truth table with just the 1 output shown. This has an output given by:

$$A \cdot \overline{B} + A \cdot B$$

This can be simplified to:

$$A \cdot \overline{B} + A \cdot B = A \cdot (\overline{B} + B) = A$$

Thus we have a rule for the map that: *when two cells containing a 1 have a common edge then we can simplify them to just the common variable, the variable that appears in the complemented and uncomplemented form is eliminated.* To help with such simplifications, we draw loops round 1s in adjacent cells. Note that in looping, adjacent cells can be considered to be those in the left- and right-hand columns. Think of the map as though it is wrapped round a vertical cylinder and the left- and right-hand edges of the map are joined together. There are other rules we can develop, namely: *looping a quad of adjacent 1s eliminates the two variables that appear in complemented and uncomplemented form* and *looping an octet of adjacent 1s eliminates the three variables that appear in both complemented and uncomplemented form.*

Figure 8.29(a) shows how we can draw a Karnaugh map for three inputs and Figure 8.29(b) for four inputs. Note that the cells are labelled so horizontally adjacent cells differ by just one variable, likewise adjacent vertical cells.

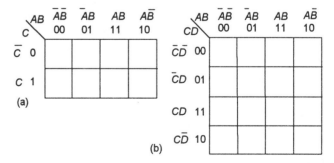

Figure 8.29 *Karnaugh maps: (a) three-input, (b) four-input*

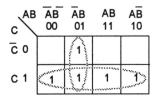

Figure 8.30 *Karnaugh map*

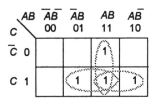

Figure 8.31 *Karnaugh map*

Example

Determine the simplified Boolean expression for the Karnaugh map shown in Figure 8.30.

We have three loops and so the outcome is:

$A \cdot B + B \cdot C + A \cdot C$

Example

Determine the simplified Boolean expression for the Karnaugh map shown in Figure 8.31.

We have a doublet loop and a quad loop and so the outcome is:

$\overline{A} \cdot B + C$

Problems 8.3

1 State a Boolean function that can be used to represent each of the switching circuits shown in Figure 8.32.
2 Give the truth tables for the switching circuits represented by the Boolean functions:

   (a) $(a + \overline{b}) + (a + \overline{c})$, (b) $\overline{a} \cdot (a \cdot b + \overline{b}) \cdot \overline{b}$

3 Determine the Boolean functions that could generate the outputs in Figure 8.33.

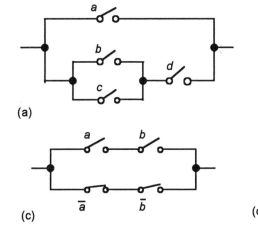

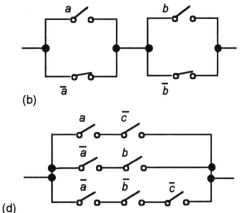

Figure 8.32 *Problem 1*

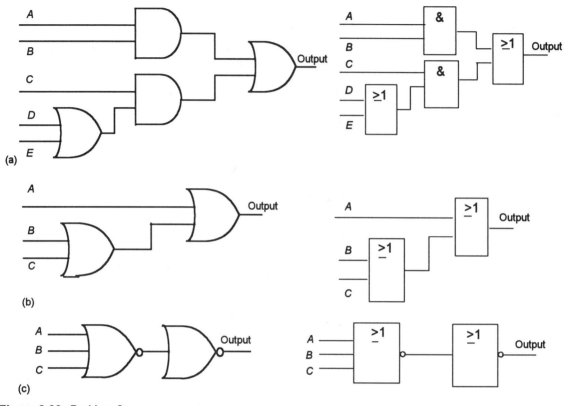

Figure 8.33 *Problem 3*

Table 8.21(a)

a	b	c	Function
0	0	0	0
0	0	1	1
0	1	0	0
0	1	1	0
1	0	0	0
1	0	1	0
1	1	0	1
1	1	1	0

Table 8.21(b)

a	b	c	Function
0	0	0	0
0	0	1	0
0	1	0	0
0	1	1	1
1	0	0	0
1	0	1	0
1	1	0	0
1	1	1	1

4 Give the truth table for the switching circuit corresponding to the Boolean function:

$$(a \cdot \overline{b}) + (\overline{a} \cdot b)$$

5 Draw switching circuits to represent the following Boolean functions:

(a) $a \cdot (a + b)$, (b) $a \cdot (a \cdot b + c)$, (c) $\overline{a} \cdot (a + \overline{b} \cdot (a + c))$

6 Determine the Boolean equations describing the logic circuits in Figure 8.34, then simplify the equations and hence obtain simplified logic circuits.

7 Draw switching circuits to represent the Boolean functions :

(a) $a \cdot b$, (b) $a \cdot b + b$, (c) $c \cdot (a \cdot b + a \cdot \overline{b})$,

(d) $a \cdot (a \cdot b \cdot \overline{c} + a \cdot (\overline{b} + c))$.

8 Derive the Boolean functions for the truth tables in Table 8.21(a) and (b).

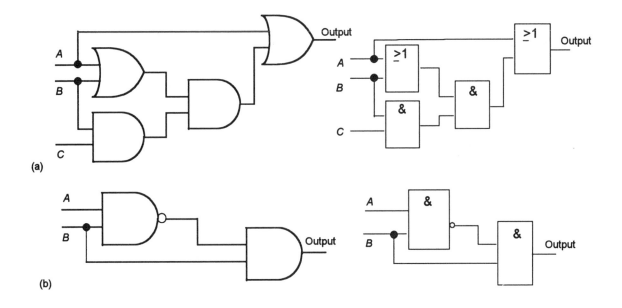

(a)

(b)

Figure 8.34 *Problem 6*

9 Determine the Boolean equations describing the logic circuits in Figure 8.35, then simplify the equations and hence obtain simplified logic circuits.

10 For the Karnaugh maps in Figure 8.36, produce the simplified Boolean expression.

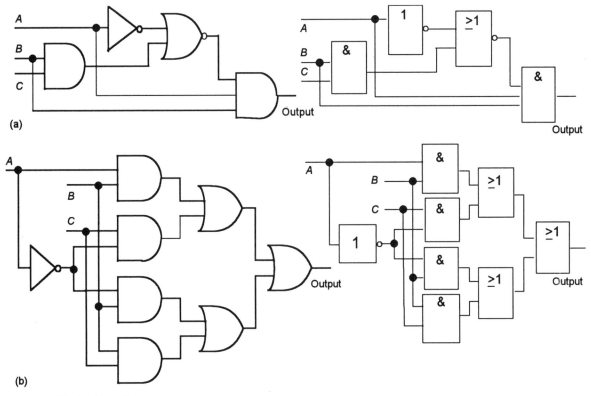

(a)

(b)

Figure 8.35 *Problem 8*

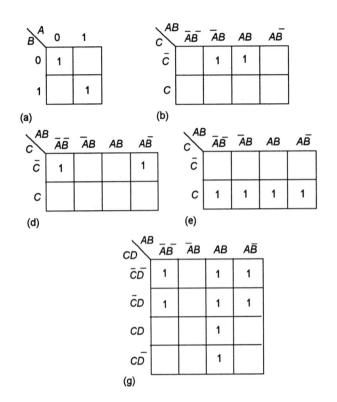
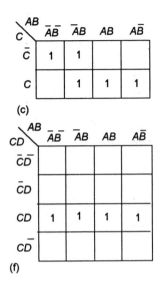

Figure 8.36 *Problem 10*

9 Probability and statistics

Summary

In any discussion of system reliability or quality control, the concept of probability plays a vital part. It is also a vital issue in the consideration of statistics when errors have to be considered in experimental measurements, all measurements being subject to some degree of uncertainty. For example, in the control of manufactured items (statistical process control SPC) control is exercised of measured variables against go/no-go criteria, the attribute, to avoid incurring scrap or reworking costs. This chapter is an introductory consideration of the principles of probability and statistics allied to such engineering issues.

Objectives

By the end of this chapter, the reader should be able to:

- understand the concept of probability;
- use probability principles in the consideration of quality control and system reliability;
- plot experimentally obtained data to show its distribution;
- use the idea of probability distributions and be able to interpret them;
- determine measures of the location and spread of distributions;
- use measures obtained from the Binomial, Poisson and Normal distributions;
- determine the errors in results obtained from experimental measurements.

9.1 Probability

What is the chance an engineering system will fail? What is the chance that a product emerging from a production line is manufactured to the required engineering tolerances, thus avoiding reworking or scrap? What is the chance that if you make a measurement in some experiment that it will be the true value of that quantity? Within what range of experimental error might you expect a measurement to be the true value? These, and many other questions in engineering and science, involve a consideration of chances of events occurring. The term *probability* is more often used in mathematics than chance and has the same meaning in the above questions. This section is about probability, its definition and determination in a number of situations.

9.1.1 Basic definitions

If you flip a coin into the air, what is the chance that it will land heads uppermost? We can try such an experiment and determine the outcomes. The result of a large number of trials leads to the result that about half the time it lands heads uppermost and half the time tails uppermost. If n is the number of trials then we can define probability P as:

$$P = \lim_{n \to \infty} \frac{\text{number of times an event occurs}}{n} \qquad [1]$$

This view of probability is *the relative frequency in the long run with which an event occurs*. In the case of the coin this leads to a probability of $\frac{1}{2} = 0.5$. If an event occurs all the time then the probability is 1. If it never occurs the probability is 0.

The result of flipping the coin might seem obvious since there are just two ways a coin can land and just one of the ways leads to heads uppermost. If there is no reason to expect one way is more likely than the other then we can define probability P as *the degree of uncertainty about the way an event can occur* and as:

$$P = \frac{\text{number of ways a particular event can occur}}{\text{total number of ways events can occur}} \qquad [2]$$

In the case of the coin, this also gives a probability of 0.5. If every possible way events can occur is the required way, then the probability is 1. If none of the possible ways are the event required, then the probability is 0.

Consider a die-tossing experiment. A die can land in six equally likely ways, with uppermost 1, 2, 3, 4, 5, or 6. Of the six possible ways the die could land, only one way is with 6 uppermost. Thus using definition [2], the probability of obtaining a 6 is 1/6. The probability of *not* obtaining a 6 is 5/6 since there are 5 ways out of the 6 possible ways we can obtain an outcome which is not a 6.

Another way the term probability is used is as *degree of belief*. Thus we might consider the probability of a particular horse winning a race as being 1 in 5 or 0.2. The probability in this case is highly subjective.

Key points

Probability can be defined as the relative frequency in the long run with which an event occurs or as the fraction of the total number of ways with which an event can occur.

Example

In the testing of products on a production line, for every 100 tested 5 were found to contain faults. What is the probability that in selecting one item from 100 on the production line that it will be faulty?

There are 100 ways the item can be selected and 5 of the ways give faulty items. Thus, using equation [2], the probability is 5/100 = 0.05.

Probability of events

If an event can occur in two possible ways, e.g. a piece of equipment can be either operating satisfactorily or have failed, then if the probability of one way is P_1 and the probability of the other way is P_2, we must have:

$$P_1 + P_2 = 1 \qquad\qquad [3]$$

The probability of either event 1 or event 2 occurring equals 1, i.e. a certainty, and is the sum of the probability of event 1 occurring, i.e. P_1, added to the probability of event 2, i.e. P_2, occurring.

A probability of 1 for an event means that the probability of it occurring is a certainty.

Suppose with the die-tossing experiment we were looking for the probability that the outcome would be an even number. Of the six possible outcomes of the experiment, three ways give the required outcome. Thus, using definition [2], the probability of obtaining an even number is $3/6 = 0.5$. This is the sum of the probabilities of 2 occurring, 4 occurring and 6 occurring, i.e. $1/6 + 1/6 + 1/6$. The 2, the 4 and the 6 are mutually exclusive events in that if the 2 occurs then 4 or 6 cannot also be occurring. Thus:

If A and B are mutually exclusive, the probability of A or B occurring is the sum of the probabilities of A occurring and of B occurring.

Example

The probability that a circuit will malfunction is 0.01. What is the probability that it will function?

The probability that it will function and the probability that it will not function must together be 1. Hence the probability that it will function is 0.99.

Example

A company manufactures two products *A* and *B*. Market research over a month showed 30% of enquiries by potential customers resulting in product *A* alone being bought, 50% buying product *B* alone, 10% buying both *A* and *B* and 10% buying neither. Determine the probability that an enquiry will result in (a) product *A* alone being bought, (b) product *A* being bought, (c) both product *A* and product *B* being bought, (d) product *B* not being bought.

(a) 30% buy product *A* alone so the probability is 0.30.
(b) 30% buy product *A* alone and 10% buy *A* in conjunction with *B*. Thus the probability of *A* being bought is 0.30 + 0.10 = 0.40.
(c) 10% buy products *A* and *B* so the probability is 0.10.
(d) 50% buy product *B* alone and 10% buy *B* in conjunction with *A*. Thus the probability of buying *B* is 0.50 + 0.10 = 0.60. The probability of not buying *B* is thus 1 − 0.60 = 0.40.

9.1.2 Ways events can occur

Suppose we flip two coins. What is the probability that we will end up with both showing heads uppermost? The ways in which the coins can land are:

HH HT TH TT

There are four possible results with just one of the ways giving HH. Thus the probability of obtaining HH is ¼ = 0.25.

There were two possible outcomes from the experiment of tossing the first coin and two possible outcomes from the experiment of tossing the second coin. For each of the outcomes from the first experiment there were two outcomes from the second experiment. Thus for the two experiments the number of possible outcomes is 2 × 2 = 4. This is an example of, what is termed, the *multiplication rule*.

Tree diagrams can be used to visualise the outcomes in such situations, Figure 9.1 showing this for the two experiments of tossing coins.

Key point

Multiplication rule: If one experiment has n_1 possible outcomes and a second experiment n_2 possible outcomes then the compound experiment of the first experiment followed by the second has $n_1 \times n_2$ possible outcomes.

Outcomes
after first
experiment

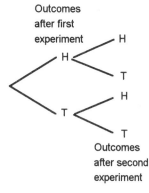

Outcomes
after second
experiment

Figure 9.1   *Tree diagram*

Example

A company is deciding to build two new factories, one of them to be in the north and one in the south. There are four potential sites in the north and two potential sites in the south. Determine the number of possible outcomes.

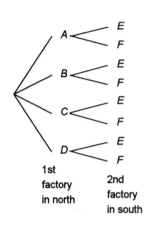

1st factory in north

2nd factory in south

Figure 9.2 *Example*

For the first experiment there are 4 possible outcomes *A*, *B*, *C* and *D* and for the second 2 possible outcomes *E* and *F*. Thus the total number of possible outcomes is given by the multiplication rule as 8. Figure 9.2 shows the tree diagram.

Permutations

Suppose we had to select two items from a possible three different items *A*, *B*, *C*. The first item can be selected in three ways. Then, since the removal of the first item leaves just two remaining, the second item can be selected in two ways. Thus the selections we can have are:

AB AC BA BC CA CB

Each of the ordered arrangements is known as a *permutation*, each representing the way distinct objects can be arranged.

If there are *n* ways of selecting the first object, there will be $(n - 1)$ ways of selecting the second object, $(n - 2)$ ways of selecting the third object and $(n - r + 1)$ ways of selecting the *r*th object. Thus, by the multiplication rule, the total number of different permutations of selecting *r* objects from *n* distinct objects is thus:

$$n(n - 1)(n - 2) \dots (n - r + 1)$$

The number $n(n - 1)(n - 2) \dots (3)(2)(1)$ is represented by $n!$. The number of permutations of *k* objects chosen from *n* distinct objects is represented by ^nP_r or $_nP_r$ or $\binom{n}{r}$ and is thus:

$$^nP_r = n(n - 1)(n - 2) \dots (n - r + 1) = \frac{n!}{(n-r)!} \qquad [4]$$

r taking values from 0 to *n*. Note that 0! is taken as having the value 1. The number of permutations of *n* objects chosen from *n* distinct objects is represented by ^nP_n or $\binom{n}{n}$ and is thus:

$$^nP_n = n! \qquad [5]$$

Example

In the wiring up of an electronic component there are four assemblies that can be wired up in any order. In how many different ways can the component be wired?

This involves determining the number of permutations of four objects from four. Thus, using equation [5]:

$^nP_n = n! = 4! = 24.$

Example

How many four-digit numbers can be formed from the digits 0 to 9 if no digit is to be repeated within any one number?

This involves determining the number of permutations of 4 objects from 10. Thus, using equation [4]:

$$^nP_n = \frac{n!}{(n-r)!} = \frac{10!}{(10-4)!}$$

$$= \frac{10 \times 9 \times 8 \times 7 \times 6 \times 5 \times 4 \times 3 \times 2 \times 1}{6 \times 5 \times 4 \times 3 \times 2 \times 1} = 5040$$

Combinations

There are often situations where we want to know the number of ways r items can be selected from n objects without being concerned with the order in which the objects are selected. Suppose we had to select two items from a possible three different items A, B, C. The selections, i.e. permutations, we can have are:

AB AC BA BC CA CB

But if we are not concerned with the sequence of the letters then we only have the three ways *AB*, *AC* and *BC*. Such an unordered set is termed a *combination*.

Consider the selection of a combination of r items from n distinct objects. In the selected r items there will be $r!$ permutations (equation [5]) of distinct objects so that the permutation of r items from n contains each group of r items $r!$ times. Since there are $n!/(n-r)!$ different permutations of r items from n we must have:

$$r! \times {}^nC_r = \frac{n!}{(n-r)!}$$

where ${}^nC_r$, ${}_nC_r$ or $\binom{n}{r}$ is used to represent the combination of r items from n. Thus:

$$^nC_r = \frac{n!}{r!(n-r)!} \qquad [6]$$

${}^nC_r$ is often termed a *binomial coefficient*. This is because numbers of this form appear in the expansion of $(x + y)^n$ by the binomial theorem (see Section 7.1.2).

When r items are selected from n distinct objects, $n - r$ items are left. The number of ways of selecting r items from n is given by equation [6] as $n!/r!(n - r)!$. The number of ways of selecting $n - r$ items from n is given by equation [6] as:

$$^nC_{n-r} = \frac{n!}{(n-r)!(n-\{n-r\})!} = \frac{n!}{r!(n-r)!}$$

Thus we can say that there are as many ways of selecting r items from n as selecting $n - r$ objects from n:

$$^nC_r = {^nC_{n-r}} \qquad\qquad [7]$$

There is just one combination of n items from n objects. Thus $^nC_n = 1$. If we select 0 items from n, then because equation [6] gives $^nC_0 = n!/0!$ and we take $1/0! = 1$, we have $^nC_0 = 1$. Evidently there are as many ways of selecting none of the items in a set of n as there are of choosing the n objects that are left.

Example

In how many ways can three objects be chosen from a sample of 20?

Using equation [6]: $^{20}C_3 = \dfrac{20!}{3!17!} = \dfrac{20 \times 19 \times 18}{1 \times 2 \times 3} = 1140$

Example

If a batch of 20 objects contains 3 with faults and a sample of 5 is chosen, what is the probability of obtaining a sample with (a) 0, (b) 1, (c) 2 faulty items?

The number of ways we choose the sample of 5 items out of 20 is, using equation [6]:

$$^{20}C_5 = \frac{20!}{5!15!}$$

(a) The number of ways we can choose a sample with 0 defective items is the number of ways we choose 5 items from 17 good items and is thus:

$$^{17}C_5 = \frac{17!}{5!12!}$$

Thus the probability of choosing a sample with 0 faulty items is:

$$\text{probability} = \frac{^{17}C_5}{^{20}C_5} = \frac{17!}{5!15!} \frac{5!15!}{20!} = \frac{91}{228}$$

(b) The number of ways we can choose a sample with 1 faulty item and 4 good items, i.e. selecting 1 faulty item from 3 faulty items and 4 good items from 17 good items, is given by the multiplication rule as $^3C_1 \times {^{17}C_4}$. Thus the probability of choosing a sample with 1 faulty item is:

$$\text{probability} = \frac{\dfrac{3!}{1!2!} \times \dfrac{17!}{4! \times 13!}}{\dfrac{20!}{5!15!}} = \frac{35}{76}$$

(c) The number of ways we can choose a sample with 2 faulty items and 3 good items, i.e. selecting 2 faulty items from 3 faulty items and 3 good items from 17 good items, is given by the multiplication rule as $^3C_2 \times {}^{17}C_3$. Thus the probability of choosing a sample with 2 faulty items is:

$$\text{probability} = \frac{\dfrac{3!}{2!1!} \times \dfrac{17!}{3! \times 14!}}{\dfrac{20!}{5!15!}} = \frac{5}{38}$$

Conditional probability

The multiplication rule is only valid when the occurrence of one event has no effect upon the probability of the second event occurring. While this can be used in many situations, there are situations where a successful occurrence of the first event affects the probability of occurrence of the second event. Suppose we have 50 objects of which 15 are faulty. What is the probability that the second object selected is faulty given that the first object selected was fault-free? This is a probability problem where the answer depends on the additional knowledge given that the first selection was fault-free. This means that there are less fault-free objects among those remaining for the second selection. Such a problem is said to involve *conditional probability*.

Example

Suppose we have 50 objects of which 15 are faulty. What is the probability that the second object selected is faulty given that the first object selected was fault-free?

Selecting the first object from 50 as fault-free has a probability of 35/50. Because the first object was fault-free we now have 34 fault-free and 15 faulty objects remaining. Now selecting a faulty object from 49 has a probability of 15/49. Using the multiplication rule gives the probability of the first object being fault-free followed by the second object faulty as (35/50)(15/49) = 0.21.

Problems 9.1

1 In a testing period of 1 year, 4 out of 50 of the items tested failed. What is the probability of finding one of the items failing?

2 In a pack of 52 cards there are 4 aces. What is the probability of selecting, at random, an ace from a pack?

3 Testing of a particular item bought for incorporation in a product shows that of 100 items tested, 4 were found to be faulty. What is the probability that one item taken at random will be (a) faulty, (b) free from faults?

4 Resistors manufactured as 10 Ω by a company are tested and 5% are found to have values below 9.5 Ω and 10% above 10.5 Ω. What is the probability that one resistor selected at random will have a resistance between 9.5 Ω and 10.5 Ω?

5 100 integrated circuits are tested and 3 are found to be faulty. What is the probability that one, taken at random, will result in a working circuit?

6 Tests of an electronic product show that 1% have defective integrated circuits alone, 2% have defective connectors alone and 1% have both defective integrated circuits and connectors. What is the probability of one of the products being found to have a (a) defective integrated circuit alone, (b) defective integrated circuit, (c) defective connector, (d) no defects?

7 Cars coming to a junction can turn to the left, to the right or go straight on. If observations indicate that all the possible outcomes are equally likely, determine the probability that a car will (a) go straight on, (b) turn from the straight-on direction.

8 In how many ways can (a) 8 items be selected from 8 distinct objects, (b) 4 items be selected from 7 distinct items, (c) 2 items be selected from 6 distinct items?

9 In how many ways can (a) 2 items be selected from 7 objects, (b) 5 items be selected from 7 objects, (c) 7 items be selected from 7 objects?

10 How many samples of 4 can be taken from 25 items?

11 A batch of 24 components includes 2 that are faulty. If a sample of 2 is taken, what is the probability that it will contain (a) no faulty components, (b) 1 faulty component, (c) 2 faulty components?

12 A batch of 10 components includes 3 that are faulty. If a sample of 2 is taken from the batch, what is the probability that it will contain (a) no faulty components, (b) 2 faulty components?

13 Of 10 items manufactured, 2 are faulty. If a sample of 3 is taken at random, what is the probability it will contain (a) both the faulty items, (b) at least 1 faulty item?

14 A security alarm system is activated and deactivated by keying-in a three-digit number in the proper sequence. What is the total number of possible code combinations if digits may be used more than once?

15 When checking on the computers used in a company it was found that the probability of one having the latest microprocessor was 0.8, the probability of having the latest software 0.6 and the probability of having the latest processor and latest software 0.3. Determine the probability that a computer selected as having the latest software will also have the latest microprocessor.

9.2 Distributions

All measurements are affected by random uncertainties and thus repeated measurements will give readings which fluctuate in a random manner from each other. This section considers the statistical approach to such variability of data, dealing with the measures of location and dispersion, i.e. mean, standard deviation and standard error, and the binomial, Poisson and Normal distributions. This statistical approach to variability is especially important in the production environment and the consideration of the variability of measurements made on products.

9.2.1 Probability distributions

Consider the collection of data on the number of cars per hour passing some point and suppose we have the following results:

10, 12, 11, 13, 11, 12, 14, 12, 12, 11

When the discrete variable is sampled 10 times the value 10 appears once. Thus the probability of 10 appearing is 1/10. The value 11 appears three times and so its probability is 3/10, 12 has a probability of 4/10, 13 has the probability 1/10, 14 has the probability 1/10. Figure 9.3 shows how we can represent these probability values as a *probability distribution*.

Consider some experiment in which repeated measurements are made of the time taken for 100 oscillations of a simple pendulum and suppose we have the following results:

20.1, 20.3, 20.8, 20.5, 21.0, 20.8, 20.3, 20.4, 20.7, 20.6,
20.5, 20.7, 20.5, 20.1, 20.6, 20.4, 20.7, 20.5, 20.6, 20.3

With a continuous variable there are an infinite number of values that can occur within a particular range so the probability of one particular value occurring is effectively zero. However, it is meaningful to consider the probability of the variable falling within a particular subinterval. The term *frequency* is used for the number of times a measurement occurs within an interval and the term *relative frequency* or *probability P(x)* for the fraction of the total number of readings in a segment. Thus if we divide the range of the above results into 0.2 intervals, we have:

values >20.0 and ≤20.2 come up twice, thus $P(x) = 2/20$
values >20.2 and ≤20.4 come up five times, thus $P(x) = 5/20$
values >20.4 and ≤20.6 come up seven times, thus $P(x) = 7/20$
values >20.6 and ≤20.8 come up five times, thus $P(x) = 5/20$
values >20.8 and ≤21.0 come up once, thus $P(x) = 1/20$

The probability always has a value less than 1 and the sum of all the probabilities is 1. Figure 9.4 shows how we can represent this graphically. The probability that x lies within a particular interval is thus the height of the rectangle for that strip divided by the sum

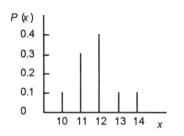

Figure 9.3 *Probability distribution with a discrete variable*

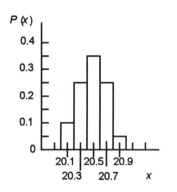

Figure 9.4   *Probability distribution with a continuous variable*

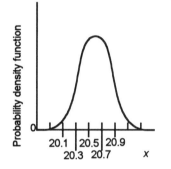

Figure 9.5   *Probability distribution function*

Key point

The *probability density function* is a function that allocates probabilities to all of the range of values that the random variable can take. The probability that the variable will be in any particular interval is obtained by integrating the probability density function over that interval.

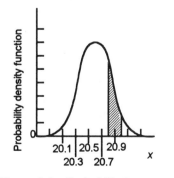

Figure 9.6   *Probability for interval 20.8 to 21.0*

---

of the heights of all the rectangles. Since each strip has the same width w:

$$\text{probability of } x \text{ being in an interval} = \frac{\text{area of strip}}{\text{total area}} \qquad [8]$$

The histogram shown in Figure 9.4 has a jagged appearance. This is because it represents only a few values. If we had taken a very large number of readings then we could have divided the range into smaller segments and still had an appreciable number of values in each segment. The result of plotting the histogram would now be to give one with a much smoother appearance. When the probability distribution graph is a smooth curve, with the area under the curve scaled to have the value 1, then it is referred to as the *probability density function $f(x)$* (Figure 9.5). Then equation [8] gives:

probability of x being in the interval $a < x \le b$
  = area under the function of that interval

$$= \int_a^b f(x)\, dx \qquad [9]$$

Consider the probability, with a very large number of readings, of obtaining a value between 20.8 and 21.0 with the probability distribution function shown in Figure 9.6. If we take a segment 20.8 to 21.0 then the area of that segment is the probability. If, say, the area is 0.30, the probability of taking a single measurement and finding it in that interval is 0.30, i.e. the measurement occurs on average 30 times in every 100 values taken.

Example

The following readings, in metres, were made for a measurement of the distance travelled by an object in 10 s. Plot the results as a distribution with segments of width 0.01 m.

13.478, 13.509, 13.502, 13.457, 13.492, 13.512, 13.475, 13.504, 13.473, 13.482, 13.492, 13.500, 13.493, 13.501, 13.472, 13.477

With segments of width 0.01 m we have:

Segment 13.45 to 13.46, frequency 1, so probability 1/16
Segment 13.46 to 13.47, frequency 0, so probability 0
Segment 13.47 to 13.48, frequency 5, so probability 5/16
Segment 13.48 to 13.49, frequency 1, so probability 1/16
Segment 13.49 to 13.50, frequency 4, so probability 4/16
Segment 13.50 to 13.51, frequency 4, so probability 4/16
Segment 13.51 to 13.52, frequency 1, so probability 1/16

Figure 9.7 shows the resulting distribution.

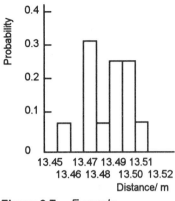

Figure 9.7 *Example*

9.2.2 Measures of location and spread of a distribution

Parameters which can be specified for distributions to give an indication of location and a measure of the dispersion or spread of the distribution about that value are the *mean* for the location and the *standard deviation* for the measure of dispersion.

Mean

The mean value of a set of readings can be obtained in a number of ways, depending on the form with which the data is presented:

- For a list of discrete readings, sum all the readings and divide by the number N of readings, i.e.:

$$\bar{x} = \frac{x_1 + x_2 + x_3 + \dots + x_j}{N} = \frac{\Sigma x_j}{N} \qquad [10]$$

- For a distribution of discrete readings, if we have n_1 readings with value x_1, n_2 readings with value x_2, n_3 readings with value x_3, etc., then the above equation for the mean becomes:

$$\bar{x} = \frac{n_1 x_1 + n_2 x_2 + n_3 x_3 + \dots + n_j x_j}{N} \qquad [11]$$

But n_1/N is the relative frequency or probability of value x_1, n_2/N is the relative frequency or probability of value x_2, etc. Thus, to obtain the mean, multiply each reading by its relative frequency or probability P and sum over all the values:

$$\bar{x} = \sum_{j=1}^{n_j} P_j x_j \qquad [12]$$

- For readings presented as a continuous distribution curve, we can consider that we have a discrete-value distribution with very large numbers of very thin segments. Thus if $f(x)$ represents the probability distribution and x the measurement values, the probability that x will lie in a small segment of width δx is $f(x)\delta x$. Thus the rule given above for discrete-value distributions translates into:

$$\bar{x} = \int_{-\infty}^{\infty} x f(x)\,dx \qquad [13]$$

With a very large number of readings, the mean value is taken as being the *true value* about which the random fluctuations occur. The mean value of a probability distribution function is often termed the *expected value*.

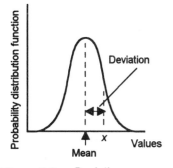

Figure 9.8 *Deviation*

Key point

Accuracy and precision
Consider two marksmen firing 5 shots, from the same gun under the same conditions, at a target with the aim of hitting the central bulls-eye. If the results obtained are as shown in Figure 9.10, then marksman A is accurate in that his shots have a mean which coincides with the bulls-eye. Marksman B has a smaller scatter of results but his mean is not centres on the bulls-eye. His shots have greater precision but are less accurate. Dispersion measurement is thus extremely important when designing manufacturing processes and machining to a set target mean value, i.e. the bulls-eye, with attainable tolerances.

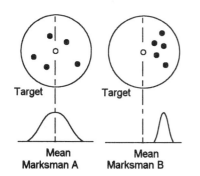

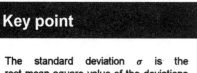

Figure 9.10 *Accuracy and precision*

Key point

The standard deviation σ is the root-mean-square value of the deviations for all the measurements in the distribution.

Standard deviation

Any single reading x in a distribution (Figure 9.8) will deviate from the mean of that distribution by:

$$\text{deviation} = x - \overline{x} \qquad [14]$$

With one distribution we might have a series of values which is widely scattered around the mean while another has readings closely grouped round the mean. Figure 9.9 shows the type of curves that might occur.

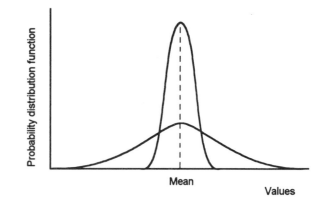

Figure 9.9 *Distributions with different spreads but the same mean*

A measure of the spread of a distribution *cannot* be obtained by taking the mean deviation from the mean, since for every positive value of a deviation there will be a corresponding negative deviation and so the sum will be zero. The measure used is the *standard deviation*.

The standard deviation σ is the root-mean-square value of the deviations for all the measurements in the distribution. The quantity σ^2 is known as the variance of the distribution.

Thus, for a number of discrete values, x_1, x_2, x_3, ..., etc., we can write for the mean value of the sum of the squares of their deviations from the mean of the set of results:

$$\text{sum of squares of deviation} = \frac{\left(x_1 - \overline{x}\right)^2 + \left(x_2 - \overline{x}\right)^2 + \left(x_3 - \overline{x}\right)^2 + ...}{N}$$

Hence the mean of the square root of this sum of the squares of the deviations, i.e. the standard deviation, is:

$$\sigma = \sqrt{\frac{\left(\left(x_1 - \overline{x}\right)^2 + \left(x_2 - \overline{x}\right)^2 + \left(x_3 - \overline{x}\right)^2 + ...\right)}{N}} \qquad [15]$$

However, we need to distinguish between the standard deviation s of a sample and the standard deviation σ of the entire population of readings that are possible and from which we have only considered a sample (many statistics textbooks adopt the convention of using Greek letters when referring to the entire population and Roman for samples). When we are dealing with a sample we need to write:

$$s = \sqrt{\frac{\left(\left(x_1 - \overline{x}_s\right)^2 + \left(x_2 - \overline{x}_s\right)^2 + \left(x_3 - \overline{x}_s\right)^2 + ...\right)}{N-1}} \qquad [16]$$

with $\overline{x}_s$ being the mean of the sample. The reason for using $N - 1$ rather than N is that the root-mean-square of the deviations of the readings in a sample around the sample mean is less than around any other figure. Hence, if the true mean of the entire population were known, the estimate of the standard deviation of the sample data about it would be greater than that about the sample mean. Therefore, by using the sample mean, an underestimate of the population standard deviation is given. This bias can be corrected by using one less than the number of observations in the sample in order to give the sample mean.

For a continuous probability density distribution, since $(n_i/N)\,\delta x$ is the probability for that interval δx, i.e. $f(x)\,\delta x$ where $f(x)$ is the probability distribution function, the standard deviation becomes:

$$\sigma = \int_{-\infty}^{\infty} \left(x - \overline{x}\right)^2 f(x)\,dx \qquad [17]$$

We can write this equation in a more useful form for calculation:

$$\sigma^2 = \int_{-\infty}^{\infty} x^2 f(x)\,dx - 2\overline{x}\int_{-\infty}^{\infty} x f(x)\,dx + \overline{x}^2 \int_{-\infty}^{\infty} f(x)\,dx$$

Since the total area under the probability density function curve is 1, the third integral has the value 1 and so the third term is the square of the means. The second integral is the mean. The first integral is the mean value of x^2. Thus:

$$\sigma^2 = \overline{x^2} - \overline{x}^2 \qquad [18]$$

i.e. the mean value of x^2 minus the square of the mean value.

Example

Determine the mean value and the standard deviation of the sample of 10 readings 8, 6, 8, 4, 7, 5, 7, 6, 6, 4.

The mean value is (8 + 6 + 8 + 4 + 7 + 5 + 7 + 6 + 6 + 4)/10 = 6.1.

The standard deviation of the sample can be calculated by considering the deviations of each reading from the mean, these being:

1.9, –0.1, 1.9, –2.1, 1.9, –1.1, 0.9, –0.1, –0.1, –2.1

The squares of these deviations are:

3.61, 0.01, 3.61, 4.41, 3.61, 1.21, 0.81, 0.01, 0.01, 4.41

The sum of these squares is 21.7. If we consider we do not have the entire population but just a sample, then the standard deviation is $\sqrt{(21.7/9)} = 1.6$.

Example

In an experiment involving the counting of the number of events that occurred in equal size time intervals the following data was obtained:

0 events 13 times, 1 event 12 times, 2 events 9 times
3 events 5 times, 4 events once

Determine the mean number of events occurring in the time interval and the standard deviation.

The total number of measurements is $13 + 12 + 9 + 5 + 1 = 40$ and so the mean value is $(0 \times 13 + 1 \times 12 + 2 \times 9 + 3 \times 5 + 4 \times 1)/40 = 1.25$.

We have 13 measurements with deviation –1.25, 12 with deviation –0.25, 9 with deviation +0.75, 5 with deviation +1.75 and 1 event with deviation 2.75. We can take the squares of these deviations, sum them and divide by (40 – 1). Hence the standard deviation is 1.1.

Example

A probability density function has $f(x) = 1$ for $0 \leq x < 1$ and elsewhere 0. Determine its mean and standard deviation.

The mean value is given by equation [13] as:

$$\bar{x} = \int_{-\infty}^{\infty} x f(x)\,dx = \int_{0}^{1} x\,dx = \left[\frac{x^2}{2} \right]_{0}^{1} = \frac{1}{2}$$

The standard deviation is given by equation [17]:

$$\sigma^2 = \int_{-\infty}^{\infty}(x-\bar{x})^2 f(x)\,dx = \int_0^1\left(x-\tfrac{1}{2}\right)^2 dx = \int_0^1\left(x^2 - x + \tfrac{1}{4}\right) d$$

$$= \left[\frac{x^3}{3} - \frac{x^2}{2} + \frac{x}{4}\right]_0^1 = \tfrac{1}{3} - \tfrac{1}{2} + \tfrac{1}{4} = \tfrac{1}{12} = 0.29$$

Alternatively, using equation [18]:

$$\sigma^2 = \overline{x^2} - \bar{x}^2$$

since we have:

$$\overline{x^2} = \int_{-\infty}^{\infty} x^2 f(x)\,dx = \int_0^1 x^2\,dx = \left[\frac{x^3}{3}\right]_0^1 = \tfrac{1}{3}$$

then:

$$\sigma^2 = \tfrac{1}{3} - \left(\tfrac{1}{2}\right)^2 = \tfrac{1}{3} - \tfrac{1}{4} = \tfrac{1}{12} = 0.29$$

Standard error of the mean

With a sample set of readings taken from a large population we can determine its mean, but what is generally required is an estimate of the error of that mean from the true value, i.e. the mean of an infinitely large number of readings. We can consider any set of readings as being just a sample taken from the very large set.

Consider one sample of readings with n values being taken: x_1, x_2, x_3, ... x_n. The mean of this sample is:

$$\bar{x} = \frac{1}{n}\sum x_j$$

This mean will have a deviation or error E from the true mean value X of :

$$E = \bar{x} - X.$$

Hence we can write:

$$E = \left(\frac{1}{n}\sum x_j\right) - X = \frac{1}{n}\sum\left(x_j - X\right)$$

If we write e_1 for the error of the first reading from the true mean, e_2 for the error of the second, etc. we obtain:

$$E = \frac{1}{n}\left(e_1 + e_2 + e_3 + \dots e_j\right)$$

Thus:

$$E^2 = \frac{1}{n^2}\left(e_1^2 + e_2^2 + e_3^2 + \dots + \text{products such as } e_1 e_2, \text{ etc.}\right)$$

E is the error from the mean for a single sample of readings. Now, consider a large number of such samples with each set having the same number n of readings. We can write such an equation as above for each sample. If we add together the equations for all the samples and divide by the number of samples considered, we obtain an average value over all the samples of E^2. Thus E is the standard deviation of the means and is known as the *standard error of the means* e_m (more usually the symbol σ). Adding together all the error product terms will give a total value of zero, since as many of the error values will be negative as well as positive. The average of all the Σe_j^2 terms is $n e_s^2$, where e_s is, what can be termed, the *standard error of the sample*. Thus:

$$e_m = \frac{e_s}{\sqrt{n}}$$

But how can we obtain a measure of the standard error of the sample? The standard error is measured from the true value X, which is not known. What we can measure is the standard deviation of the sample from its mean value. The best estimate of the standard error for a sample turns out to be the standard deviation s of a sample when we define it as:

$$s^2 = \frac{1}{n-1} \Sigma (x_j - \overline{x})^2$$

i.e. with a denominator of $N - 1$, rather than just N. Thus the best estimate of the standard error of the mean can be written as:

$$\text{standard error of the mean} = \frac{s}{\sqrt{n}} \tag{19}$$

Key point

The standard error of the mean obtained from a sample is the standard deviation of the sample divided by the square root of the sample size.

Example

Measurements are to be made of the percentage of an element in a chemical by making measurements on a number of samples. The standard deviation of any one sample is found to be 2%. How many measurements must be made to give a standard error of 0.5% in the estimated percentage of the element.

If n measurements are made, then the standard error of the sample mean is given by equation [19] and so:

$$n = \frac{2^2}{0.5^2} = 16$$

9.2.3 Common distributions

There are three basic forms of distribution which are found to represent many forms of distributions commonly encountered in engineering and science. These are the binomial distribution, the

Poisson distribution and the normal distribution (sometimes called the Gaussian distribution). Binomial distributions are often approximated by the Poisson distribution. The normal distribution is a model widely used for experimental measurements when there are random errors.

Binomial distribution

In the tossing of a single coin the result is either heads or tails uppermost. We can consider this as an example of an experiment where the results might be termed as either success or failure, one result being the complement of the other. If the probability of succeeding is p then the probability of failing is $1 - p$. Such a form of experiment is termed a *Bernoulli trial*.

Suppose the trial is the throwing of a die with a 6 uppermost being success. The probability of obtaining a 6 as the result of one toss of the die is 1/6 and the probability of not obtaining a 6 is 5/6. Suppose we toss the die n times. The probability of obtaining no 6s in any of the trials is given by the product rule as $(5/6)^n$. The probability of obtaining one 6 in, say, just the first trial out of the n is $(5/6)^{n-1}$ (1/6). But we could have obtained the one 6 in any one of the n trials. Thus the probability of one 6 is $n(5/6)^{n-1}$ (1/6). The probability of obtaining two 6s in, say, just the first two trials is $(5/6)^{n-2}(1/6)^2$. But these two 6s may occur in the n trials in a number of combinations $n!/2!(n - 2)$ (see Section 9.1.2). Thus the probability of two 6s in n trials is $[n!/2!(n - 2)](5/6)^{n-2}(1/6)^2$. We can continue this for three 6s, 4s, etc.

In general, if we have n independent Bernoulli trials, each with a success probability p, and of those n trials k give successes, and $(n - k)$ failures, the probability of this occurring is given by the product rule as:

$$^nC_k p^k(1 - p)^{n-k} = \frac{n!}{k!(n - k)!} p^k(1 - p)^{n-k} \qquad [20]$$

This is termed the *binomial distribution*. This term is used because, for $k = 0, 1, 2, 3, \ldots n$, the values of the probabilities are the successive terms of the binomial expansion of $[(1 - p) + p]^n$.

For a single Bernoulli trial of a random variable x with probability of success p, the mean value is p. The standard deviation is given by equation [18] as:

$$\sigma^2 = \overline{x^2} - \overline{x}^2 = p^2 - p = p(1 - p)$$

For n such trials:

mean value $= np$ $\qquad\qquad\qquad$ [21]

standard deviation $= \sqrt{np(1 - p)}$ $\qquad\qquad$ [22]

Key point

The characteristics of a variable that gives a binomial distribution are that the experiment consists of n identical trials, there are two possible complementary outcomes, success or failure, for each trial and the probability of a success is the same for each trial, the trials are independent and the distribution variable is the number of successes in n trials.

Key point

In the example, if the probability of a sale is p and the probability of no sale is q, then $q + p = 1$. We are concerned with 6 enquiries and so if we consider the binomial expansion of $(q + p)^6$ we have:

$$(q + p)^6 = q^6 + 6q^5p + 15q^4p^2 + 20q^3p^3 + 15q^2p^2 + 6qp^5 + p^6$$

Each successive term in the expansion gives the probability of 0, 1, 2, 3, 4, 5 or 6 sales. Thus, with $p = 0.3$ we have $q = 0.7$ and so the probability of 0 sales is $0.7^6 = 0.118$.

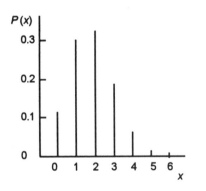

Figure 9.11 *Example*

Example

The probability that an enquiry from a potential customer will lead to a sale is 0.30. Determine the probabilities that among six enquiries there will be 0, 1, 2, 3, 4, 5, 6 sales.

Using equation [20]:

The probability of 0 is:

$$\frac{n!}{k!(n-k)!}p^k(1-p)^{n-k} = \frac{6!}{0!6!}(0.30)^0(0.70)^6 = 0.118$$

The probability of 1 is:

$$\frac{n!}{k!(n-k)!}p^k(1-p)^{n-k} = \frac{6!}{1!5!}(0.30)^1(0.70)^2 = 0.303$$

The probability of 2 is:

$$\frac{n!}{k!(n-k)!}p^k(1-p)^{n-k} = \frac{6!}{2!4!}(0.30)^2(0.70)^4 = 0.324$$

The probability of 3 is:

$$\frac{n!}{k!(n-k)!}p^k(1-p)^{n-k} = \frac{6!}{3!3!}(0.30)^3(0.70)^3 = 0.185$$

The probability of 4 is:

$$\frac{n!}{k!(n-k)!}p^k(1-p)^{n-k} = \frac{6!}{4!2!}(0.30)^4(0.70)^2 = 0.060$$

The probability of 5 is:

$$\frac{n!}{k!(n-k)!}p^k(1-p)^{n-k} = \frac{6!}{5!1!}(0.30)^5(0.70)^1 = 0.010$$

The probability of 6 is:

$$\frac{n!}{k!(n-k)!}p^k(1-p)^{n-k} = \frac{6!}{6!0!}(0.30)^6(0.70)^0 = 0.001$$

Figure 9.11 shows the distribution.

Poisson distribution

The Poisson distribution for a variable λ is:

$$P(k) = \frac{\lambda^k e^{-\lambda}}{k!} \tag{23}$$

for $k = 0, 1, 2, 3$, etc. The mean of this distribution is λ and the standard deviation is $\sqrt{\lambda}$. When the number n of trials is very large and the probability p small, e.g. $n > 25$ and $p < 0.1$, binomial

Key point

Approximating the binomial distribution to the Poisson distribution

If p is the possibility of an event occurring and q the possibility that it does not occur, then $q + p = 1$ and so if we consider n samples, we have $(q + p)^n = 1$ and the binomial expression gives:

$$(q+p)^n = q^n + nq^{n-1}p + \frac{n(n-1)q^{n-2}p^2}{2!} + \frac{n(n-1)(n-2)q^{n-3}p^3}{3!} + \dots$$

If p is small then $q = 1 - p \times 1$ and with n large the first few terms have $n - 1$ approximating to n, $n - 2$ to n, etc. The binomial expression can thus be approximated to:

$$1^n = 1^n + n1^{n-1}p + \frac{nn1^{n-2}p^2}{2!}$$

$$+ \frac{nnn1^{n-3}p^3}{3!} + \dots$$

$$1^n = 1 + np + \frac{n^2p^2}{2!} + \frac{n^3p^3}{3!} + \dots$$

If we let $np = \lambda$ then:

$$1^n = 1 + \lambda + \frac{\lambda^2}{2!} + \frac{\lambda^3}{3!} + \dots$$

But this is the series for e^λ (see Table 7.1) and so $1^n = e^\lambda$. We can thus write the binomial expression as:

$$(q+p)^n = e^{-\lambda}\left[1 + \lambda + \frac{\lambda^2}{2!} + \frac{\lambda^3}{3!} + \dots\right]$$

and so the terms in the expression are given by equation [23].

probabilities are often approximated by the *Poisson distribution*. Thus, since the mean of the binomial distribution is np (equation [21]) and the standard deviation (equation [22]) approximates to $\sqrt{np}$ when p is small, we can consider λ to represent np. Thus λ can be considered to represent the average number of successes per unit time or unit length or some other parameter.

Example

2% of the output per month of a mass produced product have faults. What is the probability that of a sample of 400 taken that 5 will have faults?

Assuming the Poisson distribution, we have $\lambda = np = 400 \times 0.02 = 8$ and so equation [23] gives for $k = 5$:

$$P(5) = \frac{8^5 e^{-8}}{5!} = 0.093$$

Example

The output from a CNC machine is inspected by taking samples of 60 items. If the probability of a defective item is 0.0015, determine the probability of the sample having (a) two defective items, (b) more than two defective items.

(a) We have $n = 60$ and $p = 0.0015$. Thus, assuming a Poisson distribution, we have $\lambda = np = 60 \times 0.0015 = 0.09$ and so equation [23] gives the probability of two defective items as $(0.092 \times e^{-0.09})/2! = 3.7 \times 10^{-3}$ or 0.37%.
(b) The probability of there being more than two defective items is $1 - \{P(0) + P(1) + P(2)\} = 1 - e^{-\lambda}\{1 + \lambda + \lambda^2/2!\} = 1 - e^{-0.09}\{1 + 0.09 + 4.5 \times 10^{-3}\} = 1.36 \times 10^{-4}$ or 0.01%.

Example

There is a 1.5% probability that a machine will produce a faulty component. What is the probability that there will be at least 2 faulty items in a batch of 100?

Assuming the Poisson distribution can be used, we have $\lambda = np = 100 \times 0.015 = 1.5$ and so the probability of at least 2 faulty items will be:

$$P(\geq 2) = 1 - P(0) - P(1) = 1 - \frac{1.5^0}{0!} e^{-1.5} - \frac{1.5^1}{1!} e^{-1.5} = 0.442$$

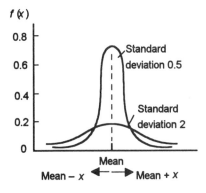

Figure 9.12   *Normal distribution*

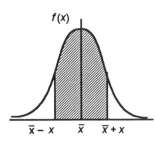

Figure 9.13   *Values within
+ or – x of the mean*

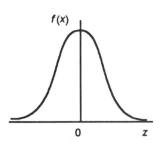

Figure 9.14   *Standard
normal distribution*

Normal distribution

A particular form of distribution, known as the *normal distribution* or *Gaussian distribution*, is very widely used and works well as a model for experimental measurements when there are random errors. This form of distribution has a characteristic bell shape (Figure 9.12). It is symmetric about its mean value, having its maximum value at that point, and tends rapidly to zero as x increases or decreases from the mean. It can be completely described in terms of its mean and its standard deviation. The following equation describes how the values are distributed about the mean:

$$f(x) = \frac{1}{\sigma\sqrt{2\pi}}\, e^{-(x-\bar{x})^2/2\sigma^2} \qquad [24]$$

The fraction of the total number of values that lies between $-x$ and $+x$ from the mean is the fraction of the total area under the curve that lies between those ordinates (Figure 9.13). We can obtain areas under the curve by integration.

To save the labour of carrying out the integration, the results have been calculated and are available in tables. As the form of the graph depends on the value of the standard deviation, as illustrated in Figure 9.12, the area depends on the value of the standard deviation σ. In order not to give tables of the areas for different values of x for each value of σ, the distribution is considered in terms of the value of:

$$(x - \bar{x})/\sigma$$

this commonly being designated by the symbol z, and areas tabulated against this quantity. z is known as the *standard normal random variable* and the distributions obtained with this as the variable are termed the *standard normal distribution* (Figure 9.14). Any other normal random variable can be obtained from the standard normal random variable by multiplying by the required standard deviation and adding the mean, i.e.

$$x = \sigma z + \bar{x}$$

Table 9.1 shows examples of the type of data given in tables for z.
When we have:

$$x - \bar{x} = 1\sigma$$

then $z = 1.0$ and the area between the ordinate at the mean and the ordinate at 1σ as a fraction of the total area is 0.341 3. The area within $\pm 1\sigma$ of the mean is thus the fraction 0.681 6 of the total area under the curve, i.e. 68.16%. This means that the chance of a value being within $\pm 1\sigma$ of the mean is 68.16%, i.e. roughly two-thirds of the values.

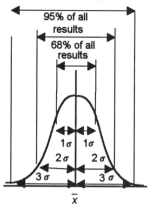

99.7% of all results

95% of all results

68% of all results

1σ 1σ

2σ 2σ

3σ 3σ

$\overline{x}$

Figure 9.15 *Normal distribution*

Table 9.1 *Areas under normal curve*

z	Area from mean	z	Area from mean
0	0.000 0	1.6	0.445 2
0.2	0.079 3	1.8	0.464 1
0.4	0.155 5	2.0	0.477 2
0.6	0.225 7	2.2	0.486 1
0.8	0.288 1	2.4	0.491 8
1.0	0.341 3	2.6	0.495 3
1.2	0.384 9	2.8	0.497 4
1.4	0.419 2	3.0	0.498 7

When we have:

$$x - \overline{x} = 2\sigma$$

then $z = 2.0$ and the area between the ordinate at the mean and the ordinate at 1σ as a fraction of the total area is 0.477 2. The area within $\pm 2\sigma$ of the mean is thus the fraction 0.954 4 of the total area under the curve, i.e. 95.44%. This means that the chance of a value being within $\pm 2\sigma$ of the mean is 95.44%.

When we have:

$$x - \overline{x} = 3\sigma$$

then $z = 3.0$ and the area between the ordinate at the mean and the ordinate at 3σ as a fraction of the total area is 0.498 7. The area within $\pm 1\sigma$ of the mean is thus the fraction 0.997 4 of the total area, i.e. 99.74%. This means that the chance of a reading being within $\pm 3\sigma$ of the mean is 99.74%. Thus, virtually all the readings will lie within $\pm 3\sigma$ of the mean. Figure 9.15 illustrates the above.

Key point

Table 9.1 is only a rough version of the more detailed tables that are needed in statistical process control. Detailed tables are readily available.

Example

Measurements are made of the tensile strengths of samples taken from a batch of steel sheet. The mean value of the strength is 800 MPa and it is observed that 8% of the samples give values that are below an acceptable level of 760 MPa. What is the standard deviation of the distribution if it is assumed to be normal?

This means that an area from the mean of 0.50 − 0.08 = 0.42. To the accuracy given in Table 9.1, this occurs when $z = 1.4$ (Figure 9.16). Thus,

$$(x - \overline{x})/\sigma = (760 - 800)/\sigma = 1.$$

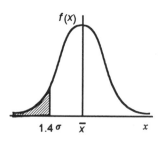

$f(x)$

1.4σ $\overline{x}$ x

Figure 9.16 *Example*

and so the standard deviation is 29 MPa. In the above analysis, it was assumed that the mean given was the true value or a good enough approximation.

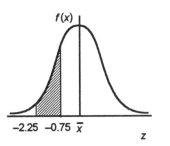

Figure 9.17 *Example*

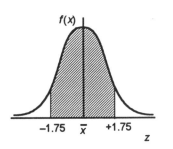

Figure 9.18 *Example*

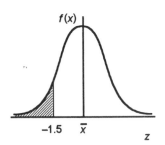

Figure 9.19 *Example*

Example

A pharmaceutical manufacturer produces tablets having a mean mass of 4.0 g and a standard deviation of 0.2 g. Assuming that the masses are normally distributed and that a table is chosen at random, determine the probability that it will (a) have a mass between 3.55 and 3.85 g, (b) will differ from the mean by less than 0.35 g, and (c) determine the number that might be expected to have a mass less than 3.7 g in a carton of 400.

(a) The probability of tablets having masses between 3.55 g and 3.85 g is the area between the normal distribution with ordinates of these masses (Figure 9.17). We have $z_1 = (3.55 - 4.0)/0.2 = -2.25$ and $z_2 = (3.85 - 4.0)/0.2 = -0.75$. Table 9.1, or better tables, gives, approximately, (area between mean and 2.25) − (area between mean and 0.75) = 0.4878 − 0.2734 = 0.2144 or about 21.4%.

(b) To determine the probability of tables differing from the mean of 4.0 g by less than 0.35 g we consider the area between the ordinates for masses between −3.65 g and 4.35 g. These give z values of −1.75 and +1.75 and the area is thus as indicated in Figure 9.18. This is $2 \times 0.4599 = 0.9198$ or about 92%.

(c) The probability of a mass less than 3.7 g is for a z value less than $(3.7 - 4.0)/0.2 = -1.5$ (Figure 9.19). The area within +1.5 and −1.5 of the mean is $2 \times 0.433 = 0.866$. The total area under the curve is 1 and so the area outside these limits is $1 - 0.866 = 0.134$. Just half of this area will be for less than 3.7 g and so the area is 0.134. For 400 tablets this means 0.134×400 or about 27 tablets.

Problems 9.2

1 Determine the mean value of a variable which can have the discrete values of 2, 3, 4 and 5 and for which 2 occurs twice, 3 occurs three times, 4 occurs three times and 5 occurs once.
2 The probability density function of a random variable x is given by $\frac{1}{2}x$ for $0 \leq x < 2$ and 0 for all other values. Determine the mean value of the variable.
3 Determine the mean and the standard deviation for the following data: 10, 20, 30, 40, 50.

4 Determine the standard deviation of the resistance values for a sample of 12 resistors taken from a batch if the values are: 98, 95, 109, 99, 102, 99, 106, 96, 101, 108, 94, 102 Ω.

5 Determine the standard deviation of the six values: 1.3, 1.4, 0.8, 0.9, 1.2, 1,0.

6 Determine the standard deviation of the probability distribution function $f(x) = 2x$ for $0 \le x < 1$ and 0 elsewhere.

7 The following are the results of 100 measurements of the times for 50 oscillations of a simple pendulum:

Between 58.5 and 61.5 s, 2 measurements
Between 61.5 and 64.5 s, 6 measurements
Between 64.5 and 67.5 s, 22 measurements
Between 67.5 and 70.5 s, 32 measurements
Between 70.5 and 73.5 s, 28 measurements
Between 73.5 and 76.5 s, 8 measurements
Between 76.5 and 79.5 s, 2 measurements

(a) Determine the relative frequencies of each segment.
(b) Determine the mean and the standard deviation.

8 A random sample of 25 items is taken and found to have a standard deviation of 2.0. (a) What is the standard error of the sample? (b) What sample size would have been required if a standard error of 0.5 was acceptable?

9 It has been found that 10% of the screws produced are defective. Determine the probabilities that a random sample of 20 will contain 0, 1, 2, 3, 4, 5, 6, 7, 8, 9, 10 defectives.

10 The probability that any one item from a production line will be accepted is 0.70. What is the probability that when 5 items are randomly selected that there will be 2 unacceptable items?

11 Packets are filled automatically on a production line and, from past experience, 2% of them are expected to be underweight. If an inspector takes a random sample of 10, what will be the probability that (a) 0, (b) 1 of the packets will be underweight?

12 1% of the resistors produced by a factory are faulty. If a sample of 100 is randomly taken, what is the probability of the sample containing no faulty resistors?

13 The probability of a mass-produced item being faulty has been determined to be 0.10. What are the probabilities that a random sample of 50 will contain 0, 1, 2, 3, 4, 5, 6, 7, 8, 9, 10 faulty items?

14 A product is guaranteed not to contain more than 2% that are outside the specified tolerances. In a random sample of 10, what is the probability of getting 2 or more outside the specified tolerances?

15 A large consignment of resistors is known to have 1% outside the specified tolerances. What would be the expected number of resistors outside the specified tolerances in a batch of 10 000 and the standard deviation?

16 The number of cars that enter a car park follows a Poisson distribution with a mean of 4. If the car park can accommodate 12 cars, determine the probability that the car park is filled up by the end of the first hour it is open.

17 On average six of the cars coming per day off a production line have faults. What is the probability that four faulty cars will come off the line in one day?

18 The number of breakdowns per month for a machine averages 1.8. Determine the probability that the machine will function for a month with only one breakdown.

19 Measurements of the resistances of resistors in a batch gave a mean of 12 Ω with a standard deviation of 2 Ω. If the resistances can be assumed to have a normal distribution about this mean, how many from a batch of 300 resistors are likely to have resistances more than 15 Ω?

20 Measurements made of the lengths of components as they come off the production line have a mean value of 12 mm with a standard deviation of 2 mm. If a normal distribution can be assumed, in a sample of 100 how many might be expected to have (a) lengths of 15 mm or more, (b) lengths between 13.7 and 16.1 mm?

21 Measurements of the times taken for workers to complete a particular job have a mean of 29 minutes and a standard deviation of 2.5. Assuming a normal distribution, what percentage of the times will be (a) between 31 and 32 minutes, (b) less than 26 minutes?

22 Inspection of the lengths of components yields a normal distribution with a mean of 102 mm and a standard deviation of 1.5 mm. Determine the probability that if a component is selected at random it will have a length (a) less than 100 mm, (b) more than 104 mm, (c) between 100 and 104 mm.

23 A machine makes resistors with a mean value of 50 Ω and a standard deviation of 2 Ω. Assuming a normal distribution, what limits should be used on the values of the resistance if there are to be not more than 1 reject in 1000.

24 A series of measurements was made of the periodic time of a simple pendulum and gave a mean of 1.23 s with a standard deviation of 0.01 s. What is the chance that, when a measurement is made, it will lie between 1.23 and 1.24 s?

25 The measured resistance per metre of samples of a wire have a mean resistance of 0.13 Ω and a standard deviation of 0.005 W. Determine the probability that a randomly selected wire will have a resistance per metre of between 0.12 and 0.14 Ω.

26 A set of measurements has a mean of 10 and a standard deviation of 5. Determine the probability that a measurement will lie between 12 and 15.

27 A set of measurements has a normal distribution with a mean of 10 and a standard deviation of 2.1. Determine the probability of a reading having a value (a) greater than 11 and (b) between 7.6 and 12.2.

9.3 Experimental errors

Experimental *error* is defined as the difference between the result of a measurement and the true value:

error = measured value – true value   [25]

Errors can arise from such causes as instrument imperfections, human imprecision in making measurements and random fluctuations in the quantity being measured. This section is a brief consideration of the estimation of errors and their determination in a quantity which is a function of more than one measured variable.

With measurements made with an instrument, errors can arise from fluctuations in readings of the instrument scale due to perhaps the settings not being exactly reproducible and operating errors because of human imprecision in making the observations. The term *random error* is used for those errors which vary in a random manner between successive readings of the same quantity. The term *systematic error* is used for errors which do not vary from one reading to another, e.g. those arising from a wrongly set zero. Random errors can be determined and minimised by the use of statistical analysis, systematic errors require the use of a different instrument or measurement technique to establish them.

With random errors, repeated measurements give a distribution of values. This can be generally assumed to be a normal distribution. The standard error of the mean of the experimental values can be estimated from the spread of the values and it is this which is generally quoted as the error, there being a 68% probability that the mean will lie within plus or minus one standard error of the true mean. Note that the standard error does not represent the maximum possible error. Indeed there is a 32% probability of the mean being outside the plus and minus standard error interval.

With just a single measurement, say the measurement of a temperature by means of a thermometer, the error is generally quoted as being plus or minus one-half of the smallest scale division. This, termed the *reading error*, is then taken as an estimate of the standard deviation that would occur for that measurement if it had been repeated many times.

Key points

Random errors are those errors which vary in a random manner between successive readings of the same quantity. Random errors can be determined and minimised by the use of statistical analysis.

Systematic errors are those errors which do not vary from one reading to another, e.g. those arising from a wrongly set zero. Systematic errors require the use of a different instrument or measurement technique to establish them.

Example

A rule used for the measurement of a length has scale readings every 1 mm. Estimate the error to be quoted when the rule is used to make a single measurement of a length.

The error is quoted as ±0.5 mm.

> ## Example
>
> Measurements of the tensile strengths of test pieces taken from a batch of incoming material gave the following results: 40, 42, 39, 41, 45, 40, 41, 43, 45, 46 MPa. Determine the mean tensile strength and its error.
>
> The mean is given by equation [10] as (40 + 42 + 39 + 41 + 45 + 40 + 41 + 43 + 45 + 44)/10 = 42. The standard deviation can be calculated by the use of equation [12]. The deviations from the mean are –2, 0, –3, –1, 3, –1, 1, 3, 4 and their squares are 4, 0, 9, 1, 9, 1, 1, 9, 16. The standard deviation is thus $\sqrt{[50/(10 - 1)]}$ = 2.4 MPa. The standard error is thus 2.4/$\sqrt{10}$ = 0.8 MPa. Thus the result can be quoted as 42 ± 0.8 MPa.

Statistical errors

In addition to measurement errors arising from the use of instruments, there are what might be termed *statistical errors*. These are not due to any errors arising from an instrument but from statistical fluctuations in the quantity being measured, e.g. the count rate of radioactive materials. The observed values here are distributed about their mean in a Poisson distribution and so the standard deviation is the square root of the mean value.

> ## Example
>
> In an experiment, the number of alpha particles emitted over a fixed period of time is measured as 4206. Determine the standard deviation of the count.
>
> Assuming the count follows a Poisson distribution, the standard deviation will be the square root of 4206 and so 65. Thus the count can be recorded as 4206 ± 65.

9.3.1 Combining errors

An experiment might require several quantities to be measured and then the values inserted into an equation so that the required variable can be calculated. For example, a determination of the density of a material might involve a determination of its mass and volume, the density then being calculated from mass/volume. If the mass and volume each have errors, how do we combine these errors in order to determine the error in the density?

Consider a variable Z which is to be determined from sets of measurements of A and B and for which we have the relationship $Z = A + B$. If we have A with an error ΔA and B with an error ΔB

then we might consider that $Z + \Delta Z = A + \Delta A + B + \Delta B$ and so we should have

$$\Delta Z = \Delta A + \Delta B \qquad [26]$$

i.e. the error in Z is the sum of the errors in A and B. However, this ignores the fact that the error is the standard error and so is just the value at which there is a probability of ±68% that the mean value for A or B will be within that amount of the true mean. If we consider the set of measurements that were used to obtain the mean value of A and its standard error and the set of measurements to obtain the mean value of B and its standard error and consider the adding together of individual measurements of A and B then we can write $\Delta Z = \Delta A + \Delta B$ for each pair of measurements. Squaring this gives:

$$\Delta Z^2 = (\Delta A + \Delta B)^2 = (\Delta A)^2 + (\Delta B)^2 + 2\,\Delta A\,\Delta B$$

We can write such an equation for each of the possible combinations of measurements of A and B. If we add together all the possible equations and divide by the number of such equations, we would expect the $2\,\Delta A\,\Delta B$ terms to cancel out since there will be as many situations with it having a negative value as a positive value. Thus the equation we should use to find the error in Z is:

$$\Delta Z^2 = (\Delta A)^2 + (\Delta B)^2 \qquad [27]$$

The same equation is obtained for $Z = A - B$.

Now consider the error in Z when $Z = AB$. As before, we might argue that $Z + \Delta Z = (A + \Delta A)(B + \Delta B)$ and so $\Delta Z = B\,\Delta A + A\,\Delta B$, if we ignore as insignificant the $\Delta A\,\Delta B$ term. Hence:

$$\frac{\Delta Z}{Z} = \frac{B\,\Delta A + A\,\Delta B}{AB} = \frac{\Delta A}{A} + \frac{\Delta B}{B} \qquad [28]$$

i.e. the fractional error in Z is the sum of the fractional errors in A and B or the percentage error in Z is the sum of the percentage errors in A and B. However, this ignores the fact that the error is the standard error and so is just the value at which there is a probability of ±68% that the mean value for A or B will be within that amount of the true mean. If we consider the set of measurements that were used to obtain the mean value of A and its standard error and the set of measurements to obtain the mean value of B and its standard error and use equation [28] for each such combination, then we can write:

$$\left(\frac{\Delta Z}{Z}\right)^2 = \left(\frac{\Delta A}{A} + \frac{\Delta B}{B}\right)^2 = \left(\frac{\Delta A}{A}\right)^2 + \left(\frac{\Delta B}{B}\right)^2 + 2\left(\frac{\Delta A}{A}\right)\left(\frac{\Delta B}{B}\right)$$

We can write such an equation for each of the possible combinations of measurements of A and B. If we add together all the possible equations and divide by the number of such equations, the $2(\Delta A/A)(\Delta B/B)$ terms cancel out since there will be as many situations with it having a negative value as a positive value. Thus the equation we should use to find the error in Z is:

$$\left(\frac{\Delta Z}{Z}\right)^2 = \left(\frac{\Delta A}{A}\right)^2 + \left(\frac{\Delta B}{B}\right)^2 \qquad [29]$$

The same equation is obtained for $Z = A/B$. If $Z = A^2$ then this is just the product of A and A and so equation [29] gives $(\Delta Z/Z)^2 = 2(\Delta A/A)^2$. Thus for $Z = A^n$ we have:

$$\left(\frac{\Delta Z}{Z}\right)^2 = n\left(\frac{\Delta A}{A}\right)^2 \qquad [30]$$

In all the above discussion it was assumed that the mean value of Z was given when the mean values of A and B were used in the defining equation.

Example

The resistance of a resistor is determined from measurements of the potential difference across it and the current through it. If the potential difference has been measured as 2.1 ± 0.2 V and the current as 0.25 ± 0.01 A, what is the resistance and its error?

The mean resistance is 2.1/0.25 = 8.4 Ω. The fractional error in the potential difference is 0.2/2.1 = 0.095 and the fractional error in the current is 0.01/0.25 = 0.04. Hence the fractional error in the resistance is $\sqrt{(0.095^2 + 0.04^2)}$ = 0.10. Thus the resistance is 8.4 ± 0.9 Ω.

Example

If $g = 4\pi^2 L/T^2$ and L has been measured as 1.000 ± 0.005 m and T as 2.0 ± 0.1 s, determine g and its error.

The mean value of g is $4\pi^2(1.000)/2.0^2$ = 9.87 m/s^2. The fractional error in L is 0.005 m and that in T is 0.05 s. Thus:

(fractional error in g)^2 = $0.005^2 + 2 \times 0.05^2$

Thus the fractional error in g is 0.071 and so g = 9.87 ± 0.7 m/s^2.

Problems 9.3

1 Determine the mean value and standard error for the measured diameter of a wire if it is measured at a number of points and gave the following results: 2.11, 2.05, 2.15, 2.12, 2.16, 2.14, 2.16, 2.17, 2.13, 2.15 mm.

2 An ammeter has a scale with graduations at intervals of 0.1 A. Give an estimate of the standard deviation.

3 Determine the mean and the standard error for the resistance of a resistor if repeated measurements gave 51.1, 51.2, 51.0, 51.4, 50.9 Ω.

4 In an experiment the number of gamma rays emitted over a fixed period of time is measured as 5210. Determine the standard deviation of the count.

5 How big a count should be made of the gamma radiation emitted from a radioactive material if the percentage error should be less than 1%?

6 Repeated measurements of the voltage necessary to cause the breakdown of a dielectric gave the results 38.9, 39.3, 38.6, 38.8, 38.8, 39.0, 38.7, 39.4, 39.7, 38.4, 39.0, 39.1, 39.1, 39.2 kV. Determine the mean value and the standard error of the mean.

7 Determine the mean value and error for Z when (a) $Z = A - B$, (b) $Z = 2AB$, (c) $Z = A^3$, (d) $Z = B/A$ if $A = 100 \pm 3$ and $B = 50 \pm 2$.

8 The resistivity of a wire is determined from measurements of the resistance R, diameter d and length L. If the resistivity is RA/L, where A is the cross-sectional area, which measurement requires determining to the greatest accuracy if it is not to contribute the most to the overall error in the resistivity?

9 The cross-sectional area of a wire is determined from a measurement of the diameter. If the diameter measurement gives 2.5 ± 0.1 mm, determine the area of the wire and its error.

10 Determine the mean value and error for Z when (a) $Z = A + B$, (b) $Z = AB$, (c) $Z = A/B$ if $A = 100 \pm 3$ and $B = 50 \pm 2$.

Solutions to problems

Chapter 1

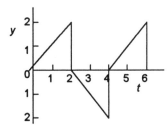

Figure S.1

1.1

1 (a) 3, (b) 5
2 (a) 0, (b) 6
3 (a)
4 (a) 2, (b) 1
5 (a) 0, (b) 2
6 $v = 0$ for $0 \le t \le 2$, $v = 10$ V for $2 \le t$
7 See Figure S.1
8 (a) 2.01 s, (b) 6.34 s
9 (a) 2 m/s, (b) 7 m/s
10 (a) $3x + 1$, (b) $2x + 2$, (c) $2x + 1$
11 (a) $x^2 + 3x + 1$, (b) $9x^2 + 3$, (c) $3x^2 + 3$, (d) $x^2 - 3x - 1$,
    (e) $9x^2 + 12x + 5$

1.2

1 (a) Straight line through origin, (b) straight line not through
   origin, (c) not straight line, (d) no straight line
2 (a) $i = 0.5t + 2$ amps, (b) $e = 1.2 \times 10^{-3} L$ m

1.3

1 (a) –3.2, 1.2, (b) –2.6, –0.38, (c) 3.8, –0.41, (d) no real roots
2 38.2°C, 261.8°C
3 0.93 m, –0.11m
4 $y = \dfrac{mgL^3}{48EI} \pm \sqrt{\left(\dfrac{mgL^3}{48EI}\right)^2 + 2h\left(\dfrac{mgL^3}{48EI}\right)}$
5 18.8 m or 1.6 m
6 5.41 cm and 8.41 cm

1.4

1 (a) $\frac{1}{5}(x+3)$, (b) $x - 4$, (c) $\sqrt[3]{x}$, (d) $\sqrt[3]{\frac{1}{2}(x+1)}$
2 No, but if domain restricted to $x \ge 0$ then yes
3 (a) $x \ge 1$, $1 + \sqrt{x}$, (b) $x \ge -1$, $-1 + \sqrt{x+4}$

1.5

1 Amplitude = 5, phase angle = +30° leading
2 Amplitude = 4, angular frequency = 3 rad/s
3 (a) 2, 1.27 s, 1 rad, (b) 6, 2.09 s, 0, (c) 5, 9.4 s, 0.33 rad, (d) 2, 6.28 s, −0.6 rad lagging
4 (a) 6, π, 1, (b) 2, $2\pi/9$, 0, (c) 6, 5π, −0.2, (d) 2, 2π, −0.2, (e) 6, $\pi/2$, $\pi/8$, (f) 0.5, 2π, $-\pi/6$
5 40, 20 Hz
6 (a) 0.87 V, (b) −0.87 V
7 (a) 100 mA, (b) 100 Hz, (c) 0.25 rad or 14.3° lagging
8 (a) 12 V, (b) 50 Hz, (c) 0.5 rad or 28.6° lagging
9 16.2 V
10 As given in the problem
11 101.3°, 355.4°
12 (a) $\sqrt{41} \sin (\theta - 5.61)$, (b) $\sqrt{41} \cos (\theta + 5.39)$
13 $R = W(1 + \mu^2)^{1/2}$, $\tan \beta = 1/\mu$
14 (a) $5 \sin (\omega t + 0.927)$, (b) $8.63 \sin (\omega t - 1.01)$, (c) $4.91 \sin (\omega t + 4.13)$ or $4.91 (\omega t - 2.15)$ for $-\pi \le a \le \pi$
15 $5.83 \sin (\theta + 59.03)$
16 (a) $15 \sin (\omega t - 0.64)$, (b) $8.06 \sin (\omega t + 2.62)$, (c) $6.71 \sin (\omega t - 2.03)$
17 $22.4 \sin (\omega t + 1.11)$ mA
18 $6.81 \sin (\omega t + 0.147)$ V
19 $126.2\sqrt{2} \sin (\omega t - 0.071)$ V
20 $8.72\sqrt{2} \sin (\omega t - 9.639)$ A
21 (a) 0.83, (b) 1.47, (c) 0.67, (d) −0.41

1.6

1 (a) The negative sign, (b) N_0, (c) λ small
2 (a) The power is positive, (b) L_0, (c) α high means a high expansion
3 (a) 0, (b) 3 A
4 (a) 2, infinite, (b) 10, 0, (c) 0, 2, (d) 2, 0, (e) −4, 0,(f) 0, 0.5, (g) 0, 4,
   (h) 10, 0, (i) 0, 0.2
5 (a) 18.10 V, (b) 7.36 V
6 9.96×10^4 Pa
7 (a) 0, (b) $0.86E/R$ A
8 (a) 0, (b) $0.86E$
9 0.95 μC
10 (a) 0, (b) 1.57 A, (c) 2.53 A, (d) 3.11 A, (e) 4 A
11 5.13 V
12 5637 Ω
13 (a) 1.26 A, (b) 1.72 A, (c) 1.90 A
14 0.03 g
15 (a) 8.61×10^4 Pa, (b) 7.41×10^4 Pa
16 (a) 198.0 V, (b) 190.2 V
17 (a) $-E$, (b) $-0.61E$
18 (a) 0, (b) $0.63E$

19 (a) 200°C, (b) 134.1°C, (c) 0°C

20 + gives growth, – decay.

21 (a) e^{8t}, (b) e^{-2t}, (c) e^{-12t}, (d) $1 + 2\,e^{2t} + e^{4t}$, (e) e^{-3t},
    (f) e^{-2t}, (g) e^{-8t}, (h) $5\,e^{3t}$, (i) $0.4\,e^{4t}$

1.7

1 (a) $4 \lg x$, (b) $5 \lg x - \lg 2$

2 (a) $\lg b + 0.5 \lg 2 - \lg a - \lg c$, (b) $3 \lg a + 3 \lg b - 1.5 \lg c$

3 (a) 5.19, (b) –0.593, (c) 0.419

4 $v = 10.2\,e^{-0.1t}$

5 $\theta = 800\,e^{-0.2t}$

6 $Q = 2.6 h^{2.5}$

7 $A = 400\,e^{-0.02t}$

8 $T = 50\,e^{0.3\theta}$

9 $I = 2430$ mA, $T = -51.3$

10 60.62×10^3

1.8

1 (a) 3.627, (b) 74.210, (c) 0.964, (d) –3.627, (e) 0.525,
    (f) 0.748

2 1.622 m

3 (a) $\sqrt{\left[\left(\dfrac{g\lambda}{2\pi} + \dfrac{2\pi\gamma}{\rho\lambda}\right)\dfrac{2\pi h}{\lambda}\right]}$,
    note when surface tension neglected $\sqrt{gh}$,

    (b) $\sqrt{\left(\dfrac{g\lambda}{2\pi} + \dfrac{2\pi\gamma}{\rho\lambda}\right)}$

Chapter 2

2.1

1 (a) 7.43 m/s, N 73° W, (b) 3.58 m/s, N 54° E, (c) 8.16 m/s, N 75° E

2 (a) 13 m, N 67° E, (b) 13 m, N 67° W, (c) 13 m, S 67° E,
    (d) 24.5 m, N 78° E

3 (a) $\overrightarrow{AC}$, (b) $\overrightarrow{BD}$, (c) $\overrightarrow{DB}$

4 (a) $\mathbf{b} - \mathbf{a}$, (b) $\mathbf{a} + \mathbf{b}$, (c) $\mathbf{a} - 3\mathbf{b}$, (d) $2\mathbf{b}$

5 7.8 N at 54° to AB

6 (a) $\overrightarrow{AD}$, (b) $\overrightarrow{AE}$, (c) 0, (d) $\overrightarrow{AC}$

7 1.36 N, 8.82 N

8 (a) 3.6, 56.3°, (b) 5.4, 21.8°, (c) 4.2, 45°

9 (a) $4\mathbf{i} + 6\mathbf{j}$, (b) $-8\mathbf{i}$, (c) $10\mathbf{i} + 9\mathbf{j}$

10 (a) $7\mathbf{i} + 5\mathbf{j}$, (b) $3\mathbf{i} - 1\mathbf{j}$, (c) $1\mathbf{i} - 4\mathbf{j}$

11 (a) $9\mathbf{i} + 2\mathbf{j}$, (b) $3\mathbf{i} + 4\mathbf{j}$, (c) $-13\mathbf{i} - 3\mathbf{j}$

12 (a) $\sqrt{74}$; $\dfrac{3}{\sqrt{74}}$, $\dfrac{7}{\sqrt{74}}$, $\dfrac{-4}{\sqrt{74}}$

    (b) $\sqrt{38}$; $\dfrac{2}{\sqrt{38}}$, $\dfrac{3}{\sqrt{38}}$, $\dfrac{5}{\sqrt{38}}$,

    (c) $\sqrt{38}$; $\dfrac{-3}{\sqrt{38}}$, $\dfrac{5}{\sqrt{38}}$, $\dfrac{2}{\sqrt{38}}$

13 $\sqrt{78}$; $\dfrac{2}{\sqrt{78}}$, $\dfrac{-5}{\sqrt{78}}$, $\dfrac{2}{\sqrt{78}}$

14 $17\mathbf{i} + 11\mathbf{j} + 5\mathbf{k}$, $\sqrt{255}$, $\sqrt{93}$, $\sqrt{2}$

15 112.6°

16 48.2°, 131.8°, 70.5°

2.3

1 (a) $-j$, (b) 1, (c) $-j$, (d) -1

2 (a) $\pm j4$, (b) $-2 \pm j2$, (c) $0.5 \pm j1.1$

3 (a) $4.12\angle166°$, (b) $5\angle233°$, (c) $3\angle0°$, (d) $6\angle270°$, (e) $1.4\angle45°$,
   (f) $3.61\angle326°$

4 (a) $-2.5 + j4.3$, (b) $7.07 + j7.07$, (c) -6,
   (d) $0.68 + j2.72$, (e) $1.73 + j1$, (f) $1.5 - j2.6$

5 (a) $1 + j6$, (b) $5 - j2$, (c) $-14 + j8$, (d) $0.23 - j0.15$, (e) $0.1 - j0.8$

6 (a) $5 - j2$, (b) $-2 - j1$, (c) $-1 + j7$, (d) 1, (e) $12 + j8$,
   (f) $-10 + j6$, (g) $11 - j2$, (h) 12, (i) $10 + j5$, (j) $0.9 + j1.2$,
   (k) $0.23 - j0.15$, (l) $j1$, (m) $-0.3 + j1.1$

7 (a) $20\angle60°$, (b) $50\angle80°$, (c) $0.1\angle(-20°)$, (d) $0.5\angle(-40°)$,
   (e) $5\angle(-20°)$, (f) $0.4\angle(-20°)$

8 (a) $10\angle(-30°)$, $8.66 - j5$, (b) $10\angle150°$, $-8.66 + j5$, (c) $22\angle45°$,
   $15.6 + j15.6$

9 (a) $5.5 + j2.6$, $6.1\angle25.3°$, (b) $-2 + j7$, $7.3\angle105.9°$,
   (c) $3.7 + j4.5$, $5.8\angle50.6°$

10 (a) $25\angle90°$, (b) $20\angle75°$, (c) $44.5\angle83.3°$, (d) $4\angle(-30°)$, (e)
   $1.25\angle15°$, (f) $0.164\angle9.2°$

11 (a) $20 + j17.32 = 26.46\angle40.9°$ V, (b) $26.46 \sin(\omega t + 40.9°)$ V

12 $25\angle(-30°)$ Ω

13 $2\angle(-36.8°)$ A

14 (a) $12 - j5$ Ω, (b) $136.6 + j136.6$ Ω, (c) $32.1 + j7.4$ Ω,
   (d) $1.88 - j6.34$ Ω, (e) $0.384 - j1.922$ Ω, (f) $j13.3$ Ω

15 (a) $5 + j2$ Ω, (b) $50 - j10$ Ω, (c) $2 + j1$ Ω, (d) $1.92 - j0.38$ Ω,
   (e) $-j125$ Ω

Chapter 3

3.1

1 You might like to consider it to be like a swinging chain
   which, in itself, is rather like a form of simple pendulum.

2 See Figure S.2, $m\dfrac{d^2x}{dt^2} + c\dfrac{dx}{dt} + kx = F$

3 (a) $k_1 x = M\dfrac{d^2y}{dt^2} + c\dfrac{dy}{dt} + (k_1 + k_2)y$,

   (b) $T = I\dfrac{d^2\theta}{dt^2} + c\dfrac{d\theta}{dt} + k\theta$,

   (c) $v = L\dfrac{di}{dt} + Ri$,

   (d) $p = c\dfrac{d\theta}{dt} + cpq\theta$, where $\theta = \theta_o - \theta_i$

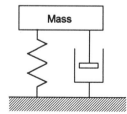

Figure S.2

3.2

1 (a) $E = 0.45L + 5$, (b) $R = 4.2L$
2 (a) T against $\sqrt{L}$, gradient $2\pi/\sqrt{g}$, intercept 0,
   (b) s/t against t, gradient $a/2$, intercept u,
   (c) e/θ against θ, gradient b, intercept a,
   (d) R against θ, gradient $R_0\alpha$, intercept R_0,
3 $R = \dfrac{1200}{V} + 50$
4 $R = \dfrac{0.16}{d^2}$
5 $V = 16p^{-1}$
6 $T = 500p^{0.28}$
7 $C = 0.001n^3 + 30$
8 $s = 0.1v^2 + 0.5v$
9 $I = 0.001V^2$
10 $v = 2090\,e^{-t/12}$

Chapter 4

4.1

1 (a) $5x^4$, (b) $-8x^{-5}$, (c) $-6x$, (d) $\frac{1}{2}$, (e) $4\pi x$, (f) $3\sec^2 3x$,
   (g) $-10\sin 2x$, (h) $8\,e^{x/2}$, (i) $-4\,e^{-2x}$, (j) $9\,e^{3x}$, (k) $-(5/6)x^{-3/2}$,
   (l) $-(12/3)x^{-3}$, (m) $-\dfrac{7}{2\sqrt{3}}x^{-3/2}$, (n) $-\dfrac{15}{8}x^{-4}$, (o) $\dfrac{\sqrt{3}}{2}x^{3/2}$,
   (p) $-24x^2 + 4x + 15$, (q) $5x\cos x + 5\sin x$, (r) $e^{x/2} + \frac{1}{2}x\,e^{x/2}$,
   (s) $(x^2 + 1)\cos x + 2x\sin x$, (t) $\dfrac{-13}{(x-6)^2}$, (u) $\dfrac{x-1}{2x^{3/2}}$,
   (v) $\dfrac{x\cos x - \sin x}{x^2}$, (w) $\dfrac{2\,e^{2x}(x^2 - x + 1)}{(x^2 + 1)^2}$,
   (x) $\dfrac{2\cosh 2x - \cosh 3x - 3\sinh 2x \sinh 3x}{(\cosh 3x)^2}$, (y) $\dfrac{7}{(2 - 7x)^2}$,
   (z) $\dfrac{1}{2(3 - x)^{3/2}}$
2 (a) 2, (b) $-4\cos 2x$, (c) $6/x^4$, (d) $36x^2 - 2 - 2/x^3$, (e) $12x^2 + 12x$,
   (f) $\dfrac{3\sqrt{x} - \sqrt{x^3}}{4x^3}$
3 7 m/s, -4 m/s^2
4 $6\cos 2t - 9\sin 3t$ m/s, $-12\sin 2t - 27\cos 3t$ m/s^2
5 $0.03\cos 5t$ A/s
6 $50\,e^{-100t}$ V/s
7 πr^2
8 $4\pi r^2$
9 $-\dfrac{f_s}{c(1 + v/c)^2}$
10 $L_0(a + 2bT)$
11 (a) $(2, -1)$ min., (b) $(1, 7)$ max., $(3, 3)$ min., (c) $(1, -4)$ min.,
    $(-1, 4)$ max., (d) $(\pi/2, 1)$ max., $(3\pi/2, -1)$ min., (e) $(-2, 23)$
    max., $(1, -4)$ min.
12 $r = h = 1$ m
13 $h = r = 4$ cm
14 $\frac{1}{2}L$
15 $0, 0.32L$
16 47.7 V

17 3.33 m from smaller source

18 10 mA/s

19 0.58L

20 $-a/2b$

21 45°

22 $x = y$

23 ½$(a - b)$

24 $r = 4/3$ m

25 6×6 cm

26 $4 \times 4 \times 2$ m

4.2

1 (a) $4x + C$, (b) $\frac{1}{2}x^4 + C$, (c) $\frac{1}{4}x^4 + \frac{5}{2}x^2 + C$,
  (d) $\frac{3}{5}x^{5/3} - 2x^{3/2} + C$, (e) $4x + \frac{1}{5}\cos 5x + C$, (f) $-\frac{2}{3}e^{-3x} + C$,
  (g) $8e^{x/2} + \frac{1}{3}x^3 + 2x + C$, (h) $4\ln|x| + C$

2 (a) 15, (b) 8, (c) 1.10, (d) –2.25, (e) –4.5, (f) 2.67, (g) 0.83

3 (a) 116/3, (b) 11/3

4 ½

5 1/12

6 1

7 4½

8 (a) Diverges, (b) 1/18, (c) 1/3, (d) diverges

9 (a) $\frac{1}{40}(4x^2 - 1)(x^2 + 1)^4 + C$, (b) $\frac{1}{2}e^{4x-1} + C$,
  (c) $\frac{1}{3}(x + 2)\sqrt{2x+1} + C$, (d) $-\frac{1}{2}\cos x + C$,
  (e) $2\sinh^{-1}\frac{x}{2} + \frac{1}{2}x\sqrt{x^2 + 4} + C$, (f) $\frac{2}{15}(3x + 2)(x - 1)^{3/2} + C$,
  (g) $\ln\left|\dfrac{1 + \tan\frac{1}{2}x}{1 - \tan\frac{1}{2}x}\right| + C$, (h) $\sin^{-1}\frac{x}{3} + C$,
  (i) $\frac{1}{6}\sin^3 2x - \frac{1}{10}\sin^5 2x + C$

10 (a) $\frac{2}{5}(x + 2)^{5/2} - \frac{4}{3}(x + 2)^{3/2} + C$, (b) $\dfrac{x}{\sqrt{x^2 + 1}} + C$,
  (c) $\frac{1}{3}\cos^3 x - \cos x + C$, (d) $\frac{1}{2}\sin^{-1}2x + C$,
  (e) $\frac{1}{3}(x^2 + 2)^{3/2} + C$, (f) $\frac{1}{10}\tan^{-1}\frac{5x}{2} + C$,
  (g) $\frac{1}{5}\sin^5 x - \frac{1}{7}\sin^7 x + C$, (h) $\frac{1}{4}\tan^4 x + C$,
  (i) $\frac{2}{3}\tan^{-1}\left(\frac{1}{3}\tan\frac{x}{2}\right) + C$

11 (a) $\ln 2$, (b) 1/90, (c) $\pi/4$, (d) $\pi/4$, (e) $\pi/8$

12 (a) $\frac{1}{3}x^3\ln|x| - \frac{1}{9}x^3 C$, (b) $\frac{1}{2}x e^{2x} - \frac{1}{4}e^{2x} + C$,
  (c) $x^3\sin x + 3x^2\cos x - 6x\sin x - 6\cos x + C$,
  (d) $\frac{1}{25}\sin 5x - \frac{1}{5}x\cos 5x + C$, (e) $\frac{1}{2}x^2\ln|3x| - \frac{1}{4}x^2 + C$,
  (f) $\frac{1}{2}x - \frac{1}{4}\sin 2x + C$

13 (a) $\frac{\pi}{2} - 1$, (b) $\frac{1}{16}\pi^2 + \frac{1}{4}$, (c) 2

14 (a) $\frac{1}{4}x^2 + \frac{3}{4}x + \frac{9}{8}\ln|2x - 3| + C$, (b) $-\frac{1}{2}x - \frac{1}{4}\ln|1 - 2x| + C$,
  (c) $\frac{1}{2}x + \frac{1}{5}\ln|x - 1| - \frac{9}{20}\ln|2x + 3| + C$,
  (d) $\frac{1}{6}\ln|x - 1| + \frac{1}{2}\ln|x + 1| - \frac{1}{6}\ln|2x + 1| + C$,
  (e) $-\frac{1}{4}\ln|x| + \frac{3}{8}\ln|x - 2| - \frac{1}{8}\ln|x + 2| + C$,
  (f) $\frac{2}{3}\ln|x - 1| + \frac{5}{6}\ln|2x + 1| + C$,
  (g) $\frac{1}{2}x^2 + 2x + \frac{2}{3}\ln|x - 1| + \frac{5}{6}\ln|2x + 1| + C$,
  (h) $\ln|x - 3| - \ln|x - 2| + C$, (i) $6\ln|x| - \ln|x + 1| - \dfrac{9}{x+1} + C$,

(j) $2\ln|x| - 2\ln|x-1| + \ln|x^2+4| + 2\tan^{-1}\frac{1}{2}x + C$,

(k) $-\frac{1}{x} - \tan^{-1}x + C$

15 (a) $Mh^2/18$, (b) $Mh^2/6$

16 $5Mr^2/2$

17 (a) $\frac{1}{12}ML^2$, (b) $M\left(\frac{1}{12}L^2 + d^2\right)$

18 (a) 1, (b) 7, (c) 16, (d) 0.5

19 $2A/\pi$

20 $0.623N_0$

21 (a) 4.92, (b) 1.15, (c) 1.23, (d) 0.707, (e) 1.35

22 $V/2$

Chapter 5

5.1

1 (a) $m\dfrac{dv}{dt} + kv^2 = mg$, (b) $m\dfrac{d^2x}{dt^2} + c\dfrac{dx}{dt} + kx = 0$,

(c) $m\dfrac{dv}{dt} + kv = 0$, (d) $m\dfrac{dv}{dt} + kv^2 = mg$, (e) $\dfrac{dI}{dx} + kx = 0$,

(f) $A\dfrac{dh}{dt} + \sqrt{2gh} = 0$, (g) $\dfrac{A}{\rho g}\dfrac{dp}{dt} + \sqrt{\dfrac{2p}{\rho}} = q_1$

2 As given in the problem

3 (a) $y = x\,e^x$, (b) $y = \dfrac{1}{\omega}\sin\omega t + 2\cos\omega t$, (c) $y = (2+x^2)\,e^{-x}$

4 $y = -\dfrac{w}{2EI}\left(\frac{1}{2}L^2x^2 - \frac{1}{3}Lx^3 + \frac{1}{12}x^4\right)$

5.2

1 (a) $y = \ln x + A$, (b) $y = 2\sin\frac{1}{2}x + A$, (c) $-1/y = x + A$,

(d) $\ln y = -2x + A$ or $y = C\,e^{-2x}$, (e) $\tan^{-1}y = x^2 + A$,

(f) $y = \ln(x^3 + A)$ or $e^y = x^3 + A$, (g) $y = 4x + x^3 + A$,

(h) $-1/y = 2x + A$, (i) $e^y = x^3 + A$, (j) $y^2 = x + A$,

(k) $y = A\,e^{-1/x}$, (l) $y = -2/(x^2 + A)$

2 $y = 1/(2 - x^2)$

3 $V = V_0\,e^{-t/RC}$

4 $N = N_0\,e^{-kt}$

5 $i = \dfrac{V}{R}(1 - e^{-Rt/L})$

6 $y = 2 - e^{-x}$

7 $v = 10 - 10\,e^{-t}$

8 $T = T_0\,e^{\mu\theta}$

9 3.41 hours

10 RC

11 51.4°C

12 As given in the problem

13 10.9 V

14 $i = I(1 - e^{-Rt/L})$

15 (a) $x = 1 - e^{-t/2}$, (b) $x = 8\,e^{-t/2} + 4t - 8$

16 $x = 5(1 - e^{-t/\tau})$

17 (a) 36.8%, (b) 13.5%

18 12 s

19 (a) $2\dfrac{dx}{dt} + x = 45$, (b) $x = 45 - 25\,e^{-t/2}$

5.3

1 (a) $y = \frac{3}{2}x^2 + 4x + 4$, (b) $y = \frac{3}{5} e^{2x} + \frac{2}{5} e^{-3x}$,
  (c) $y = e^{-2x} + 2x e^{-2x}$, (d) $y = \frac{5}{2} e^x - \frac{5}{2} e^{-x}$,
  (e) $y = 2 e^x - e^{2x}$, (f) $y = 2 e^{-x} + x e^{-x}$

2 (a) $y = A e^{x/2} + B e^{-x}$, (b) $y = (A + Bx) e^{3x}$, (c) $y = (A + Bx) e^{5x}$,
  (d) $y = e^{2x}(A \cos x + B \sin x)$, (e) $y = e^x(A \cos 2x + B \sin 2x)$,
  (f) $y = A e^{2x} + B e^{-5x}$, (g) $y = e^{2x}(A \cos x + B \sin x)$,
  (h) $y = A e^{2x} + B e^{-4x}$,

3 (a) $y = -\frac{3}{4} e^{-5x} + \frac{3}{4} e^{-x}$, (b) $y = e^{3x}(3 \cos 4x - 2 \sin 4x)$,
  (c) $y = (2 + 4x) e^{-3x/2}$, (d) $y = 2 e^{3x} - e^{4x}$, (e) $y = (1 + 2x) e^{-x}$,
  (f) $y = e^{3x}(\cos 4x - \sin 4x)$

4 (a) $y = A e^{2x} + B e^{-2x} + \frac{2}{5} e^{3x}$,
  (b) $y = e^{-3x/2}\left(\cos \sqrt{\frac{7}{2}} x + \sin \sqrt{\frac{7}{2}} x\right) + \frac{3}{4}x - \frac{1}{16}$,
  (c) $y = A e^{4x} + B e^{2x} + \frac{21}{85} \cos x - \frac{18}{85} \sin x$,
  (d) $y = A e^{2x} + B e^{3x} + \frac{1}{2}e^x$,
  (e) $y = A e^{2x} + B e^{3x} + e^x - \frac{1}{4} e^{-x}$,
  (f) $y = A e^{2x} + B e^{-x} + \frac{1}{4} \cos 2x - \frac{3}{4} \sin 2x$,
  (g) $y = A e^x + B e^{2x} - 2 - 3x - x^2$,

5 $y = e^{3x} - 3 e^{2x}$

6 $y = \cos 3x - \frac{2}{15} \sin 3x + \frac{1}{5} \sin 2x$

7 $x = A e^{-4t} + B e^{-t}$

8 (a) $x = 0.2 \cos 3t$, (b) $x = e^{-t}(0.2 \cos 2.83t + 0.070 \sin 2.83t)$

9 0.44

10 (a) 5 rad/s, (b) 1.25

11 316 rad/s, 6.3 N s/m

12 Over damped

13 6 N s/m

14 2.6 rad/s, 0.76

15 $e^{-t}(0.2 \cos 2.24t + 0.22 \sin 2.24t)$

16 $\theta = \frac{4}{9}\pi e^{-t} - \frac{1}{9}\pi e^{-4t}$

17 As given in the problem

Chapter 6

6.1

1 (a) $\frac{2}{s^3}$, (b) $\frac{6}{s^4}$, (c) $\frac{a}{s^2 - q^2}$

2 (a) $\frac{4}{s}$, (b) $\frac{3}{s^2} - \frac{1}{s}$, (c) $\frac{1}{s - 3}$, (d) $\frac{2}{s^2} + \frac{3}{s - 1}$,
  (e) $\frac{2}{s^3} + \frac{4}{s + 2}$, (f) $\frac{2}{s^3} + \frac{2}{s^2} + \frac{1}{s}$, (g) $\frac{6}{s^2 + 9}$,
  (h) $\frac{15}{s^2 - 9}$, (i) $\frac{3}{s^2 + 36}$, (j) $\frac{1}{(s + 3)^2}$, (k) $\frac{4}{s} - \frac{6}{s^2 + 9} + \frac{1}{s - 2}$,
  (l) $\frac{6}{(s + 2)^2}$, (m) $\frac{2}{s^2 - 1}$, (n) $\frac{s - 3}{s^2 - 6s + 10}$, (o) $\frac{s^2 + 4s + 5}{(s + 1)^3}$,
  (p) $\frac{4}{(s + 1)(s^2 + 2s + 5)}$, (q) $\frac{s^2 + 9}{(s^2 - 9)^2}$, (r) $\frac{2s(s^2 + 27)}{(s^2 - 9)^3}$,
  (s) $\frac{6}{(s + 3)^4}$

3 (a) $\frac{2}{s^3} + \frac{3}{s^2} + \frac{2}{s}$, (b) $\frac{2}{s} + \frac{12}{s^2 + 9}$, (c) $\frac{1}{s - 4} + \frac{s}{s^2 - 4}$,
  (d) $\frac{2}{s} + \frac{5}{s - 3}$, (e) $\frac{s}{s^2 + 4} + \frac{s}{s^2 + 9}$, (f) $\frac{6}{s^4} + \frac{4}{s + 1}$

4 (a) $\dfrac{2}{(s+3)^2+4}$, (b) $\dfrac{2}{(s-4)^3}$, (c) $\dfrac{s-2}{(s-2)^2+1}$

5 (a) $e^{-5s}\dfrac{1}{s}$, (b) $1\,e^{-4s}$ (c) $\dfrac{3\,e^{-10s}}{s^2}$

6 $\dfrac{3(1-e^{-s})}{s(1+e^{-s})}$

7 $\dfrac{1}{s^2}-\dfrac{e^{-s}}{s(1-e^{-s})}$

8 (a) $\dfrac{1}{s(1+e^{-s})}$, (b) $\dfrac{1}{s^2(1+e^{-s})}-\dfrac{e^{-s}}{s(1-e^{-2s})}$, (c) $\dfrac{1-e^{-s}}{s^2(1+e^{-s})}$

9 (a) e^{2t}, (b) 5, (c) $\cos 4t$, (d) $\sinh 3t$, (e) $e^{2t}\sin 5t$,
   (f) $\tfrac{1}{6}t^3\,e^{-3t}$, (g) $(t-2)u(t-2)$, (h) $(t-3)\,e^{-2(t-3)}u(t-3)$

10 (a) $e^{-2t}+2\,e^{3t}$, (b) $2\,e^t+e^{-2t}$, (c) $\cos 2t-e^{-t}$, (d) $e^{-2t}+2t\,e^{-2t}$

11 (a) 0, (b) 1

12 (a) 2, (b) 0

6.2

1 (a) $x=3\,e^{2t}-3\,e^{-t/2}$, (b) $x=\tfrac{2}{5}-\tfrac{2}{5}e^{-5t}$, (c) $x=\tfrac{1}{4}-\tfrac{1}{4}\cos 2t$,
   (d) $x=e^{-t}-e^{-t}\cos t$, (e) $x=\tfrac{1}{2}e^t-\tfrac{1}{2}t\,e^t+\tfrac{1}{2}\cos t$,
   (f) $x=2\,e^t-1$, (g) $x=\tfrac{4}{17}\cos t+\tfrac{2}{17}\sin t-\tfrac{4}{17}e^{-4t}$,
   (h) $x=3-3\cos t+\sin t$,
   (i) $x=\tfrac{1}{5}e^{-3t}+t\,e^{2t}-\tfrac{1}{5}e^{2t}$,
   (j) $x=\tfrac{3}{65}e^{3t}+\tfrac{1}{20}e^{-2t}-\tfrac{1}{52}\sin 2t-\tfrac{5}{52}\cos 2t$,
   (k) $x=\tfrac{5}{36}-\tfrac{1}{6}t-\tfrac{1}{12}e^{3t}+\tfrac{121}{252}e^{6t}+\tfrac{369}{252}e^{-t}$

6.3

1 $3/s$

2 2

3 $5/s^2$

4 e^{-5t} V

5 (a) $(5/3)(1-e^{-3t})$ V, (b) $5e^{-3t}$ V

6 (a) $6(1-e^{-t})$ V, (b) $6e^{-t}$ V

7 (a) $2(1-e^{-2t})$ V, (b) $\tfrac{1}{2}[t-\tfrac{1}{2}(1-e^{-2t})]$ V

8 (a) $5/(s-1)-4/(s-2)$, (b) $4/(s+1)-3/(s+2)$,
   (c) $2/(s+1)-3/(s+1)^2$

9 $24-12e^{-2t}-4e^{-4t}$

10 $8e^{-2t}-8t\,e^{-2t}$

11 (a) Critical, (b) overdamped, (c) underdamped

12 $-1.5+3.0t+1.5e^{-2t}$

13 $0.5-e^{-t}+0.5e^{-2t}$

14 $0.5(e^{-t}-e^{-3t})$

15 $0.5(1-e^{-10t})$

16 10, 0.05

17 Underdamped

18 Critically damped

19 $1/53$ s

20 (a) $0.01s\,\Omega$ in series with -0.002 V, in parallel with $0.2/s$ A,
   (b) $0.5/s$ MΩ in series with $5/s$ V, in parallel with 10 μA.

21 (a) $10+0.002s\,\Omega$, (b) $\dfrac{0.02s}{10+0.002s}\,\Omega$

22 $i(t)=\dfrac{v_0}{R}\,e^{-t/RC}$

23 $i(t) = \dfrac{V}{R} - \dfrac{V}{R}\, e^{-Rt/L}$

24 $i(t) = \dfrac{1}{R}\left(\delta(t) - \dfrac{1}{RC}\, e^{-t/RC}\right)$

Chapter 7

7.1

1   0, 1, 0, −1, 0, 1, ...
2   (a) 0, 1, 2, 3, 4, (b) 1, 0.37, 0.13, 0.05, 0.007
3   (a) 116, (b) 0.75
4   (a) $(-1)^{k-1}$, (b) $5k$, (c) $2.5 - 0.5k$
5   (a) 0, 1, 4, 9, 16, (b) 1, 2.72, 7.39, 20.09, 54.60,
    (c) 0, 2.5, 6, 10.5, 16
6   13
7   0.5
8   (a) 0.25^k, (b) $2(-1)^k$, (c) $3 + 0.1^k$
9   (a) 0.1, 0.01, 0.001, (b) 5.1, 5.01, 5.001, (c) −1, +1, −1
10 (a) 222, (b) 9.998, (c) 28.70
11 (a) 7.5, (b) 23.98, (c) 1023
12 (a) 12, (b) 16, (c) 48
13 (a) Convergent, (b) divergent
14 (a) Convergent, (b) divergent
15 (a) Divergent, (b) convergent, (c) convergent, (d) convergent,
    (e) convergent, (f) divergent
16 (a) $1 + 4x + 6x^2 + 4x^3 + x^4$, (b) $1 + \dfrac{3x}{2} + \dfrac{3x^2}{8} - \dfrac{x^3}{16}$,
    (c) $1 + \dfrac{5x}{2} + \dfrac{35x^2}{8} + \dfrac{105x^3}{16}$ (d) $1 - 0.25 + 0.062 - 0.015$,
    (e) $2 + \dfrac{x}{4} - \dfrac{x^2}{64} + \dfrac{x^3}{512}$
17 (a) $1 + 12x + 66x^2 + 220x^3 + ...$, (b) $1 + 4x + 12x^2 + 32x^3 + ...$,
    (c) $3^{2/5}\left(1 - \tfrac{4}{15}x - \tfrac{4}{75}x^2 - \tfrac{64}{3375}x^3 + ...\right)$, (d) $1 + x + x^2 + x^3 + ...$,
    (e) $1 - \tfrac{3}{2}x + \tfrac{27}{8}x^2 - \tfrac{135}{16}x^3 + ...$, (f) $1 + 2x^3 + 3x^6 + 4x^9 + ...$
18 1.013 2
19 As given in the problem
20 $1 - \dfrac{x^2}{2!} + \dfrac{x^4}{4!} + ...$
21 $x + \tfrac{1}{3}x^3 + \tfrac{2}{5}x^5 + ...$
22 (a) $1 + 2x + 2x^2 + \tfrac{4}{3}x^3 + ...$ (b) $1 + x - \tfrac{1}{2}x^3 + ...$,
    (c) $1 - \tfrac{1}{2}x + \tfrac{1}{4}x^2 - \tfrac{5}{8}x^3 + ...$, (d) $x + \dfrac{x^2}{2!} + \dfrac{2x^3}{3!} + ...$,
    (e) $1 + \tfrac{1}{2}x^2 + \tfrac{5}{24}x^4 + ...$, (f) $1 - x^2 + \tfrac{1}{3}x^4 - ...$
23 As given in the problem
24 As given in the problem
25 As given in the problem
26 Reduced by 6%
27 Increased by 1%

7.2

1   Second and fourth (and a d.c. term)
2   $\dfrac{4A}{\pi}\left(\sin \omega t + \tfrac{1}{3}\sin 3\omega t + \tfrac{1}{5}\sin 5\omega t + ...\right)$

3 $\frac{A}{2} - \frac{4A}{\pi^2}\left(\sin\omega t + \frac{1}{9}\sin 3\omega t + \frac{1}{25}\sin 5\omega t + ...\right)$

4 $\frac{2A}{\pi} - \frac{4A}{\pi}\left(\frac{1}{3}\cos 2\omega t + \frac{1}{15}\cos 4\omega t + \frac{1}{35}\cos 6\omega t + ...\right)$

5 (a) Only even harmonics, (b) only odd harmonics

6 $\frac{4A}{\pi}\left(\cos\omega t - \frac{1}{3}\cos 3\omega t + \frac{1}{5}\cos 5\omega t - ...\right)$

7 (a) Odd sines, (b) a_0, even sines and cosines

8 (a) 10/6, (b) 0

9 (a), (b), (d) odd, (c), (e) even

10 (a) sine, (b) a_0 and cosine, (c) a_0 and cosine

11 $2\left(\sin t - \frac{1}{2}\sin 2t + \frac{1}{3}\sin 3t + ...\right)$

12 (a) 0.5, 0.31, 0.16, 0.11, 180°, 180°, 180°,
    (b) 1.57, 1.19, 0.5, 0.13, −32°, 180°, −12°

13 $0.32 + 0.5\cos 100t + 0.21\cos 200t$ mA

14 $0.5\cos(100t - 90°) + 0.21\cos(200t - 90°)$ A

15 $3.2\sin(100t + 90°) + 3.2\sin(200t + 90°)$ mA

Chapter 8

8.2

1 (a) 1, (b) 1, (c) 0

2 (a) $a \cdot b$, (b) $a + b + c$, (c) b, (d) $a \cdot \overline{b}$, (e) $a + \overline{b}$

8.3

1 (a) $(b + c) \cdot d + a$, (b) $(a + \overline{a}) \cdot (b + \overline{b})$, (c) $a \cdot b + \overline{a} \cdot \overline{b}$,
    (d) $a \cdot \overline{c} + \overline{a} \cdot b + \overline{a} \cdot \overline{b} \cdot \overline{c}$

2 (a)

a	b	c	$(a + \overline{b}) + (a + \overline{c})$
0	0	0	1
0	0	1	0
0	1	0	0
0	1	1	0
1	0	0	1
1	0	1	1
1	1	0	1
1	1	1	1

(b)

a	b	$b \cdot a$	$\overline{b}$	$a \cdot b + \overline{b}$	$\overline{a}$	$\overline{a} \cdot (a \cdot b + \overline{b}) \cdot \overline{b}$
0	0	0	1	1	1	1
0	1	0	0	0	1	0
1	0	0	1	1	0	0
1	1	1	0	1	0	0

3 (a) $A \cdot B + C \cdot (D + E)$, (b) $A + (B + C)$, (c) $A + B + C$

4

a	b	$\overline{a}$	$\overline{b}$	$a \cdot \overline{b}$	$\overline{a} \cdot b$	$(a \cdot \overline{b}) + (\overline{a} \cdot b)$
0	0	1	1	0	0	0
0	1	1	0	0	1	1
1	0	0	1	1	0	1
1	1	0	0	0	0	0

5 (a) See Figure S.3(a), (b) see Figure S.3(b),
   (c) see Figure S.3(c)

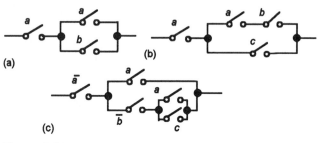

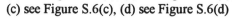

(a)   (b)

(c)

Figure S.3

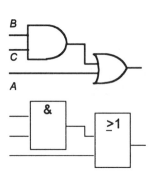

Figure S.4

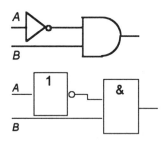

Figure S.5

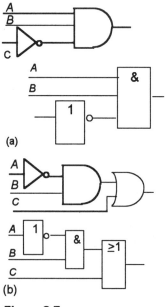

(a)

(b)

Figure S.7

6 (a) $(A + B) \cdot B \cdot C + A, B \cdot C + A$, see Figure S.4,
   (b) $\overline{A \cdot B \cdot B}, \ \overline{A} \cdot B$, see Figure S.5

7 (a) See Figure S.6(a), (b) see Figure S.6(b),
   (c) see Figure S.6(c), (d) see Figure S.6(d)

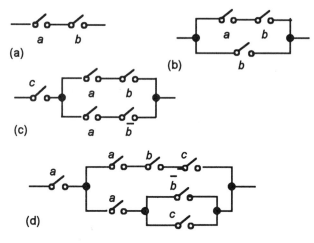

(a)

(b)

(c)

(d)

Figure S.6

8 (a) $\bar{a} \cdot \overline{b} \cdot c + a \cdot b \cdot \overline{c}$, (b) $c \cdot (a \cdot b + \overline{a} \cdot b)$

9 (a) $A \cdot B \cdot (A + B + C), A \cdot B \cdot C$,
   (b) $(A \cdot B + \overline{A} \cdot C) + (\overline{A} \cdot B + B \cdot C), B + \overline{A} \cdot C$
   See Figure S.7

10 (a) $\overline{A} \cdot \overline{B} + A \cdot B$, (b) $B \cdot \overline{C}$, (c) $\overline{A} \cdot \overline{C} + A \cdot B + A \cdot C$, o
   $\overline{A} \cdot \overline{C} + \overline{A} \cdot B + A \cdot C$, (d) $\overline{B} \cdot \overline{C}$, (e) C, (f) $C \cdot D$,
   (g) $A \cdot B + \overline{B} \cdot \overline{C}$

Chapter 9

9.1

1   4/50
2   4/52
3   (a) 0.004, (b) 0.96
4   0.85
5   0.97
6   (a) 0.01, (b) 0.02, (c) 0.03, (d) 0.96
7   (a) 1/3, (b) 2/3
8   (a) 40 320, (b) 840, (c) 30
9   1260
10   (a) 21, (b) 21, (c) 1
11   (a) 77/92, (b) 11/69, (c) 1/276
12   (a) 1/15, (b) 7/15
13   (a) 1/15, (b) 8/15
14   1000
15   0.5

9.2

1   3.0
2   4/3
3   40, 31.6
4   5
5   0.237
6   0.24
7   (a) 0.02, 0.06, 0.22, 0.32, 0.28, 0.08, 0.02, (b) 69.3, 2.3
8   (a) 0.4, (b) 7
9   0.122, 0.270, 0.285, 0.190, 0.090, 0.032, 0.009, 0.002,
     0.000 4, 0.000 1
10   0.132
11   (a) 0.817, (b) 0.016
12   0.366
13   0.005, 0.029, 0.078, 0.138, 0.181, 0.185, 0.154, 0.108, 0.064,
     0.033
14   0.016
15   100, 9.9
16   0.001
17   0.135
18   0.297
19   20
20   (a) 6.68, (b) 17.75
21   (a) 9.68, (b) 11.5
22   (a) 0.091 3, (b) 0.091 3, (c) 0.817 4
23   $50 \pm 6.6 \, \Omega$
24   0.34
25   0.954
26   0.185 9
27   (a) 0.315 6, (b) 0.726 0

9.3

1 2.134 mm, 0.011 mm
2 0.05 A
3 51.12 Ω, 0.08 Ω
4 72
5 4.9 ± 0.3 mm²
6 10 000
7 39.0 kV, 0.11 kV
8 (a) 50 ± 3.6, (b) 10 000 ± 500, (c) 1 000 000 ± 52 000, (d) 2 ± 0.1
9 Diameter
10 (a) 150 ± 3.6, (b) 5000 ± 250, (c) 2 ± 0.1

Index